THE PHYSICS HANDBOOK
Fundamentals and Key Equations

CHARLES P. POOLE, JR.
Department of Physics and Astronomy
University of South Carolina

A Wiley-Interscience Publication
JOHN WILEY & SONS, INC.
New York / Chichester / Weinheim / Brisbane / Singapore / Toronto

This book is printed on acid-free paper.$^{\infty}$

Published simultaneously in Canada.

Library of Congress Cataloging-in-Publication Data:

Poole, Charles P.
The physics handbook : fundamentals and key equations / by Charles P. Poole, Jr.
p. cm.
Includes bibliographical references and index.
1. Physics–Handbooks, manuals, etc. 2. Mathematical physics–Handbooks, manuals, etc. I. Title.
QC61.P65 1998
530′.02′02–dc21 97-45863
ISBN 0-471-18173-0 (cloth: alk. paper)
ISBN 0-471-31460-9 (pbk: alk. paper)

Printed in the United States of America

10 9 8 7 6 5 4 3 2 1

THE PHYSICS HANDBOOK

To Oswald Schuette,
friend and colleague
throughout my professional life

CONTENTS

PREFACE

This volume is an attempt to bring together a compendium of physics at the graduate level. It is at the same time a summary of the basic subject matter that a physicist learns during the first two years of a doctoral program, and a précis of much of what he or she frequently has occasion to look up during his or her subsequent academic and/or industrial career.

Most of the chapters contain material that graduate students must know before taking the PhD qualifying (admission to candidacy) examination. The first draft of the manuscript was assembled from notes I had accumulated during twenty years of teaching a qualifying examination preparation course. During these years the students asked many questions and provided comments on the lecture notes that were very helpful in improving the work. Each student enrolled in the course was required to teach two of the sessions, and this gave me the opportunity to listen to the subject matter of the chapters from a student's viewpoint. I am indebted to these students for their insights.

The handbook aspects of this volume are the result of my keeping track of the various things that I have looked up in reference books during my 40-year career as both an industrial and an academic physicist. Some types of information are needed quite often, such as matrix algebra and vector identities, together with special functions like Legendre polynomials, and these are found in the last two chapters. Other types of information are needed more occasionally such as clarifications of concepts involving Lagrangians, parity, dispersion relations, chaos, free energies, statistical mechanical ensembles, and elementary particle classifications, and these materials are spread throughout the remainder of the chapters.

This volume should be useful for its original intent, namely providing a convenient vehicle for reviewing graduate-level physics in preparation for the qualifying examination. However its greatest value will doubtless be to the community of working physicists who can benefit from a one-volume compendium of fundamental concepts and key equations. There are frequent occasions during the working life of a physicist when it is necessary to refresh one's memory concerning basic questions of theory or practice that were learned years before, and this volume attempts to satisfy that need. I hope that it will serve both of these purposes well.

It has been very satisfying to spend the beginning of my formal retirement gathering together in one volume things that I initially learned as a graduate student at Fordham and the University of Maryland during the early 1950s, such as classical mechanics, electrodynamics, and quantum mechanics, together with more recently acquired knowledge on topics such as closed-form solutions to the three-body problem, elementary particles, and chaos. A number of interesting topics including many of the useful function and formula tabulations in the final two chapters were added at the recommendation of several reviewers to whom I am thankful. I also wish to thank my son Michael for drawing many of the figures.

PHYSICAL CONSTANTS

Fundamental Constants

Avogadro's number	$N_A = 6.0221 \times 10^{23}\ mol^{-1}$
Boltzmann's constant	$k_B = R/N_A = 1.3807 \times 10^{-23}\ J/K$
Dielectric constant of vacuum	$\epsilon_0 = 8.8542 \times 10^{-12}\ F/m$
Electron charge	$e = 1.60218 \times 10^{-19}\ C$
Fine structure constant	$\alpha = e^2/4\pi\epsilon_0\hbar c = 7.2974 \times 10^{-3}$ $\alpha^{-1} \sim 137$
Gas constant	$R = N_A k_B = 8.3145\ J/moleK$
Gravitation constant	$G = 6.6726 \times 10^{-11} m^3/Kgs^2$
Light speed in vacuum	$c = 2.9979 \times 10^8 m/s$
Molar volume ($T_0 = 273.15K,\ P_0 = 10^5 Pa$)	$V_m = RT_0/P_0 = 2.2711 \times 10^{-2} m^3/mol$
Permeability of vacuum	$\mu_0 = 4\pi \times 10^{-7} N/A^2 = 1.25664 \times 10^{-6} N/A^2$
Planck's constant	$h = 6.6261 \times 10^{-34} Js$; $(4.1357 \times 10^{-15} eVs)$
Planck's reduced constant	$\hbar = h/2\pi = 1.0546 \times 10^{-34} Js$
Quantum of circulation	$h/m_e = 7.2739 \times 10^{-4} m^2/s$
Rydberg (H-atom) energy	$e^2/8\pi\epsilon_0 a_0 = 2.17987 \times 10^{-18} J$ (13.606 eV)

Stefan Boltzmann constant	$\sigma = 5.6705 \times 10^{-8}\ \mathrm{w/m^2K^4}$
Wien displacement law	$b = \lambda_{max} T = 2.8978 \times 10^{-3}\mathrm{mK}$

Electromagnetic Constants

Bohr magneton	$\mu_B = e\hbar/2m_e = 9.27402 \times 10^{-24}\mathrm{J/T}$
Nuclear magneton	$\mu_N = 5.0508 \times 10^{-27}\mathrm{J/T}$
Faraday constant	$F = N_A e = 96485\mathrm{C/mol}$
Flux quantum, magnetic	$\Phi_0 = h/2e = 2.0678 \times 10^{-15}\mathrm{Wb}$
***g*-factor, electron**	$g_e = 2.00232$
Hall resistance	$R_H = h/e^2 = 25,813\Omega$
Josephson frequency	$\omega_J = 2\pi(2eV/h)\ [1\mu V = 483.60\mathrm{MHz}]$

Conversion Factors

Energy: $1\ \mathrm{eV} = 1.6022 \times 10^{-19}\mathrm{J} = 1.7827 \times 10^{-36}\mathrm{kg} = 2.4180 \times 10^{14}\mathrm{Hz}$
$= 8065.5\mathrm{cm}^{-1} = 1.0735 \times^{-9}\ \mathrm{u} = 1.1604 \times 10^4\mathrm{K}$

Temperature: $1\mathrm{K} = 1.3807 \times 10^{-23}\mathrm{J} = 1.5362 \times 10^{-40}\mathrm{kg}$
$= 2.0837 \times 10^{10}\mathrm{Hz} = 0.69504\mathrm{cm}^{-1} = 9.2511 \times 10^{-14}\mathrm{u}$
$= 8.6174 \times 10^{-5}\mathrm{eV}$

Length: $1\mathrm{m} = 100\ \mathrm{cm} = 10^{10}\mathrm{\AA}$; 1 marathon = 42.195 km

Magnetic field: $1\ \mathrm{T} = 1\ \mathrm{Wb/m^2} = 10^4\mathrm{G}$

Pressure: $1\ \mathrm{GPa} = 10\mathrm{kBar} = 7.5 \times 10^6\mathrm{Torr} = 0.987 \times 10^4\mathrm{atm}$

Submultiples: m(milli, 10^{-3}); μ(micro, 10^{-6}); n(nano, 10^{-9}); p(pico, 10^{-12}); f(femto, 10^{-15}); a (atto, 10^{-18}); z(zepto, 10^{-21}); y(yocto, 10^{-24})

Multiples: k(kilo, 10^3); M(mega, 10^6); G(giga, 10^9); T(tera, 10^{12}); P(peta, 10^{15}); E(exa, 10^{18}); z(zetta, 10^{21}); y(yotta, 10^{24})

Length

Planck length	$\ell_p = (\hbar G/c^3)^{\frac{1}{2}} = 1.6161 \times 10^{-35}\mathrm{m}$
Classical electron radius	$r_e = \alpha^2 a_0 = e^2/4\pi\epsilon_0 m_e c^2 = 2.8179 \times 10^{-15}\mathrm{m}$
Thomson cross section	$\sigma = (8\pi/3)\ r_e^2 = 6.6525 \times 10^{-29}\mathrm{m}^2$
Compton wavelength (electron)	$\lambda_c = h/m_e c = 2.4263 \times 10^{-12}\mathrm{m}$
Bohr radius	$a_0 = r_e/\alpha^2 = 4\pi\epsilon_0\hbar^2/me^2 = 5.2918 \times 10^{-11}\mathrm{m}$
Sun radius	$R_s = 6.96 \times 10^8\mathrm{m}$

Earth–sun distance	$R_{es} = 1.496 \times 10^{11}$m
Light year	$ly = 9.4607 \times 10^{15}$m
Parsec	$pc = 3.0857 \times 10^{16}$m
Earth to nearest star	4.1×10^{16}m
Earth to center of galaxy	$\sim 3 \times 10^{20}$m
Earth to Andromeda	2.1×10^{22}m
Approximate radius of universe	$\sim 10^{26}$m

Time

Planck time	$t_p = \left(\hbar G/c^5\right)^{\frac{1}{2}} = 5.3906 \times 10^{-44}$sec
Strong force interaction time	$\sim 10^{-23}$sec
Electromagnetic force interaction time	$\sim 10^{-19}$sec
Weak force interaction time	$\sim 10^{-11}$sec
Sun's light to reach earth	499 sec
Revolution of earth (1 year)	3.1557×10^{7}sec
Light to reach nearest star	1.4×10^{8}sec
Light to reach Andromeda	7×10^{13}sec
Age of universe	$\sim 10^{17}$sec

Particle Properties

Particle	Mass ($\times 10^{-31}$kg)	Rest energy ($\times 10^{-13}$J)	Rest energy (MeV)	Magnetic moment ($\times 10^{-26}$J/T)
Electron	9.10939	0.81871	0.51100	928.48
Muon	1883.53	169.29	105.66	4.4905
Proton	16726.2	1503.3	938.27	1.4106
Neutron	16749.3	1505.4	939.57	0.99624
Deuteron	33435.9	3005.0	1875.6	0.43307
Atomic constant (1/12 of ^{12}C)	16605.4	1492.4	931.494	—

THE PHYSICS HANDBOOK

CHAPTER 1

LAGRANGIAN AND HAMILTONIAN FORMULATIONS

1 INTRODUCTION

In this chapter we outline several approaches to treating problems of dynamics in classical mechanics, i.e., problems concerning motion in which forces are involved. We examine the force and the energy approaches, then discuss the Lagrangian and Hamiltonian formulations, and finally point out some variational approaches. Explicit expressions are given for the Lagrangians, Hamiltonians, and canonical momenta of various commonly encountered systems.

2 NEWTON'S LAW APPROACH

The force approach to non-relativistic dynamics makes use of Newton's second law which states that the force **F** on a body equals its rate of change of momentum $\dot{\mathbf{p}}$

$$\mathbf{F} = \dot{\mathbf{p}} = \frac{d\mathbf{p}}{dt} \tag{1}$$

Any reference frame in which this law holds is called an inertial frame. In one dimension the momentum for a particle of mass m and velocity v is $p = mv$, and Eq. (1) becomes

$$F = ma = m\frac{dv}{dt} = m\frac{d^2x}{dt^2} \tag{2}$$

where x is the position coordinate and a is the acceleration. The more general expression (1) must be used when the mass changes, as in nuclear decay problems involving the creation and annihilation of particles.

In dynamics we encounter several types of forces, such as:

$F = mg$	gravity near the earth's surface	(3a)
$\mathbf{F} = Gmm'\hat{\mathbf{r}}/r^2$	Newton's law of gravity	(3b)
$\mathbf{F} = qq'\hat{\mathbf{r}}/4\pi\varepsilon_0 r^2$	Coulomb's law	(3c)
$F = -k(x - x_0)$	harmonic restoring force from the equilibrium position x_0	(3d)
$F = -\mu N$	static or kinetic friction	(3e)
$F = -k\|v\|^n$	frictional force, often $n = 1$ as in Stokes' law $F = 6\pi\eta rv$ for streamline flow, or $n = 2$ with turbulence	(3f)
$\mathbf{F} = q(\mathbf{E} + \mathbf{v} \times \mathbf{B})$	Lorentz force on the charge q	(3g)

Ordinarily it is easier to solve non-relativistic elementary mechanics problems by balancing the energy at the beginning (b) and at the end (e) of the interaction. For a many-particle system in which each particle i has the kinetic energy $\frac{1}{2}m_i v_i^2$ and the potential energy V_i, we can write

$$\sum \tfrac{1}{2}m_i v_{ib}^2 + \sum V_{ib} = \sum \tfrac{1}{2}m_i v_{ie}^2 + \sum V_{ie} \tag{4}$$

where the summations are over the particles of the system. If the particles interact with each other, then we can add double summations over the initial and final interaction energies V_{ij}^b and V_{ij}^e to this equation. This energy approach is valid for conservative systems for which there is no dissipation.

3 LAGRANGIAN FORMULATION

In the Langrangian approach the kinetic and potential energies are expressed in terms of generalized coordinates q_i and the velocities $\dot{q}_i$ of each particle i. The Lagrangian itself, $L(q_1, \ldots, q_N, \dot{q}_1, \ldots, \dot{q}_n, t)$, is the difference between the total kinetic energy T and the total potential energy V of the system

$$L = T - V \tag{5}$$

Examples of some Lagrangians are:

$$L = \tfrac{1}{2}m\dot{x}^2 - \tfrac{1}{2}k(x - x_0)^2 \qquad \text{harmonic oscillator} \tag{6a}$$

$$\left.\begin{aligned} L &= -mc^2/\gamma - q\phi + q\mathbf{A}\cdot\mathbf{v} \\ L &\approx \tfrac{1}{2}mv^2 - q\phi + q\mathbf{A}\cdot\mathbf{v} - mc^2 \end{aligned}\right\} \quad \text{charge } q \text{ in electromagnetic fields } [\gamma = (1-\beta^2)^{-1/2}] \tag{6b}$$

$$L = \tfrac{1}{2}m(\dot{r}^2 + r^2\dot{\theta}^2) + k/r \qquad \text{Kepler problem of the earth in orbit around the sun} \tag{6c}$$

$$L = \tfrac{1}{2}I_\perp(\dot{\theta}^2 + \dot{\phi}^2 \sin^2\theta) + \tfrac{1}{2}I_\parallel(\dot{\psi} + \dot{\phi}\cos\theta)^2 - mgl\cos\theta \tag{6d}$$

Symmetric top with one point fixed using Euler angles
$\psi =$ rotation about top axis
$\phi =$ precession about vertical
$\theta =$ angle of inclination from vertical

In writing Lagrangians it is good to keep in mind the following details about three commonly used coordinate systems with their respective differential lengths $ds_i = h_i dq_i$.

$$\text{cartesian:} \quad \begin{matrix} x, y, \\ dx, dy, dz \end{matrix} \tag{7a}$$

$$\text{cylindrical:} \quad \begin{matrix} \rho, \phi, z \\ d\rho, \rho d\phi, dz \end{matrix} \tag{7b}$$

$$\text{spherical:} \quad \begin{matrix} r, \theta, \phi \\ dr, rd\theta, r\sin\theta d\phi \end{matrix} \tag{7c}$$

Lagrangians are important because, by Hamilton's principle, the motion from time t_1 to time t_2 follows a path that makes the line integral

$$I = \int_{t_1}^{t_2} Ldt \tag{8}$$

a stationary value. Using this principle we can show that Lagrange's equation

$$\frac{d}{dt}\left(\frac{\partial L}{\partial \dot{q}_i}\right) - \frac{\partial L}{\partial q_i} = 0 \quad i = 1, 2, \ldots, N \tag{9}$$

is satisfied for each of the N coordinates q_i and their velocities $\dot{q}_i$.

The motion is often limited by constraints. Of particular interest is a holonomic constraint, which can be expressed in terms of equations involving the positions $\mathbf{r}_i$ of the particles, but not the velocities

$$f(\mathbf{r}_1, \ldots, \mathbf{r}_N, t) = 0 \tag{10}$$

If the constraint contains an inequality rather than an equal sign, such as the condition $r^2 - a^2 \geq 0$ for being outside a sphere of radius a, then it is not holonomic. The differential df of the function $f(\mathbf{q}_i, t)$

$$df = \sum \frac{\partial f}{\partial q_i} dq_i + \frac{\partial f}{\partial t} dt = 0 \tag{11}$$

where the summation is over the N particles, may be written in the form

$$\sum a_i dq_i + a_t dt = 0 \tag{12}$$

where $a_i = \partial f / \partial q_i$. If there is more than one constraint equation, then we add an index, writing a_{ji} and a_{jt}. The coefficients a_{ji} of the n constraint equations written as first-order differential equations

$$\sum a_{ji}\dot{q}_i + a_{jt} = 0 \quad j = 1, 2, \ldots, n \tag{13}$$

enter the N Lagrangian equations

$$\frac{d}{dt}\left(\frac{\partial L}{\partial \dot{q}_i}\right) - \frac{\partial L}{\partial q_i} = \sum a_{ji}\lambda_j \quad i = 1, 2, \ldots, N \tag{14}$$

through the n Lagrange multipliers λ_j, where the summation over j is from 1 to n. These are now $N + n$ equations to solve and n unknown Lagrange multipliers to be determined. Each λ_j usually has the dimension of force or torque, and in a typical case it is a reaction force.

Sometimes the constraint can be written in the form of Eq. (12) or (13) even though no function $f(q, t)$ exists. Such a constraint might be called pseudoholonomic; it is not holonomic but it is adequate for the application of the Lagrange multiplier method (14).

Sometimes friction can be taken into account with the aid of a velocity dependent Rayleigh dissipation function D

$$D = \tfrac{1}{2}\sum (k_x v_{ix}^2 + k_y v_{iy}^2 + k_z v_{iz}^2) \tag{15}$$

associated with the frictional force $\mathbf{F}_f$

$$\mathbf{F}_f = -\nabla_v D \tag{16}$$

where the gradient is with respect to the velocities. The associated Lagrange equations are

$$\frac{d}{dt}\left(\frac{\partial L}{\partial \dot{q}_i}\right) - \frac{\partial L}{\partial q_i} + \frac{\partial D}{\partial \dot{q}_i} = 0 \quad i = 1, 2, \ldots, N \tag{17}$$

The Rayleigh dissipation function corresponds to the force of Eq. (3f) with $n = 1$.

Each coordinate q_j has a conjugate momentum p_j defined by the expression

$$p_j = \frac{\partial L}{\partial \dot{q}_j} \tag{18}$$

Sometimes the conjugate momentum is a linear momentum, as $m\dot{x}$, or an angular momentum, as $I\dot{\theta}$, where I is the moment of inertia, but it can also be more complicated than that, as in the case of a charge q moving in the presence of a magnetic field $\mathbf{B} = \nabla \times \mathbf{A}$. Several conjugate momenta are:

$$p_i = \gamma m v_i + qA_i \qquad \text{charge } q \text{ in an electromagnetic field} \tag{19a}$$

$$\left.\begin{aligned} p_r &= m\dot{r} \\ p_\theta &= mr^2\dot{\theta} \end{aligned}\right\} \qquad \text{Kepler problem} \tag{19b}$$

$$\left.\begin{aligned} p_\theta &= I_\perp \dot{\theta} \\ p_\psi &= I_\parallel(\dot{\psi} + \dot{\phi}\cos\theta) \\ p_\phi &= (I_\parallel \cos^2\theta + I_\perp \sin^2\theta)\dot{\phi} + I_\parallel \dot{\psi}\cos\theta \end{aligned}\right\} \quad \text{symmetric top with one one point fixed in Euler angles} \tag{19c}$$

4 HAMILTONIAN FORMULATION

We have discussed the properties of the Lagrangian $L(q_i, \dot{q}_i, t)$, which is a function of the generalized coordinates and velocities. There is another function, called the Hamiltonian $\mathcal{H}(q_i, p_i, t)$, which depends on the coordinates q_i and their associated conjugate momenta p_i defined by Eq. (18), and it is formed from the Lagrangian through the following Legendre transformation

$$\mathcal{H}(q_i, p_i, t) = \sum \dot{q}_i p_i - L(q_i, \dot{q}_i, t) \tag{20}$$

where the summation is over the coordinates. To carry out this transformation each conjugate momentum p_i is determined by differentiating the Lagrangian using Eq. (18). Then the resulting equations are solved for the generalized velocities $\dot{q}_i$ and the latter are put into Eq. (20). This provides the Hamiltonian $\mathcal{H}$ expressed as a function of the q_i, p_i and t variables.

Sometimes this procedure is easy, as with the angle of inclination θ of the symmetric top (19c) for which we have

$$\dot{\theta} = p_\theta / I_\parallel \tag{21}$$

It is more complicated to determine the other two symmetric top angles ψ and ϕ which require the solution of two simultaneous equations.

The equations of motion are found from the canonical equations of Hamilton

$$\dot{q}_i = \frac{\partial \mathcal{H}}{\partial p_i} \tag{22a}$$

$$\dot{p}_i = -\frac{\partial \mathcal{H}}{\partial q_i} \tag{22b}$$

$$\frac{\partial L}{\partial t} = -\frac{\partial \mathcal{H}}{\partial t} \tag{22c}$$

These $2N$ first-order Hamilton differential equations replace the N second-order Lagrange equations (8).

Examples of some Hamiltonians are:

$$\mathcal{H} = \frac{p^2}{2m} + \tfrac{1}{2}k(x - x_0)^2 \quad \text{harmonic oscillator} \tag{23a}$$

$$\left.\begin{aligned} \mathcal{H} &= [(\mathbf{p} - q\mathbf{A})^2 c^2 + m^2 c^4]^{1/2} + q\phi \\ &\approx \frac{(\mathbf{p} - q\mathbf{A})^2}{2m} + q\phi + mc^2 \end{aligned}\right\} \quad \begin{array}{l}\text{charge } q \text{ in an} \\ \text{electromagnetic field}\end{array} \tag{23b}$$

$$\mathcal{H} = \frac{p_r^2}{2m} + \frac{p_\theta^2}{2mr^2} - \frac{k}{r} \quad \text{Kepler problem} \tag{23c}$$

$$\mathcal{H} = \frac{p_\theta^2}{2I_\perp} + \frac{(p_\phi - p_\psi \cos\theta)^2}{2I_\perp \sin^2\theta} + \mathrm{M}g\ell \cos\theta \quad \text{symmetric top} \tag{23d}$$

Coordinates that do not appear explicitly in the Lagrangian and Hamiltonian are called cyclic or ignorable. We see from Eqs. (22b) that the momentum p_i conjugate to such a coordinate is a constant of the motion, independent of the time, which we will denote by α_i

$$\dot{p}_i = 0 \tag{24}$$

$$p_i = \alpha_i \tag{25}$$

Because of this it is ordinarily easier to use the Hamiltonian formulation for writing the equations of motion for cyclic coordinates, and the Lagrangian approach for the non-cyclic coordinates.

When some coordinates $q_i, \ldots, q_c$ are present explicitly in the Lagrangian and the remainder $q_{c+1}, \ldots, q_N$ are cyclic, then it is convenient to restrict the term $\sum q_i p_i$ of Eq. (20) to a summation over the cyclic coordinates and thereby form the Routhian R

$$R = R(q_1, \ldots, q_N, \dot{q}_i, \ldots, \dot{q}_c, p_{c+1}, \ldots, p_N) \tag{26}$$

which is a function of the non-cyclic velocities $\dot{q}_i$ and the constant momenta p_i. This permits Hamilton's equations to be written for the cyclic coordinates and Lagrange's equations for the non-cyclic ones. For example, the Kepler problem has the cyclic coordinate θ and hence from Eqs. (6c) and (19b) the Routhian $R(r, \theta, \dot{r}, p_\theta)$ is given by

$$R(r, \theta, \dot{r}, p_\theta) = \tfrac{1}{2}m\dot{r}^2 + p_\theta^2/2mr^2 - k/r \tag{27}$$

where, since θ is cyclic, the angular momentum $p_\theta = mr^2\dot{\theta}$ is constant in time.

5 VARIATIONAL PRINCIPLES AND VIRTUAL DISPLACEMENTS

There are several general principles associated with the subject of mechanics. We have already encountered Hamilton's principle (8), which states that the line integral of the Lagrangian over time has a stationary value for the correct path of motion. This can be expressed differently by saying that the variation of the line integral I of Eq. (8) with fixed end points t_1 and t_2 is zero

$$\delta I = \delta \int_{t_1}^{t_2} L dt \tag{28}$$

In other words, as infinitesimal deviation in the path of integration about its stationary trajectory does not change the value of the integral.

Now let us consider infinitesimal displacements $\delta \mathbf{r}_i$ in the presence of an applied force $\mathbf{F}^{\text{app}}$ which is holding a system in equilibrium. The principle of virtual work states that the work done by the applied force, called the virtual work, is zero

$$\sum \mathbf{F}_i^{\text{app}} \cdot \delta \mathbf{r}_i = 0 \tag{29}$$

and this provides the condition for equilibrium in statics. In dynamics when the applied forces can cause accelerations $\ddot{x} = \dot{p}_x / m$, we have D'Alembert's principle

$$\sum (\mathbf{F}_i^{\text{app}} - \dot{p}_i) \cdot \delta r_i = 0 \tag{30}$$

which reduces to the principle of virtual work for static conditions.

CHAPTER 2

CENTRAL FORCES

1 INTRODUCTION

In this chapter we begin by commenting on mutual interactions between particles, then we examine the Kepler problem of closed orbit motion under an inverse square law force. This is followed by a discussion of open orbits and scattering, after which we make some concluding remarks on the three-body problem.

2 ACTION AND REACTION

When two particles exert forces on each other the weak law of action and reaction states that these forces are equal in magnitude and opposite in direction. Particle m_1 exerts the force $\mathbf{F}_{12}$ on m_2 and particle m_2 exerts $\mathbf{F}_{21}$ on m_1 such that

$$\mathbf{F}_{12} + \mathbf{F}_{21} = 0 \tag{1}$$

and from Newton's first law we have

$$m_1\mathbf{a}_1 + m_2\mathbf{a}_2 = 0 \tag{2}$$

where $\mathbf{a}$ denotes the acceleration. If this equation is integrated over a short time interval Δt, during which the forces do not change, then we obtain

$$m_1\mathbf{v}_1 + m_2\mathbf{v}_2 = \text{const} \tag{3}$$

where the constant might be, for example, the linear momentum of the center of mass of the two particles. Thus, the linear momentum of the two particles is conserved. The law of the conservation of linear momentum states that if the net external force is zero, then the linear momentum of a system of particles is conserved. When all of the internal forces between particles in a system obey Eq. (1) then, in the absence of externally applied forces, the total linear momentum of the system is constant in time.

If the forces $\mathbf{F}_{12}$ and $\mathbf{F}_{21}$ are not collinear, then the masses m_1 and m_2 acquire an angular momentum relative to each other. Hence, the weak law is not sufficient to ensure the conservation of angular momentum. For this it is necessary for the forces between particles in the system to be not only equal in magnitude and opposite in direction, but in addition to lie along the straight line connecting the particles. Such forces are called central forces. When all of the internal forces in a system are central, then in the absence of externally applied torques the total angular momentum, as well as the total linear momentum, is constant in time.

Moving charges can exert forces on each other which not only are non-central, but which differ in magnitude and direction. A charge q_1 moving at a relativistic speed carries along an electric field $\mathbf{E}_1$ that is stronger in magnitude in the transverse direction than it is in the forward or backward direction, and in addition there are encircling magnetic field lines $\mathbf{B}_1$. A second moving charge q_2 interacts with the electric and magnetic fields of the first charge through the Lorentz force $\mathbf{F}_{12}$

$$\mathbf{F}_{12} = q_2(\mathbf{E}_1 + \mathbf{v}_2 \times \mathbf{B}_1) \tag{4a}$$

and the force $\mathbf{F}_{21}$ of the second charge on the first

$$\mathbf{F}_{21} = q_1(\mathbf{E}_2 + \mathbf{v}_1 \times \mathbf{B}_2) \tag{4b}$$

is, in general, different in magnitude and direction from its counterpart of Eq. (4a). Neither linear nor angular momentum need be conserved! We see in Chapter 13 that in a volume containing charges, currents, and electromagnetic fields the rate of change of the mechanical momentum owing to the motion of the charged particles plus the rate of change of the momentum involving the fields themselves is equal to the forces exterted across the surface enclosing the volume.

3 TWO-PARTICLE CENTRAL FORCE MOTION

In mechanics we deal with central forces arising, for example, from Newton's law of gravitation acting between two masses m_1 and m_2, which we write in scalar form as follows:

$$\mathbf{F} = -G\frac{m_1 m_2}{|\mathbf{r}_1 - \mathbf{r}_2|^2} \tag{5}$$

We see from Eq. (5) that such a central force $\mathbf{F}$

$$\mathbf{F}(|\mathbf{r}_1 - \mathbf{r}_2|) = \mathbf{F}(r) \tag{6}$$

and its associated scalar potential V, where $\mathbf{F} = -\nabla V$

$$V(|\mathbf{r}_1 - \mathbf{r}_2|) = V(r) = \frac{Gmm'}{r} \tag{7}$$

is a function from Eq. (5) of the magnitude r of the vector distance $\mathbf{r}$

$$r = |\mathbf{r}_1 - \mathbf{r}_2| \tag{8}$$

between the two masses that is defined in Fig. 2-1. Each mass m_i is acted upon by the force $\mathbf{F}_j$ exterted by the other mass m_j in accordance with Newton's first law

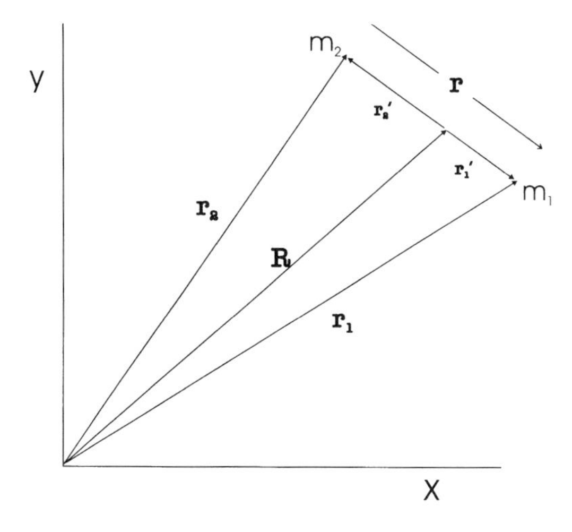

Fig. 2-1. Coordinates for the two-body problem.

$$\mathbf{F}_1(r) = m_2\ddot{\mathbf{r}}_2 \tag{9a}$$

$$\mathbf{F}_2(r) = m_1\ddot{\mathbf{r}}_1 \tag{9b}$$

where from Eq. (1) $\mathbf{F}_1(r) = -\mathbf{F}_2(r)$. These two vector force equations (9) are not very useful since we are interested in the radial and angular displacements of the two masses relative to each other, and one cannot write these expressions (9) directly in a form that will provide this information. We therefore take a different approach to this central force problem.

The locations of the masses can be expressed in terms of their position vectors $\mathbf{r}_1'$ and $\mathbf{r}_2'$ relative to the center of mass and the position vector $\mathbf{R}$ of the center of mass, as defined in Fig. 2-1. Since there are no external forces acting, and the internal force between the particles is central, the angular momentum $\mathbf{L}$ is conserved, and the motion relative to the center of mass lies in a plane perpendicular to $\mathbf{L}$.

$$\mathbf{L} = \mathbf{r} \times \mathbf{p} \tag{10}$$

We also note that

$$\mathbf{r} = \mathbf{r}_1' - \mathbf{r}_2' \tag{11}$$

since $\mathbf{r}_1'$ and $\mathbf{r}_2'$ are in opposite directions.

The Lagrangian that is given by the following expression:

$$L = \tfrac{1}{2}(m_1 + m_2)\dot{\mathbf{R}}^2 + \tfrac{1}{2}m_1\dot{r}_1'^2 + \frac{1}{2}m_2\dot{r}_2'^2 - V(|\mathbf{r}_1' - \mathbf{r}_2'|) \tag{12}$$

in terms of $\mathbf{r}_1'$ and $\mathbf{r}_2'$ is easily transformed to the more convenient position vectors $\mathbf{R}$ and $\mathbf{r}$ to give

$$L = \tfrac{1}{2}(m_1 + m_2)\dot{\mathbf{R}}^2 + \tfrac{1}{2}\frac{m_1 m_2}{m_1 + m_2}\dot{r}^2 - V(r) \tag{13}$$

We see that the center of mass coordinates are cyclic, and hence the center of mass momentum is constant, so

$$\dot{\mathbf{R}} = \mathbf{v}_{\mathrm{CM}} = \mathrm{const} \tag{14}$$

Only the relative coordinates need concern us.

Since the motion lies in a plane, the Lagrangian can be written in terms of polar coordinates r, Θ

$$L = \tfrac{1}{2}m(\dot{r}^2 + r^2\dot{\theta}^2) - V(r) \tag{15}$$

and the Hamiltonian is

$$\mathcal{H} = \frac{p_r^2}{2m} + \frac{p_\theta^2}{2mr^2} + V(r) \tag{16}$$

We see that θ is cyclic, and the canonical momenta are

$$p_r = m\dot{r} \tag{17}$$

$$p_\theta = mr^2\dot{\theta} = L = \mathrm{const} \tag{18}$$

where one should not confuse the use of the symbol L for the angular momentum and for the Lagrangian. Equation (18), the conservation of a real speed $\frac{1}{2}r^2 d\theta/dt$, is Kepler's second law of planetary motion, and also a statement that the angular momentum is conserved.

4 EQUATIONS AND ORBITS OF THE MOTION

Both Lagrange's equation and Hamilton's equation for the radial coordinate give the following equation of motion:

$$m\ddot{r} - \frac{p_\theta^2}{mr^3} = f(r) \tag{19}$$

where the radial force $f(r)$ is given by

$$f(r) = -\frac{dV}{dr} \tag{20}$$

Equation (19) can be written

$$m\ddot{r} + \frac{d}{dr}\left(V(r) + \frac{p_\theta^2}{2mr^2}\right) = 0 \tag{21}$$

and using the relation

$$\frac{dg(r)}{dt} = \frac{dg(r)}{dr}\frac{dr}{dt} = \frac{dg(r)}{dt}\dot{r} \tag{22}$$

this can be expressed in a form easy to integrate

$$\frac{d}{dt}(\tfrac{1}{2}m\dot{r}^2) + \frac{d}{dt}\left(V(r) + \frac{p_\theta^2}{2mr^2}\right) = 0 \tag{23}$$

where the term in parentheses is the effective potential energy V'

$$V' = V + \frac{L^2}{2mr^2} \tag{24}$$

which is plotted in Fig. 2-2. Using the change of variable $u = /lr$ and the expression $Ldt = mr^2 d\theta$ from Eq. (18) we can derive the equation for the orbit

$$\frac{d^2u}{d\theta^2} + u = -\frac{m}{L^2}\frac{d}{du}V(1/u) \tag{25}$$

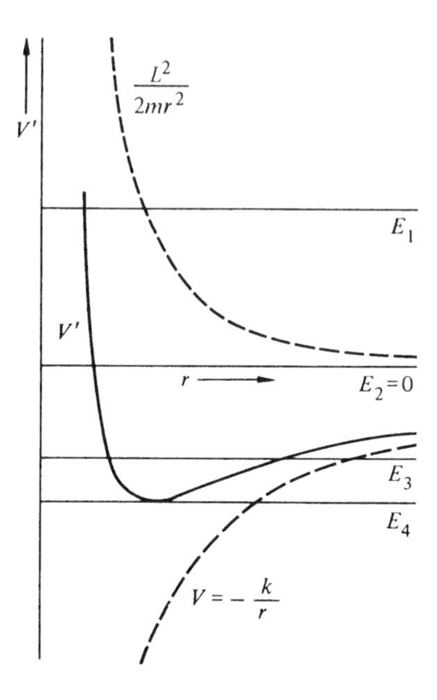

Fig. 2-2. Coulomb ($-k/r$), centrifugal ($L^2/2mr^2$) and equivalent one-dimensional ($V'(r)$) potentials for an inverse square law of attraction. The energies for hyperbolic (E_1), parabolic (E_2), elliptic (E_3), and circular (E_4) orbits are shown. (From H. Goldstein, *Classical Mechanics*, Addison-Wesley, Massachusetts, 1980, p. 77.)

The system is conservative so the energy E is a constant of the motion

$$E = \tfrac{1}{2}m\dot{r}^2 + \frac{L^2}{2mr^2} + V(r) \tag{26}$$

and this expression can be solved for $\dot{r}$ and integrated to find the position as a function of the time.

The orbit obtained from solving Eq. (26) is a plot of θ versus r, and constitutes the trajectory followed by the particle in coordinate space. For the gravitational case (7), $V = -k/r$, the orbits are conic sections. For a negative total energy the orbit is closed with an elliptical or circular shape, the former shown in Fig. 2-3. For positive energies it is a hyperbola, and for zero energy it is a parabola, both open orbit cases. There is a minimum energy $E_{\text{min}} = -mk^2/2L^2$ below which no solution exists. The orbits are distinguished by their eccentricity e which depends on the energy and angular momentum as follows:

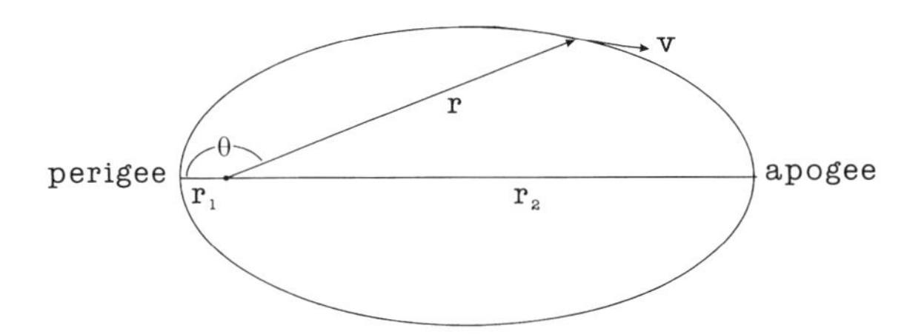

Fig. 2-3. Elliptic orbit expressed in polar coordinates (r, θ) showing the point of closest approach (perigee, $r = r_1$) and furthest approach (apogee, $r = r_2$).

$$e = 1 + \frac{2EL^2}{mk^2} \tag{27}$$

and the orbit classification is

$$e > 1 \qquad E > 0 \quad \text{hyperbola} \tag{28a}$$
$$e = 1 \qquad E = 0 \quad \text{parabola} \tag{28b}$$
$$0 < e < 1 \qquad E < 0 \quad \text{ellipse} \tag{28c}$$
$$e = 0 \qquad E = \frac{mk^2}{-2L^2} \quad \text{circle} \tag{28d}$$

The eccentricity e can be expressed in terms of the semimajor and semiminor axes a and b respectively, and the shortest distance r_1 (perihelon) and longest distance r_2 (aphelion) from a focal point, as follows:

$$e = 1 - \frac{b^2}{a^2} \tag{29a}$$
$$= 1 - (r_1/a) \tag{29b}$$
$$= (r_2/a) - 1 \tag{29c}$$

where the semimajor axis depends only on the energy

$$a = \tfrac{1}{2}(r_1 + r_2) = k/(-2E) \tag{30}$$

An elliptic orbit in polar coordinates is given by

$$r = \frac{a(1 - e^2)}{1 + e\cos\theta} \tag{31}$$

where r is the distance from a focal point, and in cartesian coordinates it has the form

$$\frac{x^2}{a^2} + \frac{y^2}{b^2} = 1 \tag{32}$$

where the center of the coordinate system is now at the center of the ellipse. Figures 2-3 and 2-4 define the quantities r, θ, a, b, r_1, r_2. A circle is the special case of an ellipse with

$$a = b = r_1 = r_2 \tag{33}$$

and eccentricity $e = 0$.

Expressions similar to those for an ellipse can be written for the other conic sections, and for a hyperbola we have

$$r = \frac{a(e^2 - 1)}{e\cos\theta + 1} \tag{34}$$

This has the nearest approach distance $r_0 = a(e - 1)$ at the angle $\theta = 0$, as shown in Fig. 2-5, and the angle θ_0 given by

$$\cos\theta_0 = 1/e \tag{35}$$

provides the asymptote for $r \Rightarrow \infty$.

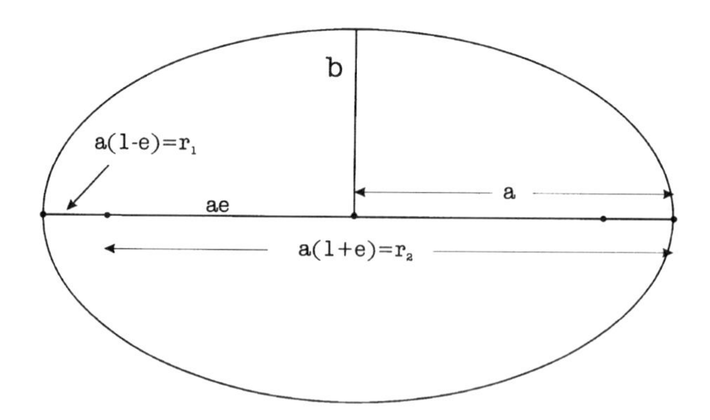

Fig. 2-4. Parameters of the elliptic orbit of Fig. 2-3 expressed in terms of the eccentricity e and the semimajor and semiminor axes a and b, respectively.

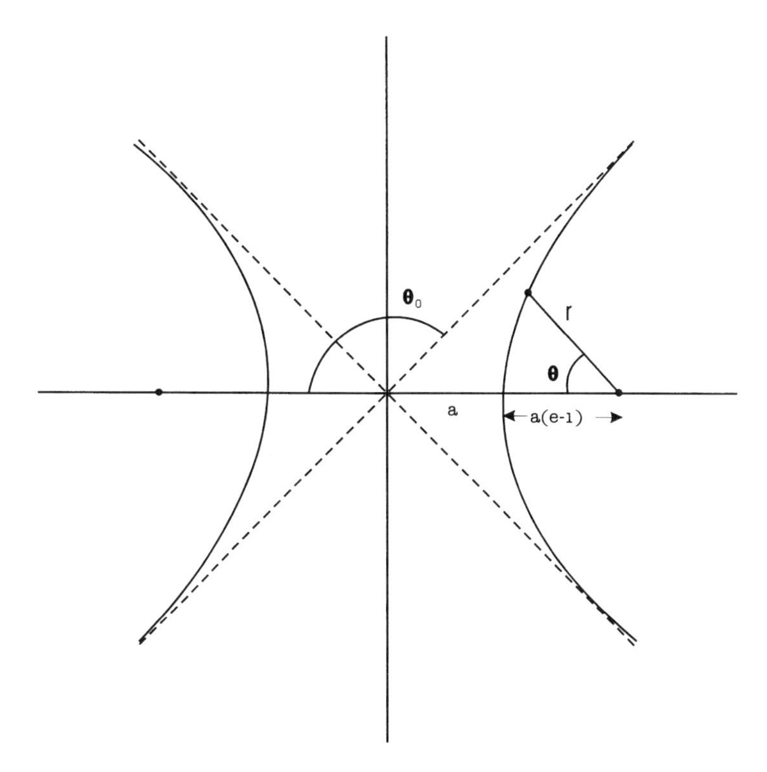

Fig. 2-5. Hyperbolic orbit (—— on the right) showing the asymptotic behavior (- - - -).

5 BOUND AND CLOSED ORBITS

For many attractive force laws there are ranges of energy over which the orbits are bounded, i.e., r remains within a range of values

$$r_1 \leq r \leq r_2 \tag{36}$$

as in the elliptic case. If the frequencies of the radial and angular motion are commensurate, which means that their ratio ω_r/ω_θ is a ratio of integers, then the orbit will be closed, i.e., it will repeat itself or retrace the same path continuously. Figure 2-6 shows samples of the cases $\omega_r/\omega_\theta = \frac{1}{2}$, 1 and 2. The Kepler problem with the inverse square law of force is, of course, the case $\omega_r/\omega_\theta = 1$.

If the frequencies are not commensurate, then the orbit will eventually fill the entire annular region between r_1 and r_2 without ever retracing its steps,

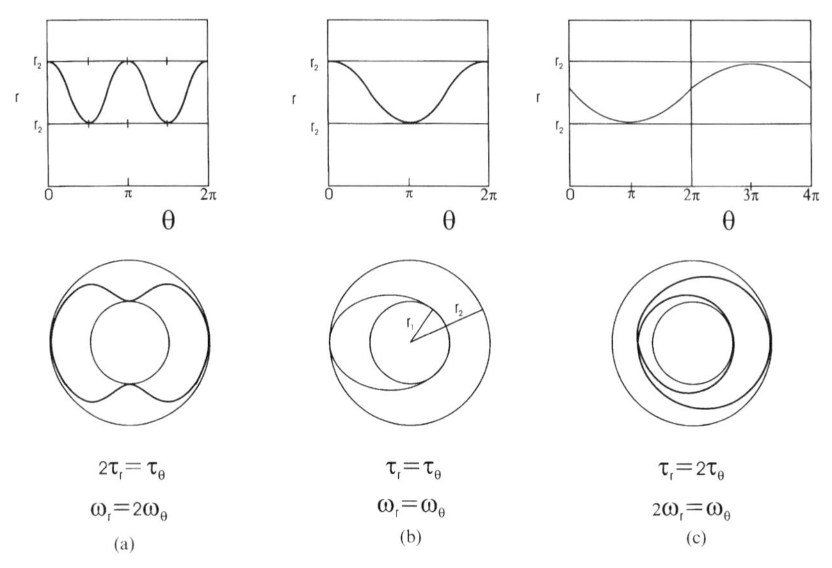

Fig. 2-6. Commensurate closed orbits plotted in polar coordinates (top) and in coordinate space (bottom). Cases are shown for the radial frequency equal to twice the azimuthal frequency (a), for equal frequencies (b) and for the azimuthal frequency equal to twice the radial frequency (c).

as illustrated in Fig. 2-7(a). Figure 2-7(b) shows an example of a slightly incommensurate case that can occur, for example, when the orbit of a planet is perturbed ($\tau_r \approx \tau_\Theta$). In 1873, J. Bertrand showed that the inverse square law $V = -k/r$ and Hooke's law for harmonic motion $V = \frac{1}{2}kr^2$ are the only cases that give closed orbits. Other force laws do give bounded orbits, i.e., ones that are closed but not commensurate.

The virial theorem is a statistical theorem involving time averages over mechanical quantities for a system of particles. We avoid its complications and begin by quoting the time average kinetic energy $\langle T \rangle$ of a system of particles

$$\langle T \rangle = -\tfrac{1}{2}\langle \sum \mathbf{F}_i \cdot \mathbf{r}_i \rangle \tag{37}$$

This is known as the virial theorem, and the summation average on the right-hand side is called the virial of Clausius. The virial theorem is applicable to periodic motion, but it can also be applied to non-periodic motion if the coordinates and velocities remain finite so there is an upper bound on

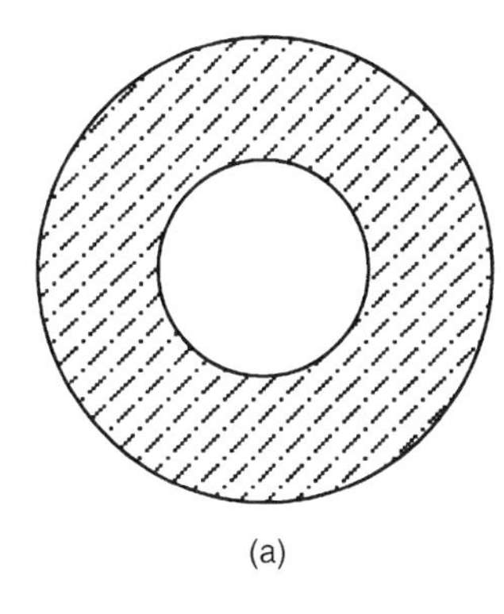

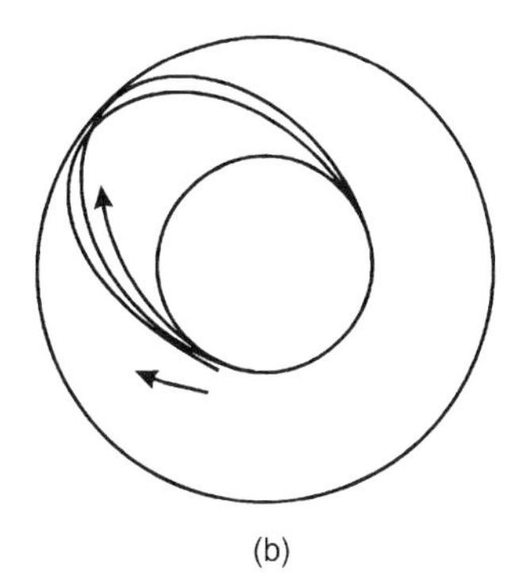

Fig. 2-7. Bounded orbits showing (a) incommensurate case with the orbit completely filling the space between r_1 and r_2, and (b) slightly perturbed case in which the orbit precesses.

the virial of Clausius. The theorem can, for example, be used to derive the ideal gas law.

For a power law potential

$$V(r) = ar^{n+1} \tag{38}$$

the viral theorem has the particularly simple form

$$\langle T \rangle = \tfrac{1}{2}(n+1)\langle V \rangle \tag{39}$$

which reduces to

$$\langle T \rangle = -\tfrac{1}{2}\langle V \rangle \qquad n = -2 \tag{40a}$$

$$\langle T \rangle = \langle V \rangle \qquad n = +1 \tag{40b}$$

for the inverse square law ($n = -2$) and harmonic oscillator ($n = +1$) cases.

6 OPEN ORBITS AND SCATTERING

Open orbits occur when incoming particles in a beam scatter off atoms in a target such as a metal foil, and are picked up far from the foil by a detector. If the detector were not in place, then the outgoing beam would recede to infinity. The kinetic energy of the incoming particles is so high that it far exceeds the potential energy and the orbits are, for the inverse square law case, hyperbolic. Very low incoming kinetic energies from sources at finite distances could lead to bound orbits.

Consider the case of an incoming beam with the intensity I particles per second per unit area perpendicular to the beam that passes by a scattering center with the impact parameter s defined in Fig. 2-8. The impact parameter is what would be the distance of closest approach if there were no deflection of the beam. The differential scattering cross-section $d\sigma/d\Omega$ in the center-of-mass coordinate system is

$$\frac{d\sigma(\Omega)}{d\Omega} = \begin{bmatrix} \text{particles scattered from } \Omega \text{ to } \Omega + d\Omega \\ \text{per unit time per incident intensity} \end{bmatrix} \tag{41}$$

where

$$d\Omega = 2\pi \sin \theta d\theta \tag{42}$$

The angular momentum L is given by

$$L = mv_0 s = s(2mE)^{1/2} \tag{43}$$

and we can write

$$2\pi Is|ds| = 2\pi I(d\sigma/d\Omega) \sin \Theta d\Theta \tag{44}$$

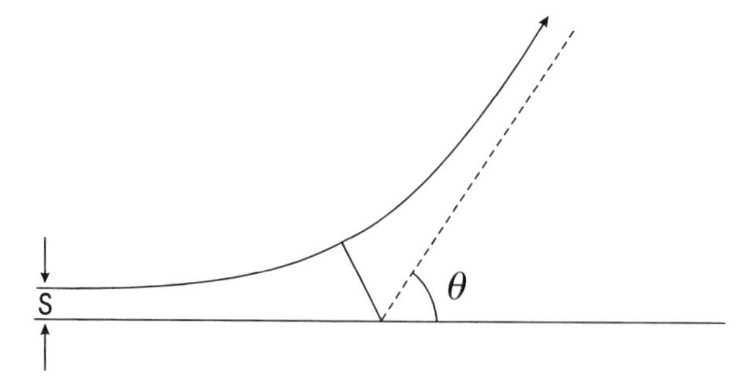

Fig. 2-8. Orbit for hyperbolic unbounded motion showing the impact parameter s and the scattering angle θ.

which gives for the differential cross-section

$$\frac{d\sigma(\theta)}{d\Omega} = \frac{s}{\sin\theta}\left|\frac{ds}{d\theta}\right| \tag{45}$$

For the inverse square law case we obtain the famous Rutherford cross-section for particles of charge Ze scattering off particles of charge $Z'e$

$$\frac{d\sigma(\theta)}{d\Omega} = \frac{1}{4}\left\{\frac{ZZ'e^2}{8\pi E_0 E}\right\}^2 \csc^4 \tfrac{1}{2}\theta \tag{46}$$

It is clear from Fig. 2-9 that the greater the impact parameter s the less the scattering angle θ. This result is for the equivalent one-body problem, corresponding to the case of a very small mass scattering off a very large one. The center of mass can be considered as located at the large particle, which remains effectively stationary.

In a more general case, the scattering angle θ_L in the laboratory system is related to the center of mass angle θ as follows

$$\tan\theta_L = \frac{\sin\theta}{\cos\theta + \rho} \tag{47}$$

where for an elastic collision we have

$$\rho = \frac{m_1}{m_2} \tag{48}$$

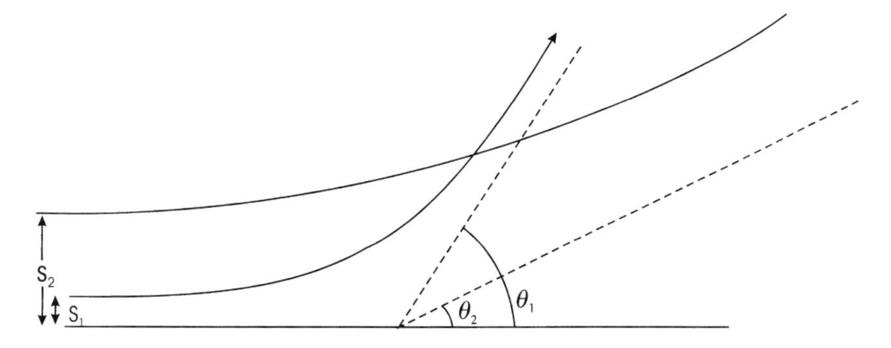

Fig. 2-9. Hyperbolic orbits with small and large impact parameters, s_1 and s_2, respectively, showing their corresponding large and small scattering angles θ_1 and θ_2.

This means that for a small incoming particle, $m_1 \ll m_2$ corresponding to the Rutherford case, the scattering angle is almost the same in both systems, $\theta_L \approx \theta$. For the opposite extreme, $m_1 \gg m_2$, we have $\theta_L \ll \theta$, so a large incoming particle m_1 undergoes very little deviation path, and hardly any measurable scattering occurs in the laboratory. When the collision is inelastic, then ρ depends on the Q-value that results from the conversion of some kinetic energy into heat.

7 THREE-BODY PROBLEM

For the two-body inverse square law case we found solutions involving motion in elliptic, parabolic and hyperbolic orbits, with closed orbits for the elliptic case. When one more mass is added we have the three-body problem, which has no known general solution. There are, however, particular alignments of the masses that simplify the equations of motion to such an extent that they can be solved in closed form. In addition, when two of the masses are large and the third is small, then a perturbation approach can be applied to obtain a solution.

The Newtonian three-body problem involves three masses, m_1, m_2 and m_3, at the respective positions $\mathbf{r}_1$, $\mathbf{r}_2$ and $\mathbf{r}_3$ interacting with each other via gravitational forces. The equation of motion of the first mass is

$$\ddot{\mathbf{r}}_1 = -Gm_2 \frac{\mathbf{r}_1 - \mathbf{r}_2}{|\mathbf{r}_1 - \mathbf{r}_2|^3} - Gm_3 \frac{\mathbf{r}_1 - \mathbf{r}_3}{|\mathbf{r}_1 - \mathbf{r}_3|^3} \tag{49}$$

and analogously for the other two masses. If we make use of the relative position vectors defined by

$$\mathbf{s}_i = \mathbf{r}_j - \mathbf{r}_k \tag{50}$$

in Fig. 2-10, where clearly

$$\mathbf{s}_1 + \mathbf{s}_2 + \mathbf{s}_3 = 0 \tag{51}$$

then the equations of motion assume the symmetrical form

$$\ddot{\mathbf{s}}_1 = -mG \frac{\mathbf{s}_1}{\mathbf{s}_1^3} + m_1 \mathbf{G} \tag{52}$$

where m is the sum of the three masses

$$m = m_1 + m_2 + m_3 \tag{53}$$

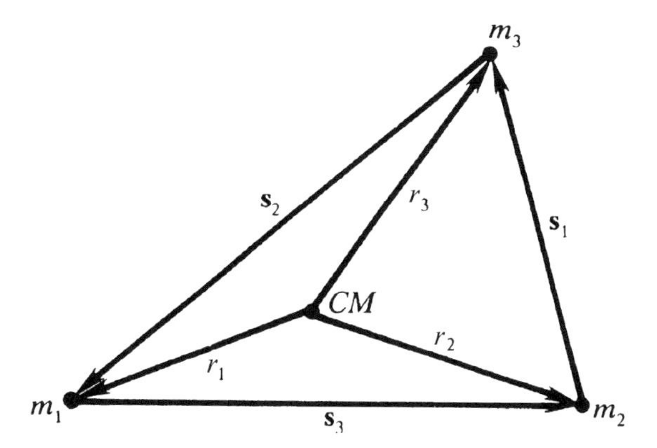

Fig. 2.10. Position coordinates r_1, r_2 and r_3, and relative coordinates $\mathbf{s}_1, \mathbf{s}_2$ and $\mathbf{s}_3$ of the masses m_1, m_2, and m_3 of the three-body problem. (From D. Hestenes, *New Foundations for Classical Mechanics*, Reidel, Dordrecht, 1986, p. 401.)

and the vector **G** is given by

$$\mathbf{G} = G\left\{\frac{\mathbf{s}_1}{s_1^3} + \frac{\mathbf{s}_2}{s_2^3} + \frac{\mathbf{s}_3}{s_3^3}\right\} \tag{54}$$

The equations in the symmetrical form (52) provide solutions to the three-body problem for some simple cases.

For example, there is a solution in which mass m_2 always lies on the straight line between the other two masses so that the four vectors $\mathbf{s}_1, \mathbf{s}_2, \mathbf{s}_3$ and **G** are collinear. Figure 2-11 shows a bound state (i.e. a negative energy, elliptic orbit) solution with the mass ratio $m_1{:}m_2{:}m_3 = 1{:}2{:}3$.

A solution can also be obtained when the vector $\mathbf{G} = 0$ so that the equations of motion decouple, and this occurs when the three masses are at the vertices of an equilateral triangle. As the motion proceeds, the equilateral triangle condition continues to be satisfied, but the triangle changes in size and orientation. Figure 2-12 shows an elliptic solution when the same mass ratio $m_1{:}m_2{:}m_3 = 1{:}2{:}3$ as previously.

Various cases can occur in the three-body problem. If the total energy is positive, then all three masses can move away from each other, or one can escape and leave the other two behind bound in elliptic orbits. If the energy is negative, one can escape and leave the other two in a bound state, or all three can be confined to bound orbits as in Figs. 2-11 and 2-12.

The restricted three-body problem is one in which two of the masses are bound and the third perturbs the motion of the other two. Examples are a

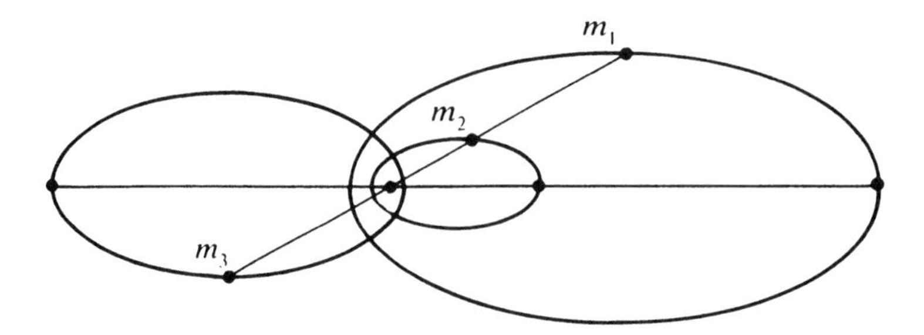

Fig. 2-11. Elliptic orbits for the Euler case three-body problem with collinear masses in the ratios $m_1{:}m_2{:}m_3 = 1{:}2{:}3$. (From D. Hestenes, *New Foundations for Classical Mechanics*, Reidel, Dordrecht, 1986, p. 403.)

spacecraft orbiting between the earth and the moon, or the perturbation of the sun on the moon's orbit. In the spacecraft case the first approach is to assume that the earth and moon move in their unperturbed orbits, and the

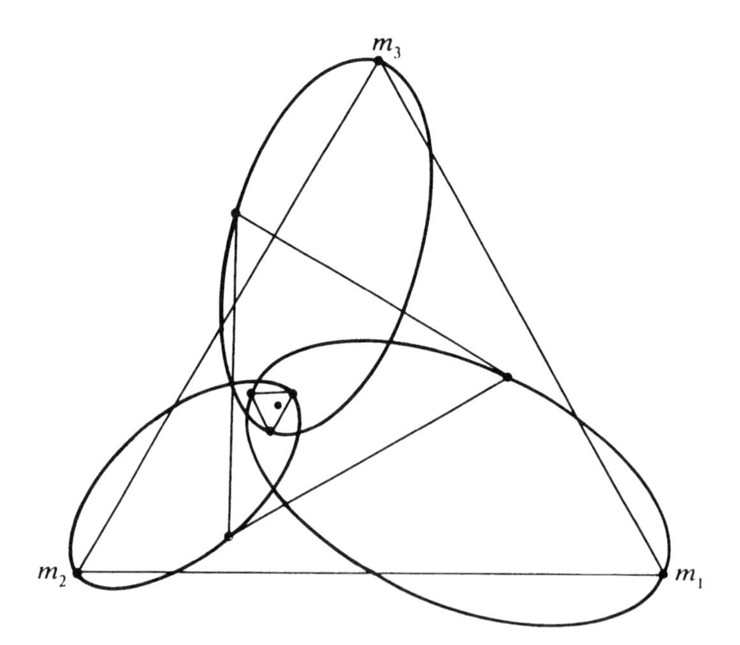

Fig. 2-12. Elliptic orbits for the Lagrange case three-body problem with the masses in an equilateral triangle configuration with the ratios $m_1{:}m_2{:}m_3 = 1{:}2{:}3$. (From D. Hestenes, *New Foundations for Classical Mechanics*, Reidel, Dordrecht, 1986, p. 404.)

satellite interacts with them through their respective inverse square gravitational forces. We should also note that satellites orbiting the earth at altitudes of 90 miles or 150 kilometers have their orbits perturbed by the nonspherical mass distribution of the earth.

CHAPTER 3

RIGID BODIES

1 INTRODUCTION

In this chapter we discuss rigid bodies, rotations, and rotational motion. This includes rotation matrices, rotations of vectors and tensors, parity, the two-dimensional representation of rotations, rotating bodies, rotating coordinate systems, and Coriolis forces. The next chapter will be devoted to vibrational motions that occur when the bodies are not so rigid, and other aspects of vibrational motion.

2 NATURE AND ORIENTATION OF A RIGID BODY

A rigid body is made up of component parts that remain fixed in position relative to each other. Consider a collection of N atoms, each with an x, y and z coordinate. There are $3N$ coordinates needed to specify the state of the system. If these N atoms form a rigid body, then they are subject to the rigid body constraint whereby the distance between each pair of atoms r_{ij} remains constant

$$r_{ij} = r_{ji} = \text{const} \quad \text{for all } i, j \tag{1}$$

The body can move about in space, and at any time its position is determined by the coordinates $x_{\text{CM}}, y_{\text{CM}}, z_{\text{CM}}$ of its center of mass. Its orientation can be specified by the direction cosines $\cos\theta_{ij}$ of the angles θ_{ij} defined in Fig. 3-1 of a reference frame fixed in the body relative to one in the laboratory, and ordinarily both frames will be centered at the center of mass. A second way is to specify polar angles θ, ϕ of an axis and an angle of rotation ψ about this axis, which brings the two frames into coincidence. In any event, there are six independent coordinates needed to specify the configuration of a rigid body, three for the position of the center of mass and three

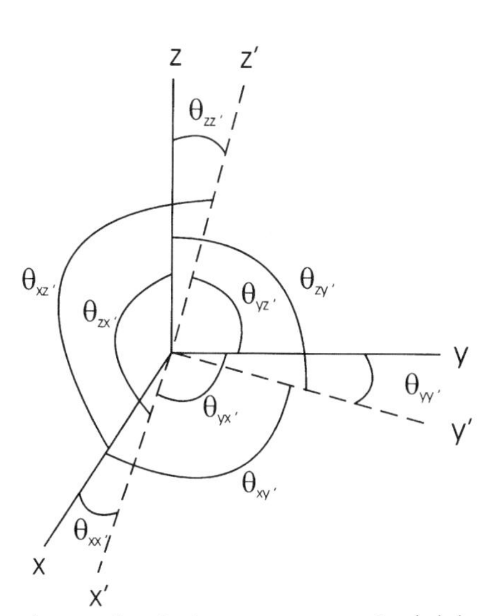

Fig. 3-1. Direction cosine angles θ_{ij} between xyz and $x'y'z'$ cartesian coordinate systems.

for the body's orientation, the first three being translation coordinates, and the other three, angular coordinates.

Another way to specify the orientation is to use the Euler angles defined in Fig. 3-2 and perform a rotation by ϕ about the z axis, then by θ about the new x axis, and finally by ψ about the new z axis, corresponding to the matrix rotation operation

$$R_{\text{Euler}} = R_z(\psi)R_x(\theta)R_z(\phi) \tag{2}$$

This sequence of rotations called the zxz convention singles out a preferred direction; namely, the z direction, and it is well adapted for treating systems with axial symmetry such as a rotating symmetrical top and the axial Zeeman effect.

Airplane pilots use an xyz convention of angles by specifying rotations about a cartesian coordinate system fixed in the airplane, with x along the

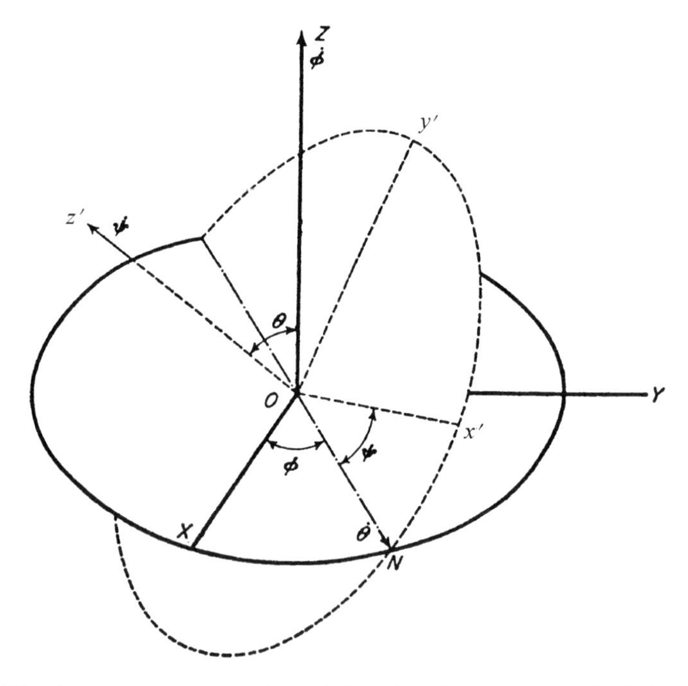

Fig. 3-2. The three successuve rotations ϕ, θ, and ψ that constitute the Euler angles for transforming from the x, y, z coordinate system to the x', y', z' coordinate system. (From H. Goldstein, *Classical Mechanics*, 2nd Ed., Addisson–Wesley, 1980, p. 146.)

so-called figure axis extending from front to back in the forward direction of motion, y along the wings and z vertical for level flight, as shown in Fig. 3-3. The direction of flight is changed by rotation through ϕ about the z axis, an operation called heading or yaw. This is accompanied by a bank or roll, a rotation by the angle ψ about x. The third operation called pitch or attitude rotates about y to begin a climb or descent. A bicycle executes a yaw by turning a corner, but the rider instinctively banks in the process, otherwise the bicycle could fall over. Starting up or down a hill constitutes a change in pitch or attitude.

There are some general theorems that concern rigid body orientations.

1. Euler's theorem states that the most general displacement of a rigid body with one point fixed is a rotation.
2. A real orthogonal matrix specifying a rotation in three dimensions has a determinant +1 and one and only one eigenvalue +1. The other two eigenvalues are complex conjugates of the form $e^{i\phi}$ and $e^{-i\phi}$.
3. Chasles' theorem states that the most general displacement of a rigid body is a translation plus a rotation, and these operations commute.
4. Rotations do not, in general, commute. Infinitesimal rotations, however, do commute to first order.

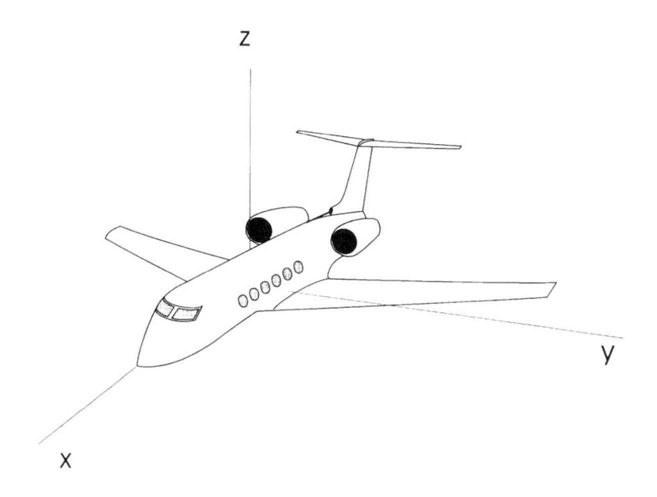

Fig. 3-3. The xyz body axis coordinate system of an airplane. The operations roll (bank), pitch (attitude) and yaw (direction change) are rotations around the x, y and z axes, respectively.

3 ROTATION MATRICES

The orthogonal rotation matrix made up of direction cosines for the angles defined in Fig. 3-1 is

$$R = \begin{pmatrix} \cos\theta_{xx} & \cos\theta_{xy} & \cos\theta_{xz} \\ \cos\theta_{yx} & \cos\theta_{yy} & \cos\theta_{yz} \\ \cos\theta_{zx} & \cos\theta_{zy} & \cos\theta_{zz} \end{pmatrix} \tag{3}$$

The direction cosines satisfy the orthogonality conditions

$$\sum \cos\theta_{ij} \cos\theta_{ik} = \sum \cos\theta_{ji} \cos\theta_{ki} = \delta_{jk} \tag{4}$$

and the determinant of R is $+1$

$$|R| = +1 \tag{5}$$

The general direction cosine matrix (3) reduces to

$$R_z(\phi) = \begin{pmatrix} \cos\phi & \sin\phi & 0 \\ -\sin\phi & \cos\phi & 0 \\ 0 & 0 & 1 \end{pmatrix} \tag{6}$$

for rotations about the z direction, and analogous expressions are easily written down for R_x and R_y.

Equations (2) and (3) involve no inversions or reflections so they are called proper rotations. It is easy to see that any two proper rotation matrices can be generated from each other by a sequence of infinitesimal rotations where, for example, ϕ becomes $d\phi$ in Eq. (6). This sequence corresponds to the gradual rotation of a vector from one orientation in space to another.

If the inversion operation I is involved, whereby $x \Rightarrow -x$, $y \Rightarrow -y$, $z \Rightarrow -z$, then the rotation is called improper. Inversion brings a vector $\mathbf{V}$ to its negative, $\mathbf{V} \Rightarrow -\mathbf{V}$ by means of the matrix I

$$I = \begin{pmatrix} -1 & 0 & 0 \\ 0 & -1 & 0 \\ 0 & 0 & -1 \end{pmatrix} \tag{7}$$

A reflection in the x, y plane ($x \Rightarrow x$, $y \Rightarrow y$, $z \Rightarrow -z$) has the matrix

$$\sigma_{xy} = \begin{pmatrix} 1 & 0 & 0 \\ 0 & 1 & 0 \\ 0 & 0 & -1 \end{pmatrix} \tag{8}$$

and analogous expressions can be written for σ_{yz} and σ_{zx}. We should note that inversion and reflection matrices have a determinant of -1.

More generally, product matrices of the type $RI = IR$ and $R\sigma = \sigma R$, where the I, σ, and the proper rotation matrix R commute, constitute improper rotations with a determinant of -1. An example is $IR_z(\phi)$, which is easily written down from Eqs. (6) and (7)

$$IR_z(\phi) = \begin{pmatrix} -\cos\phi & -\sin\phi & 0 \\ \sin\phi & -\cos\theta & 0 \\ 0 & 0 & -1 \end{pmatrix} \tag{9}$$

As in the proper rotation case, any two improper rotation matrices can be generated from each other by a sequence of infinitesimal improper rotations. The groups of proper and improper rotations are, however, disjoint in the sense that no sequence of infinitesimal proper rotations can transform a proper rotation to an improper one, and vice versa. A right hand is not converted into a left hand in a gradual way, but rather abruptly by reflection in a mirror!

4 VECTOR AND TENSOR TRANSFORMATIONS

A rotation by ϕ about the z axis transforms vector $\mathbf{V}$ to $\mathbf{V}'$ as follows

$$\begin{pmatrix} \cos\phi & \sin\phi & 0 \\ -\sin\phi & \cos\phi & 0 \\ 0 & 0 & 1 \end{pmatrix} \begin{pmatrix} V_x \\ V_y \\ V_z \end{pmatrix} = \begin{pmatrix} V'_x \\ V'_y \\ V'_z \end{pmatrix} \tag{10}$$

where

$$\begin{aligned} V'_x &= V_x \cos\phi + V_y \sin\phi \\ V'_y &= -V_x \sin\phi + V_y \cos\phi \\ V'_z &= V_z \end{aligned} \tag{11}$$

and the magnitude $|\mathbf{V}|$ is preserved

$$V_x'^2 + V_y'^2 + V_z'^2 = V_x^2 + V_y^2 + V_z^2 \tag{12}$$

When a tensor **T** is rotated to a new orientation by a similarity transformation

$$RTR^{-1} = T' \tag{13}$$

the trace is preserved

$$T'_{xx} + T'_{yy} + T'_{zz} = T_{xx} + T_{yy} + T_{zz} \tag{14}$$

If a tensor is symmetric ($T_{ij} = T_{ji}$) or antisymmetric ($T_{ij} = T_{ji}$), then this symmetry type is preserved by the similarity transformation.

The rotations that we have presented in matrix form can be written down analytically in tensor notation

$$V'_i = \sum R_{ij} V_j \tag{15}$$

$$T'_{ij} = \sum R_{im} R_{jn} T_{mn} \tag{16}$$

where $R_{ij} = \cos\theta_{ij}$ is a direction cosine, the summations $i = 1, 2, 3$ are over the coordinates x, y, z, and the matrix elements R_{ij} obey the orthogonality condition (4).

5 PARITY

A more general name for the inversion operation is parity, denoted by P, and it takes the coordinate vector **r** to its negative −**r**, corresponding to the operation $x \Rightarrow -x$, $y \Rightarrow -y$, $z \Rightarrow -z$. We noted above that parity changes the direction of a vector without affecting its magnitude, and it has no effect on a scalar. Parity also has no effect on a second-rank tensor. These facts are summarized in the Table 3-1. We see that scalars are zero-rank tensors,

TABLE 3-1 Parities of low-rank tensors and pseudotensors

Quantity	Rank	Parity	Pseudoquantity	Rank	Parity
Scalar	0	+	pseudoscalar	0	−
Vector	1	−	pseudovector (axial vector)	1	+
'Tensor'	2	+	pseudo 'tensor'	2	−

vectors are first-rank tensors, and what we ordinarily call 'tensors' are more properly called second-rank tensors. There are also higher rank tensors, and an analogous nomenclature is used for their pseudo counterparts. We see from the table that odd-rank tensors are odd parity and even-rank tensors are even parity, with the parity rules reversed for the pseudotensors.

An example of a pseudovector, which is sometimes called an axial vector, is the cross-product **C** of two vectors **A** and **B**

$$\mathbf{C} = \mathbf{A} \times \mathbf{B} \tag{17}$$

The parity operation P reverses the direction of each vector **A** and **B**, so the cross-product is unchanged

$$P(\mathbf{A} \times \mathbf{B}) = (P\mathbf{A}) \times (P\mathbf{B}) = (-\mathbf{A}) \times (-\mathbf{B}) = \mathbf{A} \times \mathbf{B} = \mathbf{C} \tag{18}$$

demonstrating that **C** is a pseudovector.

6 TWO-DIMENSIONAL REPRESENTATION OF ROTATIONS

We have discussed the usual way of handling the rotations of vectors and tensors with the aid of 3×3 rotation matrices. There is also a 2×2matrix approach to rotations that has a great deal of theoretical importance in some advanced theories, but not much practical importance in mechanics. Nevertheless, it is still worthwhile to summarize some of this material.

In this approach, vectors are written as complex 2×2 hermitian matrices

$$\mathbf{V} = \begin{pmatrix} V_z & (V_x - iV_y) \\ (V_x + iV_y) & -V_z \end{pmatrix} \tag{19}$$

and they are transformed by unitary two-dimensional rotation matrix similarity transformations

$$V' = QVQ^{-1} \tag{20}$$

where, for example, we have for rotations θ about the x, y and z axes

$$Q_x = \begin{pmatrix} \cos\frac{1}{2}\theta & i\sin\frac{1}{2}\theta \\ i\sin\frac{1}{2}\theta & \cos\frac{1}{2}\theta \end{pmatrix} \tag{21a}$$

$$Q_y = \begin{pmatrix} \cos\frac{1}{2}\theta & \sin\frac{1}{2}\theta \\ -\sin\frac{1}{2}\theta & \cos\frac{1}{2}\theta \end{pmatrix} \tag{21b}$$

$$Q_z = \begin{pmatrix} e^{i\theta/2} & 0 \\ 0 & e^{-i\theta/2} \end{pmatrix} \tag{21c}$$

These matrices can be expressed in terms of the Pauli spin matrices

$$\sigma_x = \sigma_1 = \begin{pmatrix} 0 & 1 \\ 1 & 0 \end{pmatrix} \qquad \sigma_y = \sigma_2 = \begin{pmatrix} 0 & -i \\ i & 0 \end{pmatrix} \qquad \sigma_z = \sigma_3 = \begin{pmatrix} 1 & 0 \\ 0 & -1 \end{pmatrix} \tag{22}$$

considered as forming a cartesian vector

$$\sigma = \mathbf{i}\sigma_x + \mathbf{j}\sigma_y + \mathbf{k}\sigma_z \tag{23}$$

The vector expression (19) becomes

$$\mathbf{V} \cdot \sigma = V_x\sigma_x + V_y\sigma_y + V_z\sigma_z \tag{24}$$

and the unitary rotation matrices (21) may be employed to express a rotation Q_n through an angle θ about an axis in the direction of a unit vector $\hat{\mathbf{n}}$ in the following manner

$$Q_n = I\cos\tfrac{1}{2}\theta + i\hat{\mathbf{n}} \cdot \sigma \sin\tfrac{1}{2}\theta \tag{25}$$

where I is the 2×2 unit matrix, and Eq. (25) has the exponential form

$$Q_n = \exp[i\hat{\mathbf{n}} \cdot \sigma(\theta/2)] \tag{26}$$

as can be demonstrated by expanding the exponential as a power series using the cyclic permutation and anticommutation properties of the Pauli matrices that are given in Chapter 27, in Section 6.

7 ROTATING COORDINATE SYSTEMS AND CORIOLIS FORCE

We examine relationships between a vector in a stationary coordinate system and the same vector in a coordinate system centered at the same origin of coordinates and rotating at an angular velocity ω that is constant in time. We wish to determine how the equations of motion expressed in the stationary inertial coordinate system become modified when they are expressed in the non-inertial rotating system.

Consider the velocity $\mathbf{v}_s = (d\mathbf{r}/dt)_s$ of the position vector $\mathbf{r}$ expressed in the space system. This velocity vector is related to the velocity $\mathbf{v}_r = (d\mathbf{r}/dt)_r$ expressed in the rotting system through the expression

$$\mathbf{v}_s = \mathbf{v}_r + \boldsymbol{\omega} \times \mathbf{r} \tag{27}$$

The acceleration $\mathbf{a}_s$ in the space system

$$\mathbf{a}_s = \left(\frac{d\mathbf{v}_s}{dt}\right)_s = \left(\frac{d\mathbf{v}_s}{dt}\right)_r + \boldsymbol{\omega} \times \mathbf{v}_s \tag{28}$$

obtained by substituting Eq. (27) into Eq. (28) is given by

$$\mathbf{a}_s = \mathbf{a}_r + 2(\boldsymbol{\omega} \times \mathbf{v}_r) + \boldsymbol{\omega} \times (\boldsymbol{\omega} \times \mathbf{r}) \tag{29}$$

where $\mathbf{a}_r = (d\mathbf{v}_r/dt)_r$. The true force $\mathbf{F} = m\mathbf{a}_s$ in the space inertial system and the effective force $\mathbf{F}_{\text{eff}} = m\mathbf{a}_r$ in the non-inertial rotating system are related as follows:

$$\mathbf{F}_{\text{eff}} = \mathbf{F} - 2m(\boldsymbol{\omega} \times \mathbf{v}_r) - m\boldsymbol{\omega} \times (\boldsymbol{\omega} \times \mathbf{r}) \tag{30}$$

where the second and third terms on the right-hand side are, respectively, the Coriolis force $\mathbf{F}_{\text{Cor}}$ and the centrifugal force $\mathbf{F}_{\text{cent}}$

$$\mathbf{F}_{\text{Cor}} = -2m(\boldsymbol{\omega} \times \mathbf{v}_r) \tag{31}$$

$$\mathbf{F}_{\text{cent}} = -m\boldsymbol{\omega} \times (\boldsymbol{\omega} \times \mathbf{r}) \tag{32}$$

and for the case of the earth's motion the latter has the magnitude

$$F_{\text{cent}} = m\omega^2 r \sin\theta \tag{33}$$

$$\approx 0.0035mg \sin\theta \tag{34}$$

In the northern hemisphere the Coriolis force causes winds to circulate counter clockwise about a center of low pressure, and causes freely falling bodies to be deflected to the east.

Newton's law of gravitation, Chapter 2, Eq. (5), causes one body to pull another body directly toward it, and this is what happens in an inertial frame of reference, i.e., a frame in which Newton's second law of motion $\mathbf{F} = d\mathbf{p}/dt$ is obeyed. A body in freefall toward the earth would not be deflected if the earth were not rotating. The observed deflection of such a body demonstrates that coordinate frames fixed in the rotating earth are not inertial systems.

8 ROTATING BODIES

We have discussed translational kinetic energy of rigid bodies. When a rigid body rotates it acquires rotational kinetic energy given by

$$T = \tfrac{1}{2}\boldsymbol{\omega} \cdot \mathbf{I} \cdot \boldsymbol{\omega} \tag{35}$$

$$= \tfrac{1}{2}\boldsymbol{\omega} \cdot \mathbf{L} \tag{36}$$

$$= \tfrac{1}{2}\mathrm{I}\omega^2 \tag{37}$$

where $\mathbf{L}$ is the angular momentum and $\mathbf{I}$ is the moment of inertia

$$\mathbf{L} = \mathbf{I} \cdot \boldsymbol{\omega} \tag{38}$$

The scalar moment of inertia tensor I about an axis along the direction defined by the unit vector $\hat{\mathbf{n}}$ is

$$I = \hat{\mathbf{n}} \cdot \mathbf{I} \cdot \hat{\mathbf{n}} \tag{39}$$

$$= \sum m_i[r_i^2 - (\mathbf{r}_i \cdot \hat{\mathbf{n}})^2] \tag{40}$$

$$= \sum m_i(\mathbf{r}_i \times \hat{\mathbf{n}}) \cdot (\mathbf{r}_i \times \hat{\mathbf{n}}) \tag{41}$$

and for a body of density ρ the principal moment of inertia about an axis in the x-direction is given by

$$I_{xx} = \int \rho(r^2 - x^2)d\tau \tag{42a}$$

$$= \int \rho(y^2 + z^2)d\tau \tag{42b}$$

Figure 3-4 shows moments of inertia of some common solids.

By the parallel axis theorem the moment of inertia I of a body of mass M about an arbitrary axis is

$$I = I_{\text{CM}} + Md^2 \tag{43}$$

where I_{CM} is the moment of inertia about a parallel axis through the center of mass, and d is the perpendicular distance between the two axes.

The rotational motion of a rigid body with one point fixed, subject to the torque $\mathbf{N}$, is described by the equation

$$I_{xx}\dot{\omega}_x - \omega_y\omega_z(I_{yy} - I_{zz}) = N_x \tag{44}$$

and similarly for the other two directions where the coordinate system has its origin at the fixed point, it moves with the body and it is the principal axis system with $I_{ij} = 0$ for $i \neq j$. For a symmetric top

$$I_\perp = I_{xx} = I_{yy} \tag{45a}$$

$$I_\parallel = I_{zz} \tag{45b}$$

$$\Delta I = (I_\perp - I_\parallel) \tag{45c}$$

we have

$$I_\perp\dot{\omega}_x - \omega_y\omega_z\Delta I = N_x \tag{46a}$$

$$I_\perp\dot{\omega}_y + \omega_z\omega_x\Delta I = N_y \tag{46b}$$

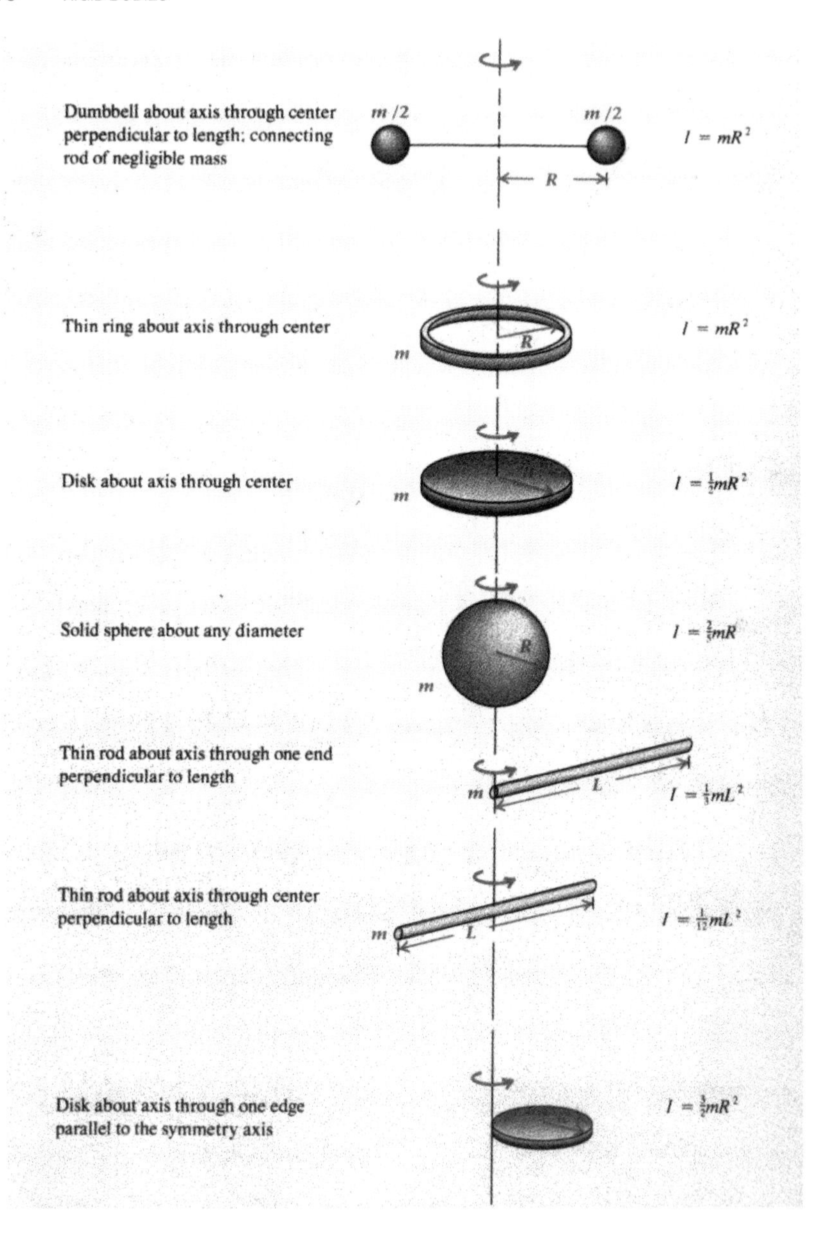

Fig. 3-4. Moments of inertia of several solids. (From E. R. Jones and R. E. Childers, *Contemporary College Physics*, Addison–Wesley, Reading, Massachusetts, 1993, p. 247.)

$$I_{\parallel}\dot{\omega}_z = N_z \tag{46c}$$

The Lagrangian and canonical momenta for this top expressed in Euler angles are given in Chapter 1. The top spins on its axis at the rate $\dot{\psi}$ and precesses around the vertical at the rate $\dot{\phi}$ as it bobs up and down, i.e., nutates, between the minimum $\theta_{\min}$ and the maximum $\theta_{\max}$ inclination angles. When θ decreases as the top rises there is an increase in the potential energy $MgL\cos\theta$, and this causes the rotation and precession angular velocities to decrease, thereby lowering the rotational kinetic energy to balance the increase in the potential energy. The overall motion is typically a fast rotation $\dot{\psi}$ about the axis of the top, and a slower precession $\dot{\phi}$ of this axis about the vertical direction accompanied by a nutation of this axis between the angles $\theta_{\max}$ and $\theta_{\min}$ with respect to the vertical.

CHAPTER 4

OSCILLATIONS AND VIBRATIONS

1 INTRODUCTION

Molecular vibrations can be treated classically or quantum mechanically. The former is done in classical mechanics, and the latter in molecular spectroscopy. In this chapter we begin with the classical description in terms of forces and Lagrangians, and then we examine the quantum mechanical formulation in terms of Hamiltonians and energy levels. The emphasis is on relatively small molecules, and this is followed by a discussion of vibrating infinite chains and solids. The thermal excitations of vibrational modes, molecular rotations, infrared and Raman spectroscopy, and the Debye theory of solids are treated.

2 MOLECULAR VIBRATIONS

Underlying much of our discussion in this chapter is the picture of nearest neighbor atoms in molecules and solids held together by connecting springs with equilibrium lengths. The force $F = -k(x - x_0)$ acts to return a chemical bond to its equilibrium length x_0 when it is compressed or stretched from this separation. The electrons, of course, are responsible for the bonding, and the electronic bonding energy is much greater than the vibrational energy of stretching and compression. For now we neglect the influence of this electric energy on the vibrations.

A molecule with N atoms has three translational and three rotational degrees of freedom, so there are $3N - 6$ vibrational modes. If the molecule is linear, then a rotation about the molecular axis does not change the atomic positions, so only rotations about the two perpendicular axes count for rotations, and there are $3N - 5$ normal vibrational modes. A diatomic molecule with $N = 2$ has $3N - 5 = 1$ so there is only one vibrational mode arising from the motion of the two atoms along their axis. A linear triatomic molecule ($N = 3$) like CO_2 has four normal vibrational modes, and non-linear H_2O has three modes.

3 RIGID BODY

We give a classical description of the vibrations of a not-so-rigid body, then we apply the treatment to small molecules, followed by a discussion of vibrations in solids. Before proceeding, however, we define a rigid body, comment on the equilibrium configuration of the not-so-rigid body, and make a couple of remarks on the rigidity of real bodies.

The masses of a rigid body are held fixed distances apart so that the distance r_{ij} between each pair of masses m_i and m_j is a constant

$$r_{ij} = (\text{const})_{ij} \tag{1}$$

This is the defining equation of a rigid body, as explained in Chapter 3. A collection of masses interacting via short-range forces, such as those involved in the chemical bonds between atoms in a molecule, has an equilibrium configuration satisfying Eq. (1). For this configuration the generalized forces Q_i acting on the masses vanish

$$Q_i = -\frac{\partial V}{\partial q_i} = 0 \tag{2}$$

and the energy is a minimum. Vibrations are treated as motions relative to this rigid body equilibrium configuration

In practice, bodies are not perfectly rigid, but they can be compressed and distorted in shape by the application of forces and torques. For a rod, Young's modulus Y is defined as the linear stress per linear strain, or, in other words, the force per unit area divided by the stretch per unit length

$$Y = \frac{F/A}{\Delta L/L} \tag{3}$$

Longitudinal waves propagate along the rod at the velocity v given by

$$v = (Y/\rho)^{1/2} \tag{4}$$

where ρ is the mass density or mass per unit volume. For a solid, the bulk modulus B, the reciprocal of its compressibility, is the pressure change per bulk strain

$$B = \frac{\Delta P}{-\Delta V/V} \tag{5}$$

These moduli describe macroscopic changes arising from applied forces. In this chapter we are more interested in the individual motions of the atoms of the body rather than its bulk response to applied forces.

When vibrations are discussed in this chapter it is assumed that there are no externally applied forces, that the otherwise rigid body is held fixed in position so there is no translation of the center of mass, and that no rotation is taking place.

4 NORMAL MODES

In this section we discuss normal modes; that is, collective oscillations of a system in which many or all of the atoms oscillate at the same frequency. These normal modes are treated in terms of generalized coordinates q_i, which might be, for example, the distance of an atom from its equilibrium position, or the angle between chemical bonds joining adjacent atoms.

Consider the deviations η_i of the generalized coordinates q_i from their equilibrium positions q_{0i}

$$q_i = q_{0i} + \eta_i \tag{6}$$

The potential energy V can be expanded about the equilibrium value of the generalized coordinates.

$$\begin{aligned} V(q_1, \ldots q_n) = & V(q_{01}, \ldots, q_{0n}) + \sum \frac{\partial V}{\partial q_i} \eta_i \\ & + \tfrac{1}{2} \sum \frac{\partial^2 V}{\partial q_i \partial q_j} \eta_i \eta_j \end{aligned} \tag{7}$$

We can select the zero of energy so the first term $V(q_{0i}, \ldots, q_{0n})$ vanishes, and we know from the equilibrium condition (2) that the second term vanishes. This leaves us with the third term

$$V(q_1, \ldots, q_n) = \tfrac{1}{2} \sum \frac{\partial^2 V}{\partial q_i \partial q_j} \eta_i \eta_j = \tfrac{1}{2} V_{ij} \eta_i \eta_j \tag{8}$$

which is quadratic in the coordinates. The kinetic energy T is quadratic in the velocities $\dot{q}_i = \dot{\eta}_i$

$$T = \tfrac{1}{2} \sum T_{ii} \dot{\eta}_i^2 \tag{9}$$

and the Lagrangian L

$$L = \sum \left[\tfrac{1}{2} T_{ii} \dot{\eta}_i^2 - \tfrac{1}{2} V_{ij} \eta_i \eta_j \right] \tag{10}$$

gives the equations of motion

$$T_{jj} \ddot{\eta}_j + \sum V_{ij} \eta_i = 0 \tag{11}$$

An oscillatory function

$$\eta_i = \eta_{i0} e^{-i\omega t} \tag{12}$$

which describes the vibrations at the normal mode frequency ω is now substituted into the equations of motion. This provides a solution for ω^2 when the determinant vanishes, and we write down the secular equation

$$\begin{vmatrix} V_{11} = \omega^2 T_{11} & V_{12} & \cdots \\ V_{21} & V_{22} - \omega^2 T_{22} & \cdots \\ \cdots & \cdots & \cdots \\ \cdots & \cdots & \cdots \end{vmatrix} = 0 \tag{13}$$

This determinant is expanded to form a polynomial equation which is solved to provide the normal mode frequencies. Then for each frequency ω the following set of simultaneous equations

$$T_{jj}\omega^2 \eta_j - \sum V_{ij}\eta_i = 0 \tag{14}$$

is solved for $j = 1, 2, \ldots, n$ to provide the normal coordinates η_i corresponding to each frequency.

5 TRIATOMIC MOLECULE

An example of a vibrational problem is a symmetric linear triatomic molecule such as CO_2 with masses m on the ends at positions x_1 and x_3, and M in the middle at position x_2, as shown in Fig. 4-1. This molecule has the following potential and kinetic energies

$$\begin{aligned} V &= \tfrac{1}{2}C(\eta_2 - \eta_1)^2 + \tfrac{1}{2}C(\eta_3 - \eta_2)^2 \\ &= \tfrac{1}{2}C(\eta_1^2 + 2\eta_2^2 + \eta_3^2 - 2\eta_1\eta_2 - 2\eta_2\eta_3) \end{aligned} \tag{15}$$

$$T = \tfrac{1}{2}m(\dot{\eta}_1^2 + \dot{\eta}_3^2) + \tfrac{1}{2}M\dot{\eta}_2^2 \tag{16}$$

where, from symmetry, $x_{03} - x_{02} = x_{02} - x_{01}$, and the symbol C is used for the spring constant to avoid confusion with the wave vector k. This gives for the secular equation

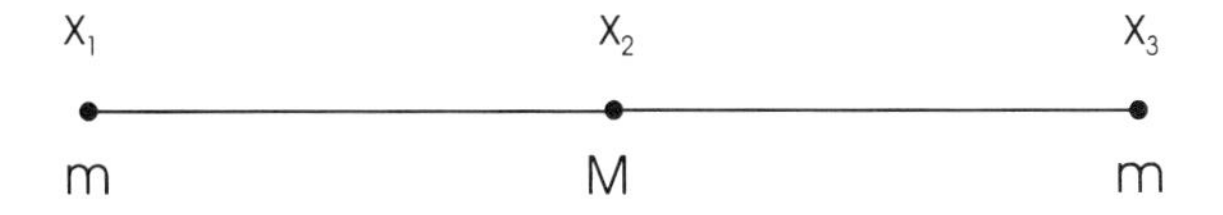

Fig. 4-1. Symmetric linear triatomic molecule m–M–m.

$$\begin{vmatrix} C-\omega^2 m & -C & 0 \\ -C & 2C-\omega^2 M & -C \\ 0 & -C & C-\omega^2 m \end{vmatrix} = 0 \tag{17}$$

which, upon multiplying out the determinant and grouping terms, gives the following equation, which is cubic in ω^2

$$\omega^2[C-\omega^2 m][C(M+2m)-\omega^2 mM]=0 \tag{18}$$

The quadratic part of the equation in square brackets has two roots. There is a symmetrical stretch normal mode vibration with the normal coordinate ζ_1 and frequency ω_1

$$\begin{aligned} \zeta_1 &= \eta_2 - \eta_1 \\ \omega_1 &= (C/m)^{1/2} \end{aligned} \tag{19}$$

and an asymmetric stretch with the parameters

$$\begin{aligned} \zeta_2 &= (\eta_1+\eta_3)(2m/M)^{1/2} - \eta_2(M/2m)^{1/2} \\ \omega_2 &= (C/m)^{1/2}[1+2(m/M)] \end{aligned} \tag{20}$$

The third root $\omega = 0$ is associated with translation of the center of mass so there is no vibration. This solution appears because we assumed that all of the normal modes correspond to vibrations in which the atoms are confined to motion along the molecular axis, and there are only two modes conforming to this stipulation. If we no longer restrict the atoms to moving along the molecular axis, then they can vibrate laterally, the equations become more complex, and we obtain a bending mode. The atom motions for these three modes are shown in Fig. 4-2. There are actually two degenerate bending modes at right angles to each other, which makes four modes in all since a linear molecule has $3N-5=4$ normal modes, where $N=3$.

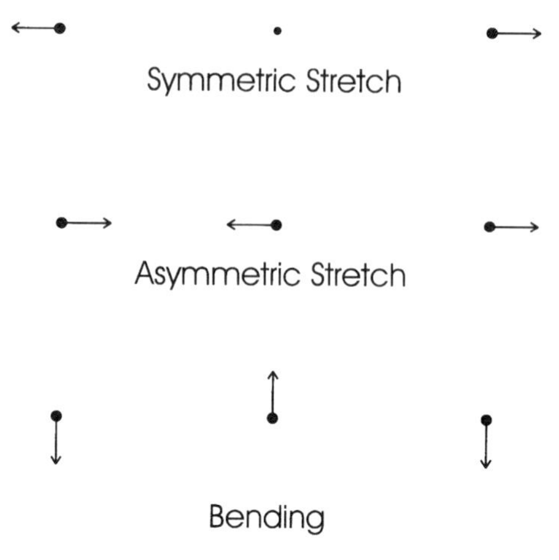

Fig. 4-2. Normal vibrational modes of the symmetric linear triatomic molecule m–M–m shown in Fig. 4-1. The bending mode is doubly degenerate with vibrations occurring in the plane of the paper and perpendicular to this plane at the same frequency.

Our treatment has been for a symmetric linear triatomic molecule. The asymmetric but linear molecule OCS can be treated in a similar manner. If the equilibrium configuration of a triatomic molecule is bent, as in the case of H_2O, then one finds a symmetric stretch, an asymmetric stretch, and a bending mode, as depicted in Fig. 4-3. There are three normal modes since now $3N - 6 = 3$.

6 LINEAR CHAIN OF *N* ATOMS

To gain some insight into vibrating solids we anlyze a linear chain of N atoms, of mass m and spacing a, with nearest-neighbor atoms coupled together through springs C, and second-nearest neighbor interactions ignored. At equilibrium the atoms are at the positions $x_n = x_{n0}$ where

$$x_{n0} = na \tag{21}$$

with the springs neither stretched nor compressed, and only longitudinal motions are allowed. We can define atom positions q_n relative to the equilibrium position x_{n0}

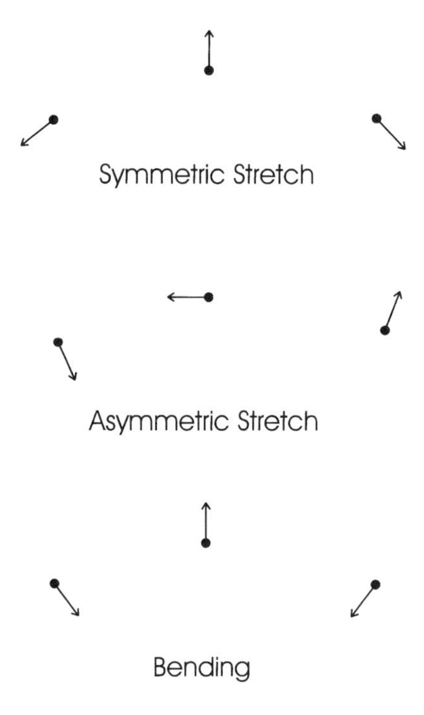

Fig. 4-3. The three normal modes of vibration of a nonlinear symmetric triatomic molecule such as H_2O.

$$q_n = x_n - na \tag{22}$$

with

$$\dot{q}_n = \dot{x}_n \tag{23}$$

for the atom velocities. The length of the spring between atoms $n - 1$ and n is

$$\text{length} = x_n - x_{n-1} = q_n - q_{n-1} + a \tag{24}$$

and the equilibrium length separating the atoms is a, so the extent to which the spring is compressed or stretched is given by $(q_n - q_{n-1})$. The ends of the chain are assumed to be fixed in place so the boundary conditions at the ends are taken to be

$$q_0 = q_{N+1} = 0 \tag{25}$$

where positions 0 and N + 1 are the remote ends of the springs attached to atoms number 1 and N, respectively.

Consider the case plotted in Fig. 4-4 in which the three atoms $n-1$, n and $n+1$ have positive relative displacements, and both springs are stretched. This can occur when

$$0 < q_{n-1} < q_n < q_{n+1} \tag{26}$$

We see from the figure that the net force on the nth mass is

$$F_{\text{net}}^{(n)} = C(q_{n+1} - q_n) - C(q_n - q_{n-1}) \tag{27}$$

and using the equation of motion $F_{\text{net}}^{(n)} = mq_n$ for the nth mass gives

$$q_n = \omega_0^2(q_{n+1} + q_{n-1} - 2q_n) \tag{28}$$

where $C = m\omega_0^2$.

In a normal mode all of the masses vibrate at the same frequency, so we look for solutions of the form

$$q_n(t) = q_n^0 e^{-i\omega t} \tag{29}$$

which gives, for each of the N atoms

$$-\omega^2 q_n^0 = \omega_0^2(q_{n+1}^0 + q_{n-1}^0 - 2q_n^0) \tag{30}$$

where q_n^0 is the maximum relative displacement of the nth atom. A traveling wave solution has the form

$$q_{n\pm 1} = A \exp[\mathrm{i}(n \pm 1)ka] \tag{31}$$

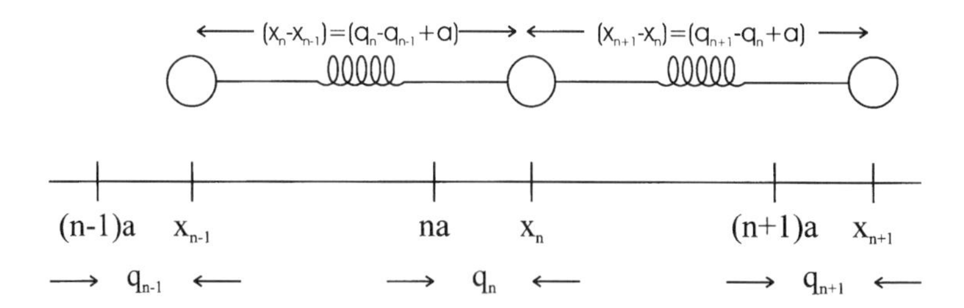

Fig. 4-4. Coordinates for describing the longitudinal vibrations of a linear chain of atoms.

and similarly for q_n, which gives

$$\omega^2 = 2\omega_0^2(1 - \cos ka) \tag{32}$$

With the aid of the trigonometric identity

$$(1 - \cos ka) = 2\sin^2 \tfrac{1}{2}ka \tag{33}$$

this expression becomes

$$\omega = 2\omega_0 \sin \tfrac{1}{2}ka \tag{34}$$

There is a shortest and a longest wavelength $\lambda = 2\pi/k$ associated with the maximum and minimum frequency, with the values

$$\begin{aligned} \lambda_{\min} &= 2L/N \qquad & \omega_{\max} &\approx 2\omega_0 \\ \lambda_{\max} &= 2L \qquad & \omega_{\min} &\approx \pi\omega_0/N \end{aligned} \tag{35}$$

where we assume a long chain so $N \gg 1$.

7 DISPERSION RELATIONS

For large N the wave number k may be regarded as a continuous variable, and the expression (34) which considers ω as a function $\omega(k)$ of k is called a dispersion relation. As a result of the dispersion relation the phase velocity $v = \omega/k$ has the following frequency dependence

$$v = \frac{a\omega}{2\sin^{-1}(\omega/2\omega_0)} \tag{36}$$

and in the low frequency limit $\omega \ll \omega_0$ this becomes

$$v = \omega_0 a \tag{37}$$

Figure 4-5 shows a plot of the dispersion relation, ω, versus k of Eq. (34).

Since frequencies cannot be negative, Fig. 4-5(a) plots the absolute value $2\omega_0|\sin\frac{1}{2}ka|$. The range of k values from $-\pi/a$ to $+\pi/a$ is called the first Brillouin zone in solid state physics texts. Another way to plot this, illustrated in Fig. 4-5(b), is in the range of k from 0 to $2\pi/a$.

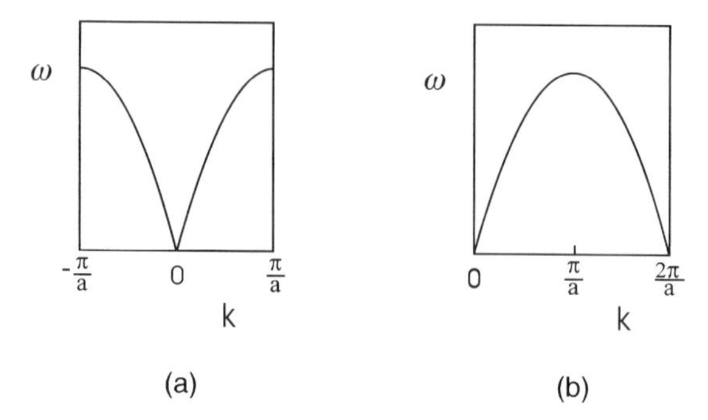

Fig. 4-5. Plots of the dispersion relation of a long linear chain of atoms (a) in the range $-\pi/a < k < \pi/a$ of the first Brillouin zone, and (b) in the equivalent range $0 < k < 2\pi/a$.

8 DIATOMIC LINEAR CHAIN

The above model can be extended to a linear chain in which the atoms alternate between small atoms of mass m and large atoms of mass M. There are two equations of motion

$$m_n q_n = C(q_{n-1} + q_{n+1} - 2q_n) \tag{38a}$$

$$M q_{n+1} = C(q_n + q_{n+2} - 2q_{n+1}) \tag{38b}$$

where even n denotes a small atom and odd n indicates the relative displacement of a large atom. Both atoms vibrate at the same frequency ω, but with different amplitudes q_n^0 for the small and large masses. The allowed frequencies are provided by the following equation, which is quadratic in ω^2:

$$mM\omega^4 - 2C(m + M)\omega^2 + 4C^2 \sin^2 \tfrac{1}{2}ka = 0 \tag{39}$$

with the k dependence of the frequencies shown in Fig. 4-6. We see that there are two types of modes, a low frequency acoustic branch with the limiting frequencies

$$\begin{aligned} \omega^2 &\approx \left(\frac{\frac{1}{2}C}{m + M}\right)k^2a^2 \quad && ka \ll \tfrac{1}{2}\pi \\ \omega^2 &\approx \frac{2C}{M} && ka \approx \tfrac{1}{2}\pi \end{aligned} \tag{40}$$

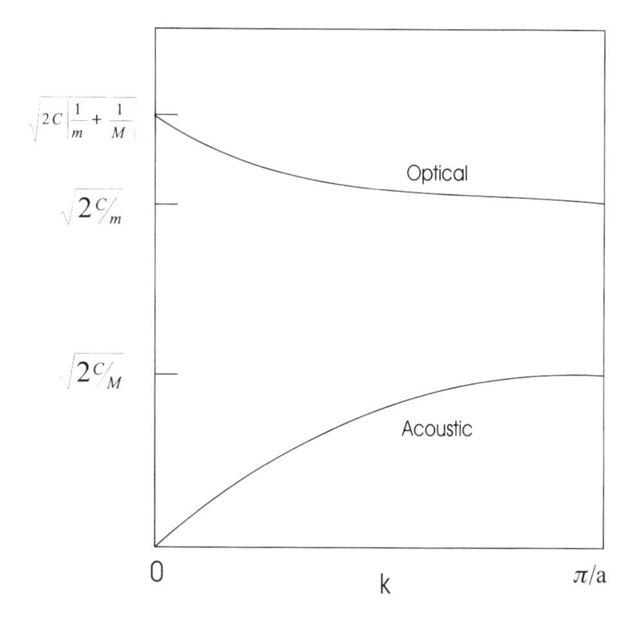

Fig. 4-6. Wave number dependence of the optic and acoustic vibrational modes of a diatomic linear chain of atoms.

and a high frequency optical branch with the limiting frequencies

$$\begin{aligned} \omega^2 &\approx 2C\left(\frac{1}{m}+\frac{1}{M}\right) \quad & ka \ll \tfrac{1}{2}\pi \\ \omega^2 &\approx \frac{2C}{m} & ka \approx \tfrac{1}{2}\pi \end{aligned} \qquad (41)$$

More generally, there can be longitudinal oscillations in which the atoms vibrate with compressions and stretchings of the springs along the direction of the chain as in Fig. 4-4, and transverse modes in which the atomic vibrations are in a direction perpendicular to the chain, as in Fig. 4-7. In three-dimensional crystals there are modes that are longitudinal as well as others that are transverse to the direction of propagation, and both of these mode types can have acoustical and optical branches.

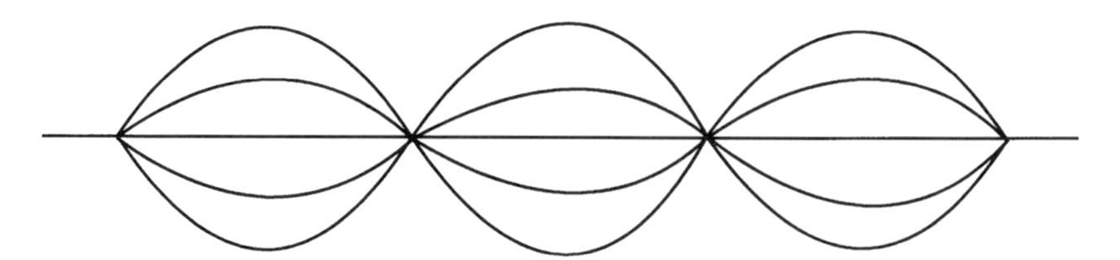

Fig. 4-7. Transverse vibration of a linear chain of atoms.

9 VIBRATING SOLIDS

Increasing the temperature of a solid produces an increase in the vibrations of the lattice. At a particular temperature T the only modes that will be appreciably populated are those with frequencies below the ratio $k_B T/\hbar$, i.e., those whose phonon energy quantum satisfies the inequality

$$\hbar\omega < k_B T \tag{42}$$

We can see from Eq. (35) that a monatomic chain of atoms has a minimum and a maximum frequency of vibration. In the Debye model for the specific heat of three-dimensional solids there is a highest vibrational frequency $\omega_{\max}$, which is called the Debye frequency ω_D, and associated with it is the Debye temperature Θ_D defined by

$$\Theta_D = \hbar\omega_D/k_B \tag{43}$$

Classically, for very low temperatures few modes are excited, the specific heat increases with the temperature, and we find in Chapter 8, Eq. (60) the expression

$$C \approx \frac{12\pi^4}{5} R\left(\frac{T}{\Theta_D}\right)^3 \quad T \ll \Theta_D \tag{44}$$

for the specific heat per mole in the low temperature range, where $R = N_A k_B$ is the gas constant. For very high tempeatures where all modes contribute the specific heat attains its classical limit

$$C = 3R \quad T \gg \Theta_D \tag{45}$$

independent of the temperature.

10 MOLECULAR ROTATIONS

The energy levels and spectroscopic transitions of rotating and vibrating molecules that are in agreement with experiment result from a quantum mechanical treatment. The rotational Hamiltonian for a molecule with three different principal moments of inertia $I_a \neq I_b \neq I_c$, called an asymmetric top molecule, is given by

$$\mathcal{H} = \frac{J_a^2 \hbar^2}{2I_a} + \frac{J_b^2 \hbar^2}{2I_b} + \frac{J_c^2 \hbar^2}{2I_c} \tag{46}$$

where the $\hbar \mathbf{J}$ is the total angular momentum

$$J^2 = J_a^2 + J_b^2 + J_c^2 \tag{47}$$

We assume that the molecule is rigid or non-vibrating for this analysis. This justified by the fact that the frequency of vibration is 10 to 100 times faster than the rotational frequency so the vibrational motion averages out, and the vibrating atoms act like they are in fixed positions during the rotation.

For an axially symmetric molecule such as chloroform $CHC1_3$ or ethane C_2H_6, called a symmetric top, $I_a = I_b$, we use the notation

$$\begin{aligned} I_{\parallel} &= I_c \\ I_{\perp} &= I_a = I_b \end{aligned} \tag{48}$$

and with the aid of Eq. (47) the Hamiltonian assumes the form

$$\mathcal{H} = \frac{J^2 \hbar^2}{2I_{\perp}} + \tfrac{1}{2} J_a^2 \hbar^2 \left(\frac{1}{I_{\parallel}} - \frac{1}{I_{\perp}} \right) \tag{49}$$

The energies are given by

$$E_{JK} = \frac{\hbar^2 J(J+1)}{2I_{\perp}} + \tfrac{1}{2} \hbar^2 K^2 \left(\frac{1}{I_{\parallel}} - \frac{1}{I_{\perp}} \right) \tag{50}$$

where K is another quantum number in addition to J. This a complicated system, so for pedagogic reasons we will confine our attention to discussing the diatomic molecule in more detail.

A diatomic or linear molecule has almost all its mass concentrated in the nuclei along the axis, so rotations around this symmetry axis involve electronic rather than nuclear rotations, and we do not take them into account. The resulting Hamiltonian for rotations around a perpendicular axis

$$\mathcal{H} = \frac{J^2 \hbar^2}{2I_\perp} \tag{51}$$

gives the following Schrödinger equation in spherical coordinates

$$\frac{\hbar^2}{2I_\perp}\left\{\frac{1}{\sin\theta}\frac{\partial}{\partial\theta}\left(\sin\theta\frac{\partial\Psi}{\partial\theta}\right)+\frac{1}{\sin^2\theta}\frac{\partial^2\Psi}{\partial\phi^2}\right\} = E\Psi \tag{52}$$

which has the solution

$$E_J = BJ(J+1) \tag{53}$$

where

$$B = \hbar^2/2I_\perp \tag{54}$$

Thus, the allowed states of rotation are quantized with the energies (53). If we take into account centrifugal distortion, or the stretching of the chemical bond by the centrifugal force arising from the rotation, there is a small correction to the energy

$$E_J = BJ(J+1) - DJ^2(J+1)^2 \tag{55}$$

where $D \ll B$. The energy spacing ΔE between two successive rotational J and $J+1$ is

$$\begin{aligned}\Delta E &= E_{J+1} - E_J \\ &= 2B(J+1) - 4D(J+1)^3\end{aligned} \tag{56}$$

The selection rules for transitions to occur between these rotational levels are

$$\Delta J = \pm 1, \qquad \Delta M = \pm 1, \tag{57}$$

although the energy itself does not depend on M.

The frequencies $\omega = \Delta E/\hbar$ of molecular rotational transitions from Eq. (56) are in the microwave region of the electromagnetic spectrum, typically in the range from 10^{10} to 10^{11} Hz. They can be measured by the direct absorption of microwave radiation, subject to selection rules (57).

11 MOLECULAR VIBRATIONAL SPECTROSCOPY

The discussion of vibrational frequencies ω earlier in the chapter was based on classical mechanics. A quantum mechanical treatment using the Hamiltonian

$$\mathcal{H} = \frac{p^2}{2m} + \tfrac{1}{2} m\omega^2 x^2 \tag{58}$$

gives the vibrational energies

$$E = \hbar\omega(n_v + \tfrac{1}{2}) \tag{59}$$

where $n_v = 0, 1, 2 \ldots$ is the vibrational quantum number, and each normal mode type has a particular fundamental vibrational frequency ω. Vibrational spectroscopy measures transitions $\Delta n_v = \pm 1$ between the energy levels of Eq. (59).

Molecular vibrations occur in the infrared region, typically from 10^{12} to 10^{14} Hz. To be able to measure directly a vibrational transition brought about by the absorption of infrared radiation the molecule must change its electric dipole moment p_{dip} during the vibration. The linear molecule CO_2 with positively charged carbon in the center and negatively charged oxygen at the ends has no dipole moment when the atoms are at their equilibrium positions. It should be clear from Fig. 4-2 that CO_2 undergoes a change in dipole moment along the molecular axis during the asymmetric stretch vibration, and undergoes a change in dipole moment in a direction perpendicular to this axis during a bending vibration. It is also clear from the figure that the dipole moment remains zero during a symmetric stretch. Because of the change in dipole moment the asymmetric stretch and bending modes can be detected by measuring this change in dipole moment, and hence they are said to be infrared active. The symmetric stretch vibration cannot be observed in this manner, and is said to be infrared inactive.

To measure indirectly a vibrational mode by Raman spectroscopy an incident optical photon is absorbed by the molecule and another optical photon is emitted, with the difference frequency $\hbar\omega$ equal to a vibrational transition frequency $\hbar\omega_{\text{vib}}$, hence

$$\omega_{\text{vib}} = \left|\omega_{\text{inc}} - \omega_{\text{emit}}\right| \tag{60}$$

where $\omega_{\text{vib}} \ll \omega_{\text{inc}}, \omega_{\text{emit}}$. For this to occur requires a change in the polarizability α, which causes the electric field vector E of the incident optical photon to polarize the molecule and induce in it an electric dipole moment p_{ind}

$$p_{\text{ind}} = \alpha E \tag{61}$$

The symmetric stretch vibration involves such a change in polarizability, and hence it is said to be Raman active.

Many large molecules have small molecular side groups such as amino (H_2N—), cyano (NC—), formyl (HCO—), hydroxy (HO—), imino (HN=), methyl (H_3C—), methylene (H_2C=), nitro (O_2N—), nitroso (ON—), and phospho (O_3P—), which have characteristic vibrational normal modes in which the remainder of the molecule does not participate actively. Hence, their presence can be detected by vibrational spectroscopy, and in addition characteristics of the molecular environment of the side groups can be inferred from small shifts in their frequencies of vibration. As a result of this both infrared and Raman spectroscopies have become major analytical tools for organic chemists.

CHAPTER 5

CANONICAL TRANSFORMATIONS

1 INTRODUCTION

In Chapter 2 we discussed Lagrangian and Hamiltonian formulations. The coordinates and canonical momenta of the Hamiltonian were determined from the coordinates and velocities of the Lagrangian. In this chapter we see how to transform the coordinates and momenta in ways that make it easier to solve problems in mechanics. Examples of transformed Hamiltonians, and some alternative ways to treat mechanical systems such as the harmonic oscillator are presented. The Hamilton–Jacobi equation, action–angle variable and Poisson bracket approaches to mechanics are delineated.

2 NATURE OF THE CANONICAL TRANSFORMATION

A canonical transformation is a coordinate transformation in phase space from initial coordinates and canonical momenta q_i, p_i to final ones Q_i, P_i that maintains the form of Hamilton's equations, and makes additive changes in the Hamiltonian from its initial configuration $\mathcal{H}(q_i, p_i, t)$ to its recast form $K(Q_i, P_i, t)$. A generating function $F(q, Q, t)$ brings about the transformation

$$\sum p_i\dot{q}_i - \mathcal{H} = \sum P_i\dot{Q}_i - K + \frac{dF}{dt} \tag{1}$$

with Hamilton's equations preserved

$$\left.\begin{aligned} \dot{q}_i &= \frac{\partial\mathcal{H}}{\partial p_i} & \dot{Q}_i &= \frac{\partial K}{\partial P_i} \\ \dot{p}_i &= -\frac{\partial\mathcal{H}}{\partial q_i} & \dot{P}_i &= -\frac{\partial K}{\partial Q_i} \end{aligned}\right\} \tag{2}$$

$$K = \mathcal{H} + \frac{\partial F}{\partial t} \tag{3}$$

and this particular generating function F, called $F_1(q_i, Q_i, t)$, may be differentiated to give the old and new canonical momenta

$$p_i = \frac{\partial F_1}{\partial q_i} \qquad P_i = -\frac{\partial F_1}{\partial Q_i} \tag{4}$$

Equations (1) and (3) are related to each other through Eq. (4),

A generating function F_j, in general, depends on one old and one new phase space coordinate, and since there are two old phase space coordinates, q and p, and two new ones, Q and P, there are four types of generating functions, as follows:

$$F = F_1(q, Q, t) \tag{5a}$$

$$F = F_2(q, P, t) - \sum Q_iP_i \tag{5b}$$

$$F = F_3(p, Q, t) + \sum q_ip_i \tag{5c}$$

$$F = F_4(p, P, t) + \sum q_ip_i - \sum Q_iP_i \tag{5d}$$

We saw from Eqs. (4) how partial derivatives of the generating function give the corresponding phase space coordinate. Another example of this is

$$p_i = \frac{\partial F_2}{\partial q_i} \qquad Q_i = -\frac{\partial F_2}{\partial P_i} \tag{6}$$

The particular generating function $F_1 = q_i Q_i$ interchanges coordinates and momenta, i.e., $p_i = Q_i$ and $q_i = -P_i$, while the generating function $F_2 = q_i P_i$ provides an identity transformation with $q_i = Q_i$ and $p_i = P_i$.

3 HARMONIC OSCILLATOR

An interesting insight into the mature of an harmonic oscillator is obtained by carrying out a canonical transformation with the aid of the following F_1 generating function:

$$F_1(q, Q) = \tfrac{1}{2} m\omega q^2 \cot Q \tag{7}$$

This converts the one dimensional harmonic oscillator Hamiltonian

$$\mathcal{H} = \frac{p^2}{2m} + \frac{kq^2}{2} \tag{8}$$

where $k = m\omega^2$, to the new Hamiltonian

$$K = \omega P \tag{9}$$

where the new momentum $P = E/\omega$ is a constant, E is the energy, and the new coordinate Q depends linearly on the time

$$Q = \omega t + \delta \tag{10}$$

This approach gives the time dependence of the original coordinate and momentum

$$q = (2E/m\omega^2)^{1/2} \sin(\omega t + \delta) \tag{11}$$

$$p = (2mE)^{1/2} \cos(\omega t + \delta) \tag{12}$$

These expressions, of course, can be found more easily by elementary methods. Plots involving Eqs. (9)–(12) presented in Fig. 5-1 illustrate two ways to picture the motion of a simple harmonic oscillator.

4 HAMILTON–JACOBI EQUATION

The Hamilton–Jacobi equation is derived by employing the generating function $F_2(q, P, t)$ to carry out a canonical transformation to a system in which the new Hamiltonian is identically zero: $K = 0$. We can see from Eq. (2) that in the new system the time derivatives of the coordinates and momenta

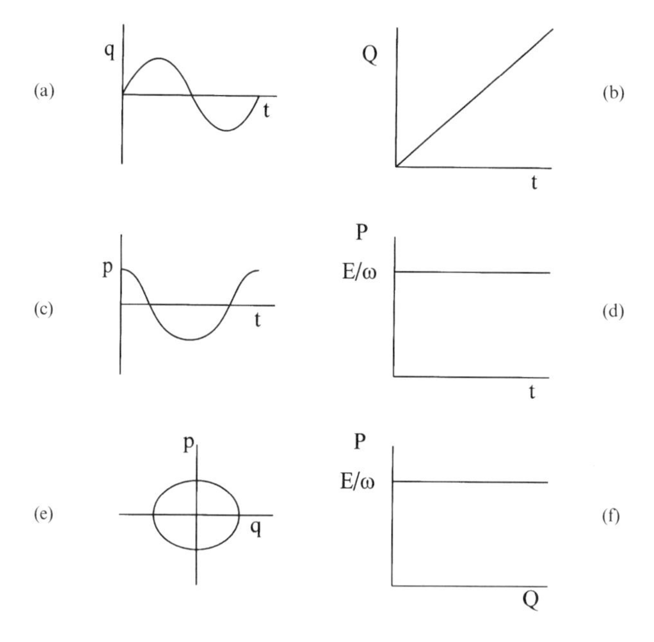

Fig. 5-1. Time dependence of the coordinate q (a) and the momentum p (c) of the simple harmonic oscillator with the Hamiltonian of Eq. (8). The p versus q phase diagram is in (e). The corresponding P, Q, t plots in the transformed coordinate system obtained from the generating function $F_1(q, Q)$ of Eq. (7) are presented in (b), (d), and (f).

vanish, i.e., $\dot{Q}_i = 0$ and $\dot{P}_i = 0$, and hence the new coordinates and momenta are constants of the motion which we designate as α_i and β_i, respectively

$$\begin{aligned} P_I &= \alpha_i \\ Q_I &= \beta_I \end{aligned} \tag{13}$$

The expression for p_i from Eq. (6)

$$p_i = \frac{\partial F_2}{\partial q_i} \tag{14}$$

can be inserted into the Hamiltonian $\mathcal{H}(q_i, p_i, t)$ of Eq. (3) to give what is called the Hamilton–Jacobi equation

$$\mathcal{H}\left(q_i, \frac{\partial S^*}{\partial q_i}, t\right) + \frac{\partial S^*}{\partial t} = 0 \tag{15}$$

where we have followed the common custom of writing S^* instead of F_2 for what is called the Hamilton principal function. If, as is often the case, the Hamiltonian does not depend explicitly on the time, then we can write

$$S^*(q_i, \alpha_i, t) = S(q_i, \alpha_i) - Et \tag{16}$$

and the Hamilton–Jacobi equation assumes the time-independent form

$$\mathcal{H}\left(q_i, \frac{\partial S}{\partial q_i}, t\right) = E \tag{17}$$

where $S(q_i, \alpha_i)$ is called the Hamilton's characteristic function, and the constant E is identified as the total energy of the system.

As an example of the above we write down the Hamilton–Jacobi equation for the one-dimensional harmonic oscillator (8)

$$\frac{1}{2m}\left(\frac{\partial S}{\partial q}\right)^2 + \tfrac{1}{2}m\omega^2 q^2 = E \tag{18}$$

The integral for Hamilton's principal function is easily written down

$$S^* = (2m)^{1/2} \int [E - \tfrac{1}{2}m\omega^2 q^2] \, dq - Et \tag{19}$$

From this expression we can derive Eq. (11) and (12), respectively, that were found earlier for the time dependence of the original coordinate q and momentum p (see Goldstein, 1980, pp. 443–444).

5 ACTION–ANGLE VARIABLES

Periodic motion can be treated in terms of the generating function $F_2(q, P)$ where the transformed momentum P, customarily referred to as the action variable J, is given by the integral

$$P = J = \oint p dq \tag{20}$$

over a closed region of phase space. This integral is the area enclosed by the curve of q, p in the two-dimensional phase space, and is constant in time. It has the units of angular momentum. The variable Q conjugate to J, called the angle variable and denoted by w, is given by

$$Q = w = \frac{\partial}{\partial J} F_2(q, P) \tag{21}$$

The Hamiltonian $\mathcal{H}(J)$ is a function of J only, so form Hamilton's equation we obtain

$$\dot{w} = \frac{\partial}{\partial J} \mathcal{H}(J) = \nu(J) \tag{22}$$

where ν is a frequency dependent only on J, and Eq (22) has the immediate solution

$$w = \nu t + \beta \tag{23}$$

where β is a phase constant.

We can show that the angle variable w changes by unity

$$\Delta w = 1 = \nu \tau \tag{24}$$

in the time τ that q goes through one period, which means that $\nu = 1/\tau$ is the reciprocal of the period. This method has the advantage that it can provide the frequency of a periodic motion without completely solving the problem.

An example of the utility of the action–angle approach is provided by the one-dimensional harmonic oscillator whose Hamiltonian (8) can be written in the form

$$\frac{p^2}{2m} + \tfrac{1}{2} m\omega^2 q^2 = E \tag{25}$$

The action variable J

$$J = (2m)^{1/2} \oint [E - \tfrac{1}{2} m\omega^2 q^2] dq \tag{26}$$

can be integrated directly to give

$$J = \frac{2\pi E}{\omega} \tag{27}$$

so

$$E = \mathcal{H} = \frac{J\omega}{2\pi} \tag{28}$$

and

$$\nu = \frac{\partial \mathcal{H}}{\partial J} = \frac{\omega}{2\pi} = \frac{1}{2\pi}(k/m)^{1/2} \tag{29}$$

The result is

$$q = (J/\pi m\omega)^{1/2} \sin 2\pi w \tag{30}$$

$$p = (mJ\omega/\pi)^{1/2} \cos 2\pi w \tag{31}$$

which, through Eqs. (23) and (28), agrees with what was found above in Eqs. (11) and (12) with $\delta = 2\pi\beta$.

6 POISSON BRACKETS

Another way to approach mechanics is through the Poisson bracket formalism. A Poisson bracket of two functions u, v with respect to two conjugate canonical variables q_i, p_i is defined as

$$[u, v] = \frac{\partial u}{\partial q_i}\frac{\partial v}{\partial p_i} - \frac{\partial u}{\partial p_i}\frac{\partial v}{\partial q_i} \tag{32}$$

It is easy to show that

$$[q_i, q_j] = [p_i, p_j] = 0 \tag{33}$$

$$[q_i, p_j] = -[p_i, q_j] = \delta_{ij} \tag{34}$$

The equation of motion for a function f expressed in terms of Poisson brackets is

$$\frac{df}{dt} = [f, \mathcal{H}] + \frac{\partial f}{\partial t} \tag{35}$$

where $\mathcal{H}$ is the Hamiltonian. In particular, for a coordinate x we have

$$v_x = \dot{x} = [x, \mathcal{H}] \tag{36}$$

Poisson brackets are the classical analogs of the commutators of quantum mechanics.

We can show, for angular momenta L_i, that

$$[L_i, L_j] = \varepsilon_{ijk} L_k \tag{37}$$
$$[p_i, L_j] = p_k \tag{38}$$

where the dimensionless Levi–Civita symbol ε_{ijk} is zero if any two of its indices are the same, it is +1 for a cyclic permutation of the three indices ijk, and it is −1 for an anticyclic permutation. Thus, we have, for example

$$
\begin{aligned}
\varepsilon_{123} &= \varepsilon_{231} = +1 \\
\varepsilon_{213} &= \varepsilon_{132} = -1 \\
\varepsilon_{113} &= \varepsilon_{232} = 0
\end{aligned}
\tag{39}
$$

The Levi–Civita symbol is also called the permutation symbol, the alternating tensor, and the isotropic tensor of rank 3.

CHAPTER 6

NON-LINEAR DYNAMICS AND CHAOS

1 INTRODUCTION

We have been treating integrable problems, ones in which the equations of motion can be integrated to give a closed-form solution. For the two-body inverse square law problem we found solutions involving motion in elliptic, parabolic and hyperbolic orbits, the former of which constitute closed orbits. However, even for the two-body problem there are many potentials for which the equations of motion are not integrable. In this chapter we investigate what happens when no well-behaved solution exists. We begin with cases in which the non-integrability arises from a perturbation on the main interactions, and we find that the orbits of the motion are close to those without a perturbation. We also find that if the perturbation becomes large enough, then the motion can become chaotic. We examine the characteristic features of chaos, and even find regions of normal behavior embedded within those of chaotic behavior. The logistic equation and several other mathematical and physical systems that exhibit chaos are discussed.

2 PERTURBATION THEORY

Perturbation theory is often applied to problems in which the dominant interaction involves an integrable Hamiltonian $\mathcal{H}_0(p, q)$ for which the solution is known, and an additional interaction that can be taken into account by adding a perturbation term $\Delta\mathcal{H}(p, q)$ to the overall Hamiltonian $\mathcal{H}(p, q)$

$$\mathcal{H} = \mathcal{H}_0 + \Delta\mathcal{H} \tag{1}$$

One convenient approach involves using the generating function $S(q, P_0, t) = F_2(q, P, t)$ to transform the dominant Hamiltonian term $\mathcal{H}_0(p, q)$ from the phase space coordinates p, q to new coordinates P_0, Q_0 of a transformed Hamiltonian $K_0(Q, P)$ that is identically zero

$$K_0(Q_0, P_0) = 0 \tag{2}$$

as in the Hamilton–Jacobi approach. The Hamilton equations

$$\frac{\partial K}{\partial P_0} = \dot{Q}_0 \qquad \frac{\partial K}{\partial Q_0} = -\dot{P}_0 \tag{3}$$

for $K = 0$ provide new coordinates and momenta Q_0 and P_0, which are constants of the motion.

$$Q = Q_0 \qquad P = P_0 \tag{4}$$

The subscripts indicate that they are constants for the zero-order Hamiltonian, not taking into account the perturbation $\Delta\mathcal{H}$.

The same transformation is now carried out for the total Hamiltonian

$$\mathcal{H}(P_0, Q_0) = \mathcal{H}(P_0, Q_0) + \Delta\mathcal{H}_0(P_0, Q_0) \tag{5}$$

and this provides us with a transformed Hamiltonian $\Delta K_0(P_1, Q_1)$, which can be used to obtain first-order corrections to the time derivatives of the coordinates and momenta

$$\frac{\partial \Delta K_0}{\partial P_1} = \dot{Q}_1 \qquad \frac{\partial \Delta K_0}{\partial Q_1} = -\dot{P}_1 \tag{6}$$

These expressions can be integrated to give the first-order determination of Q_1 and P_1

$$Q_1 = \int \frac{\partial \Delta K_0}{\partial P} \, dt \tag{7}$$

$$P_1 = -\int \frac{\partial \Delta K_0}{\partial Q} \, dt \tag{8}$$

The procedure provides us with a new generating function $S(Q_0, P_1, t)$ and hence a new perturbed Hamiltonian $\Delta K_1(Q_2, P_2)$, which can be iterated to determine the next higher order terms Q_2 and P_2, etc. Thus, we have a systematic iteration technique for calculating better and better approximations to the solution when the known perturbation $\Delta\mathcal{H}$ is present.

3 HARMONIC OSCILLATOR IN PHASE SPACE

In Chapter 1 we discussed how cyclic coordinates have associated conjugate momenta that are constants of the motion. If all of the coordinates are not cyclic, then we apply one or a sequence of canonical transformations to new variables Q_i, which are cyclic so the Hamiltonian has the form $\mathcal{H}(P_1, P_2, \ldots P_N)$. For the present we take into account only the unperturbed system, and then afterwards we consider the effect of adding perturbations. Hamilton's equations for the transformed cyclic Hamiltonian

$$\dot{Q}_i = \frac{\partial \mathcal{H}}{\partial P_i} \tag{9}$$

$$\dot{P}_i = \frac{\partial \mathcal{H}}{\partial Q_i} \tag{10}$$

have the solutions

$$Q_i(t) = \omega_i t + Q_i(0) \tag{11}$$

$$P_i(t) = P_i(0) + E_i/\omega_i \tag{12}$$

for each of the N variables, where we take $P_1(0) = 0$, and E_i is the conserved energy.

We examine the case $N = 1$ and discuss higher N values in the next section. This $N = 1$ solution was found earlier [Chapter 5, Eqs. (11) and (12)] for the harmonic oscillator, and using Eq. (9) from Chapter 5 it provides the following well-known expression for the initial variables p' and q':

$$q' = (2P/m\omega)^{1/2} \sin \omega t \tag{13}$$

$$p' = (2mP\omega)^{1/2} \cos \omega t \tag{14}$$

The motion in the q', p' plane is confined to an ellipse, and the motion of the dimensionless, normalized phase space coordinates q and p

$$q = q'/(2P/m\omega)^{1/2} \qquad p = p'/(2mP\omega)^{1/2} \tag{15}$$

is uniform along a unit circle.

4 SYSTEM TRAJECTORIES ON *N*-TORI

For the case $N = 2$ there are two simple harmonic solutions in the four-dimensional phase space q_1, p_1, q_2, p_2. If we deal with normalized variables, then one is a circle in the q_1, p_1 plane, and the other is a circle in the q_2, p_2 plane. For illustrative purposes we consider the radius of the q_1, p_1 circle to be greater, and the frequency ω_1 of the first oscillator to be less than their counterparts associated with the other oscillator. The motion of q_2, p_2 is on a circle centered on the q_i, p_1 circle and oriented at right angles to it. The joint motion is a spiraling of the system point around the q_1, p_1 path, as illustrated in Fig. 6-1.

If the frequency ω_2 is a multiple of ω_1, then the system point retraces the same spirals, so the orbit is closed. If the ratio ω_2/ω_1 is a rational fraction,

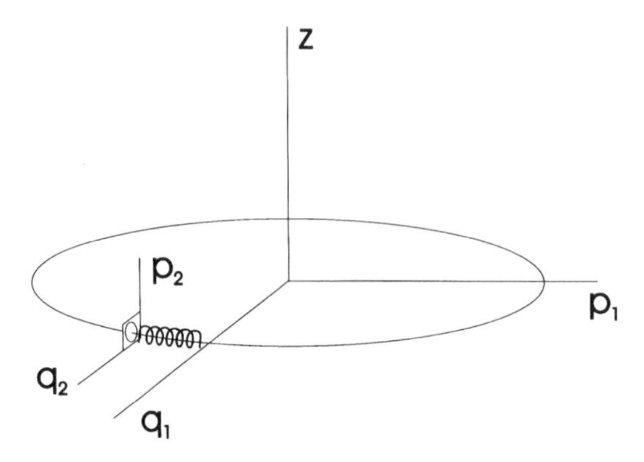

Fig. 6-1. Orbits of the system points for two harmonic oscillators in their respective q_1, p_1 and q_2, p_2 phase spaces for the case $\omega_1 \ll \omega_2$. The combined motion is a spiral that generates a torus centered along the circular parth of the first oscillator trajectory.

i.e., a ratio of integers, then the orbit will still be closed, but will consist of more than one spiral around the q_1, p_1 circle. These are cases of commensurate frequencies that lead to closed orbits. If the frequencies are not commensurate, i.e., the ratio ω_2/ω_1 is an irrational number, then the spiral will fill the surface of the torus without ever passing through the same point twice, but eventually it will pass arbitrarily close to every point on the surface. This is a bounded orbit, confined to a surface, but not closed.

This description is easily generalized to the $N = 3$ case in which the motion is confined to a three-dimensional surface called a 3-torus in the six-dimensional $q_1, q_2, q_3, p_1, p_2, p_3$ phase space, and so on for N-tori in $2N$ dimensional phase spaces.

5 STABILITY OF PERTURBED ORBITS AND CHAOS

When a mechanical system is integrable and it is subjected to a small perturbation that makes it non-integrable the question arises as to whether or not the perturbed solution will be stable, and whether or not the orbits will remain close to the unperturbed ones over long periods of time. The Kolmogorov–Arnold–Moser (KAM) theorem tells us that (i) if the perturbation causing non-integrability is small, and (ii) if the frequencies ω_i of the unperturbed motion are uncorrelated, then the motion is confined to an N-torus so the perturbed orbits are stable and localized in the same region as the unperturbed ones. The theorem also states that for $N > 2$ there may be a negligibly small set of initial conditions that lead to wandering trajectories on the energy surface. The N-tori are called KAM surfaces, or KAM curves when pictured in cross-sections of the surface.

The KAM theorem applies to small perturbations. If the perturbation becomes sufficiently large then the behavior may change to that of chaos whereby successively calculated orbits move further and further away from the original one. In fact, the extent to which they recede can increase exponentially with the number of iterations. We examine these regions of chaotic behavior.

An example of how linear and chaotic motion differs is turbulence in water. During streamline flow two nearby points in the water stream stay close to each other as they move along. After the onset of turbulence the same two points keep moving further and further apart. Another example is a spaceship in an earth orbit. A small rocket boost will move it to a nearby orbit whereas a strong boost could throw it out of orbit, or even send it into outer space.

The phenomenon of chaos has three fundamental properties; namely, mixing dense periodic points, and a sensitivity to initial conditions.

1. Mixing means that if we can choose two arbitrary small intervals or regions, I_1 and I_2 then an orbit starting in region I_1 will eventually pass through I_2. We already mentioned how incommensurate orbits pass arbitrarily close to all points on the surface of a torus.
2. Dense periodic orbits are orbits that pass repeatedly very close to the same points of the domain. The moon follows closely the same path around the earth despite perturbations from the sun and other planets.
3. An example of sensitivity to initial conditions is how these conditions influence the rate at which successive orbits on a torus move away from each other.

6 LOGISTIC EQUATION OR QUADRATIC ITERATOR

To gain some insight into the nature of chaos it is helpful to analyze a simple mathematical example that exhibits the regularities of ordinary repetitive behavior, as well as the irregularities of chaotic demeanor. The example chosen is the logistic equation, sometimes called the quadratic iterator, which is defined by the expression

$$x_{n+1} = ax_n(1 - x_n) \tag{16}$$

for the domain

$$0 \le x \le 1 \tag{17}$$

where the control parameter a is positive. It is found that the behavior of this equation changes from being regular to being chaotic near the value $a = 3.67$ so it is ordinarily studied over the range

$$0 \le a \le 4 \tag{18}$$

of control parameter a.

To illustrate how this equation can converge in the limit $n \Rightarrow \infty$ to a single solution x_∞ for which $x_{n+1} = x_n$ irrespective of the starting value x_0 of the iteration, we consider the case $a = 2$ and the three initial values of $x_0 = 0.2000$, 0.3000, and 0.9000. For these starting points successive iterations give the values shown in Table 6-1 and two of these iterations are plotted in Fig. 6-2. In fact, every initial value in the range (17) proceeds toward this same final value $x_n = \frac{1}{2}$ for large n. It appears that the values of x_{n+1} for successive iterations are attracted to the value $x_\infty = \frac{1}{2}$. This limiting value x_∞ for $n \Rightarrow \infty$ is called an attractor because arbitrary initial values proceed by iteration toward the attractor.

To illustrate how more than one solution can exist we start with the control parameter $a = 3.2$ and the previous initial value $x_0 = 0.3000$. For

TABLE 6-1. Example of how iterations of the logistic equation from three different initial values x_0 converge to the same final value for control parameter a = 2

n	x_n	n	x_n	n	x_n
0	0.2000	0	0.3000	0	0.9000
1	0.3200	1	0.4200	1	0.1800
2	0.4352	2	0.4872	2	0.2952
3	0.4916	3	0.4997	3	0.4161
4	0.4999	4	0.5000	4	0.4899
5	0.5000	5	0.5000	5	0.4996
.	.	.	.	.	.
.	.	.	.	.	.
.	.	.	.	.	.
25	0.5000	25	0.5000	25	0.5000

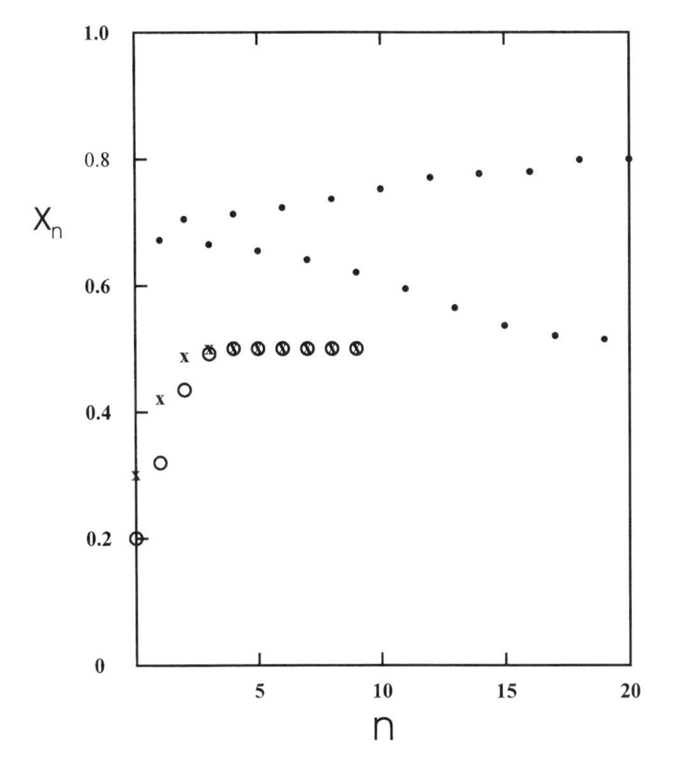

Fig. 6-2. Plot of the values x_n obtained for successive iterations of the logistic equation $N = 1$ cycle for control parameter $a = 2$ with the initial values $x_0 = 0.2$ (O) and 0.3 (X), and for successive iterations of the $N = 2$ cycle ($\cdot$) for control parameter $a = 3.2$ with the initial value $x_0 = 0.3$. Note the rapid convergence for the 1-cycle and the much slower convergence for $N = 2$.

successive iterations we have the values given in Table 6-2. It is clear from these data and their plot in Fig. 6-2 that the convergence is now slower, and after many iterations the value of x alternates between two final attractor values as follows:

$$\begin{aligned} x_n &= 0.5131 \\ x_{n+1} &= 0.7994 \end{aligned} \tag{19}$$

It is easy to show that the control parameter $a = 3.5$ gives a fourfold cycle with four attractors

TABLE 6-2. Illustration of how successive iterations of the logistic equation (16) alternate between two final values x = 0.5131 and x = 0.7994 when the control parameter a = 3.2

n	x_n
0	0.3000
1	0.6720
2	0.7053
3	0.6651
4	0.7128
5	0.6551
6	0.7230
7	0.6408
8	0.7365
9	0.6210
10	0.7531
11	0.5950
12	0.7711
13	0.5647
14	0.7866
15	0.5372
.	.
.	.
.	.
24	0.7994
25	0.5131

$$\begin{aligned} x_n &= 0.501 \\ x_{n+1} &= 0.875 \\ x_{n+2} &= 0.383 \\ x_{n+3} &= 0.827 \end{aligned} \tag{20}$$

$a = 3.55$ gives an eightfold cycle, and $a = 3.567$ gives a sixteenfold cycle.

Figure 6-3, a sequence of Feigenbaum diagrams, presents a plot of the limiting values x_n obtained by letting $n \Rightarrow \infty$ as a function of the control parameter a. We see from the graph that as a increases the system undergoes successive bifurcations at which the number of solutions to the logistic equation doubles with increasing a. This doubling continues until the

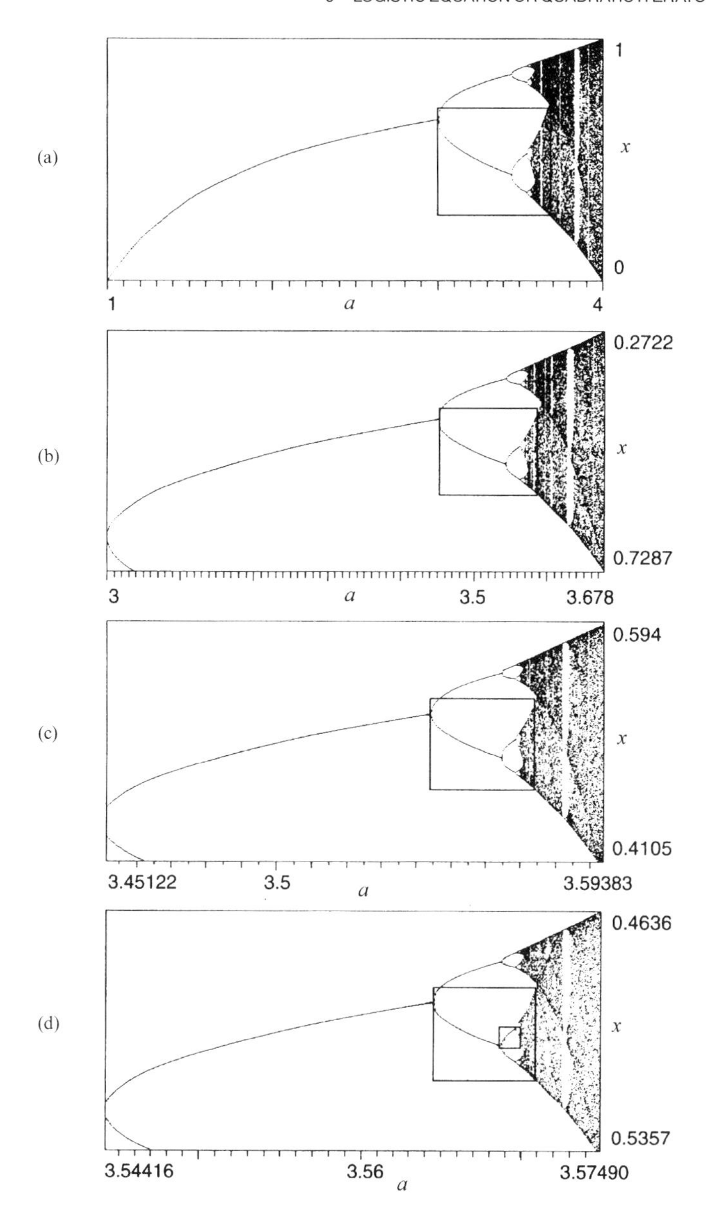

Fig. 6-3. Feigenbaum diagrams (x plotted vs. a) of the logistic equation showing the overall range for $1 \leq a \leq 4$ and $0 \leq x \leq 1$ (a), and expanded views from top to bottom (b–d) of the first bifurcation ($\Delta a = 0.68$, $\Delta x = 0.46$), the second bifurcation ($\Delta a = 0.14$, $\Delta x = 0.18$) and the third bifurcation ($\Delta a = 0.03$, $\Delta x = 0.07$). (From H. O. Peitgen et al., *Chaos and Fractals*, Springer-Verlag, Berlin, 1992, p. 589.)

control parameter a reaches a critical value $a_\infty \sim 3.5699$, which will be given to greater accuracy in the next section.

For the choice of a control parameter a greater than a_∞ the system becomes chaotic, and successive x_n terms generate a sequence of what seems like random numbers. To see this chaotic behavior, consider selecting the control parameter $a = 4.0$, which exceeds a_∞ and two initial values of x_0 that are close together, such as $x_0 = 0.3000$ and $x_0 = 3.100$. The calculations give Table 6-3, where we see that for the first couple of iterations the successive x_n values of the two lists remain close together, but they soon become randomized relative to each other. The closer the initial values are to each other the longer it takes for successive iterations to become uncorrelated.

7 CHARACTERISTICS AND CONSTANTS OF CHAOS

An analysis of the results of the previous section shows that the logistic equation exhibits the three characteristic properties of chaotic systems that were mentioned above:

1. The property of mixing is present because if we select a small interval such as from $x = 0.366$ to $x = 0.367$ and we start with any initial

TABLE 6-3. Generation of random values by the logistic equation for control parameter a > a_∞

n	x_n	n	x_n
0	0.3000	0	0.3100
1	0.8400	1	0.8556
2	0.5376	2	0.4942
3	0.9943	3	0.9999
4	0.0225	4	0.0005
5	0.0879	5	0.0020
6	0.3208	6	0.0086
7	0.8716	7	0.0340
8	0.4476	8	0.1318
9	0.9890	9	0.4576
10	0.0434	10	0.9928
11	0.1661	11	0.0029
12	0.5542	12	0.1109
13	0.7034	13	0.3945

value x_0 and a control parameter $a > a_\infty$, then the succession of iterations $x_2, x_3, x_4, \ldots$ is sufficiently random so that it will eventually yield a value of x_n that is within the chosen interval; namely, $0.366 < x_n < 0.367$.

2. Dense periodic orbits, meaning very closely spaced points in this case, are obtained as the successive iterated points x_n approach the values of the attractors.
3. The sensitivity to initial conditions is shown by how the iteration proceeds for different choices of starting values x_0.

The sequence of Feigenbaum diagrams presented in Fig. 6-3 shows the series of bifurcations in which the number of attractors a successively doubles: 1, 2, 4, 8, ..., until the control parameter a reaches the point

$$a_\infty = 3.5699456\ldots \tag{21}$$

beyond which the x_n values generated by successive iterations become random, and we say that the behavior is chaotic. When the region near each bifurcation is enlarged we see that the ensuing bifurcations appear self similar to each other, but on successively enlarged scales. The ratio of the horizontal spacings between successive bifurcations converges for $n \Rightarrow \infty$ to a limit called the Feigenbaum number δ, and δ has been evaluated to many significant figures

$$\delta = \lim \frac{a_n - a_{n-1}}{a_{n+1} - a_n} = 4.6692016\ldots \tag{22}$$

The vertical spacings at the bifurcations called Δ_n also decrease with increasing n, and as $n \Rightarrow \infty$ the ratio Δ_n/Δ_{n-1} of successive vertical spacings converges to a limit α which has the value

$$\alpha = \lim \frac{\Delta_n}{\Delta_{n+1}} = 2.50290787\ldots \tag{23}$$

The Feigenbaum number δ is a universal constant, found with many chaotic systems, but the values of the Feigenbaum point a_∞ and the number α are specific to the logistic equation.

Another curious thing is that beyond the Feigenbaum limit a_∞ there are intervals of control parameter values, which appear white on Figs. 6-3 and 6-4, where regions of order with attractors and period doublings are imbedded in the chaotic zone. Figure 6-4 shows the region of order near $a = 3.84$, and presents enlargements of two bifurcations found there.

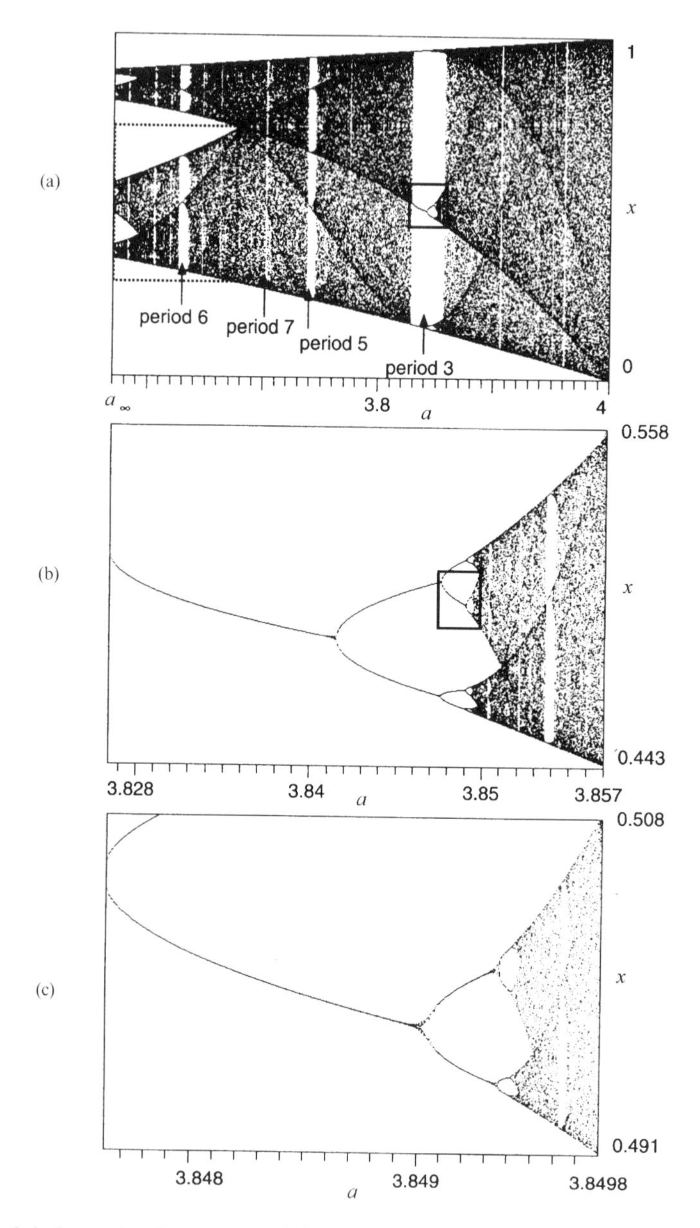

Fig. 6-4. Lateral enlargement of the Feigenbaum plot (x vs. a) of Fig. 6-3 in the region of chaos $a_\infty \leq a \leq 4$ (a) and expanded views of the first bifurcation (b) and the second bifurcation (c) embedded in the region of chaos near the control parameter value $a = 3.84$. (From H. O. Peitgen et al., *Chaos and Fractals*, Springer-Verlag, Berlin, 1992, p. 637.)

The period doubling behavior that we have described is typical of many chaotic systems, as is the universal Feigenbaum number. This Feigenbaum number δ has been determined experimentally by measurements on several physical systems that exhibit period doubling and the onset of chaos, such as: hydrodynamics (H_2O, He, Hg); electronics (diodes, transistors); laser feedback, Josephson junctions and acoustics (He).

8 CHAOS IN DIFFERENT SYSTEMS

Until now the discussion of chaos has been in terms of the one-dimensional logistic equation. There are many other systems that exhibit chaos, and we discuss briefly several of them.

The first example is the Henon–Heiles coupled harmonic oscillator system, which has a Hamiltonian that we write in a dimensionless form

$$\mathcal{H} = \tfrac{1}{2}(p_x^2 + p_y^2 + x^2 + y^2) + x^2 - \tfrac{1}{3}y^2)y \tag{24}$$

where the first term on the right is the standard two-dimensional harmonic oscillator part and the coupling term $\mathcal{H}_{\mathrm{COUP}} = x^2 - \frac{1}{3}y^2)$ may be treated as a perturbation to carry out calculations. This Hamiltonian is of interest because astronomers have used it to model the motion of a star moving in a cylindrically symmetric potential in the galactic disc.

The energy is a constant of the motion so the orbits lie on a three-dimensional constant energy hypersurface in the four-dimensional phase space p_x, p_y, x, y. Figure 6-5 presents slices through this energy surface in the p_y versus y plane, called Poincaré maps, for the three (dimensionless) energies $E = \frac{1}{12}, \frac{1}{8}$, and $\frac{1}{6}$. The Poincaré maps (b) calculated by canonical perturbation theory show how the constant energy surface slices through the p_y versus y plane. Such perturbation calculations do not take into account the possibility of chaotic behavior. The more realistic computer calculations (a) approximate normal behavior for the lowest energy $E = \frac{1}{12}$, and exhibit widespread chaos for the highest energy $E = \frac{1}{6}$. The intermediate energy $E = \frac{1}{8}$ is in the neighborhood of the onset of chaos. Analogous Poincaré maps of the p_x versus x plane, not shown, have orbits that differ from those of the p_y versus y plane because of the lack of symmetry of the coupling term $\mathcal{H}_{\mathrm{COUP}}$, but they exhibit the same degree of chaos for each energy.

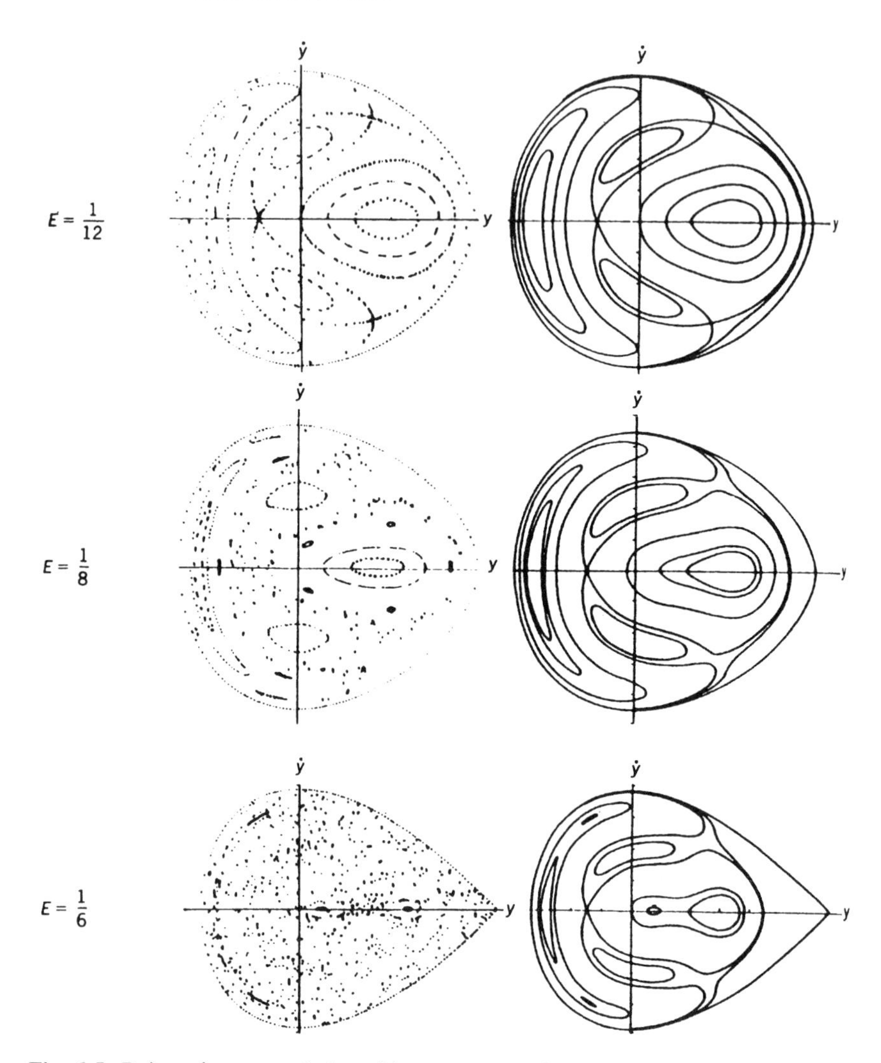

Fig. 6-5. Poincaré map made by taking a cross-section of the Henon–Heiles energy surface in the y, p_y plane for three sets of energies; namely, $E = \frac{1}{12}$ in the non-chaotic region, $E = \frac{1}{8}$ near the transition to chaos, and $E = \frac{1}{6}$ in the chaotic region. Each dot on the left-side figures is made by an orbit as it passes through the y, p_y plane, and the right-side figures were calculated by perturbation theory. In the non-chaotic region the dots are on the regular trajectories of the pertubation calculations, while in the chaotic region they exhibit a great deal of randomness. (See F. Gustavson, *Astron. J.*, **71**, 1996, 670; J. Moser, *Amer. Math. Soc.* **81**, 1968; and R. J. Creswick, H. A. Farach, and C. P. Poole, Jr., *Inroduction to Renormalization Group Methods in Physics*, Wiley, New York, 1992, p. 36.)

Several sets of coupled equations have been found that exhibit chaos. Otto E. Rössler proposed the following set, which we write as a function of time:

$$\begin{aligned} x'(t) &= -[y(t) + z(t)] \\ y'(t) &= x(t) + ay(t) \\ z'(t) &= b + z(t)[x(t) - c] \end{aligned} \tag{25}$$

and Figure 6-6 plots a typical behavior of the system in three dimensions. The motion is confined to spend most of its time close to the attractor in the x, y plane where it spirals outward from the origin until it reaches a threshold value where it rises upward along the z direction toward a maximum height. It is then thrust downward and reinserted into the dense orbit region

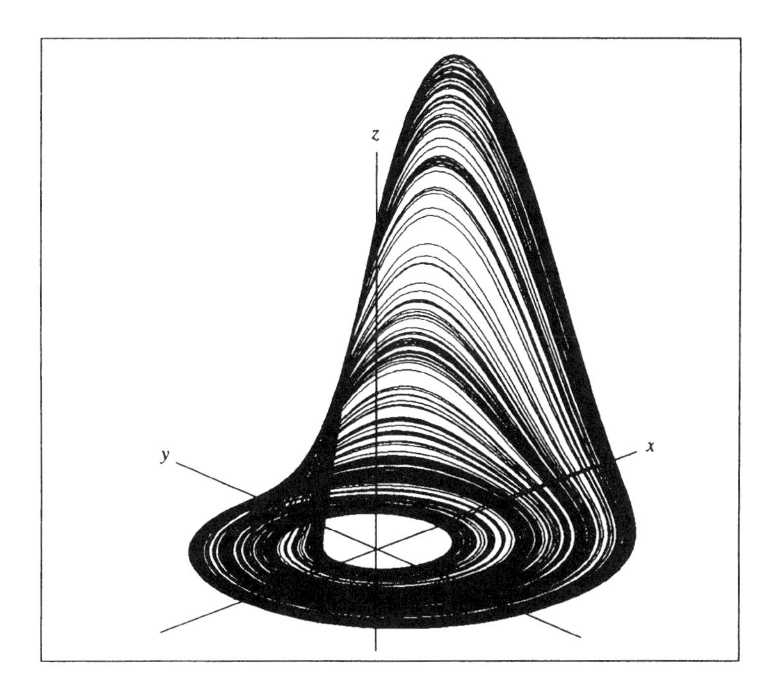

Fig. 6-6. Rössler attractor showing the dense orbits spiraling outward around the attractor near the x, y plane, and the less dense excursion trajectories along z that constitute ejections from an outer orbit and reinsertion into an inner orbit near the plane. (From H. O. Peitgen et al., *Chaos and Fractals*, Springer-Verlag, Berlin, 1992, p. 688.)

close to the x, y plane to repeat the spiraling outwards motion. The overall sequence of spiraling outward, ejection, and then reinsertion repeats itself continuously, as shown.

Figure 6-7 shows a Feigenbaum diagram of the Rössler system with minimal $|x|$ values plotted against the parameter c in the range $2.5 < c < 10$ when the other two parameters are set at $a = b = 0.2$. For

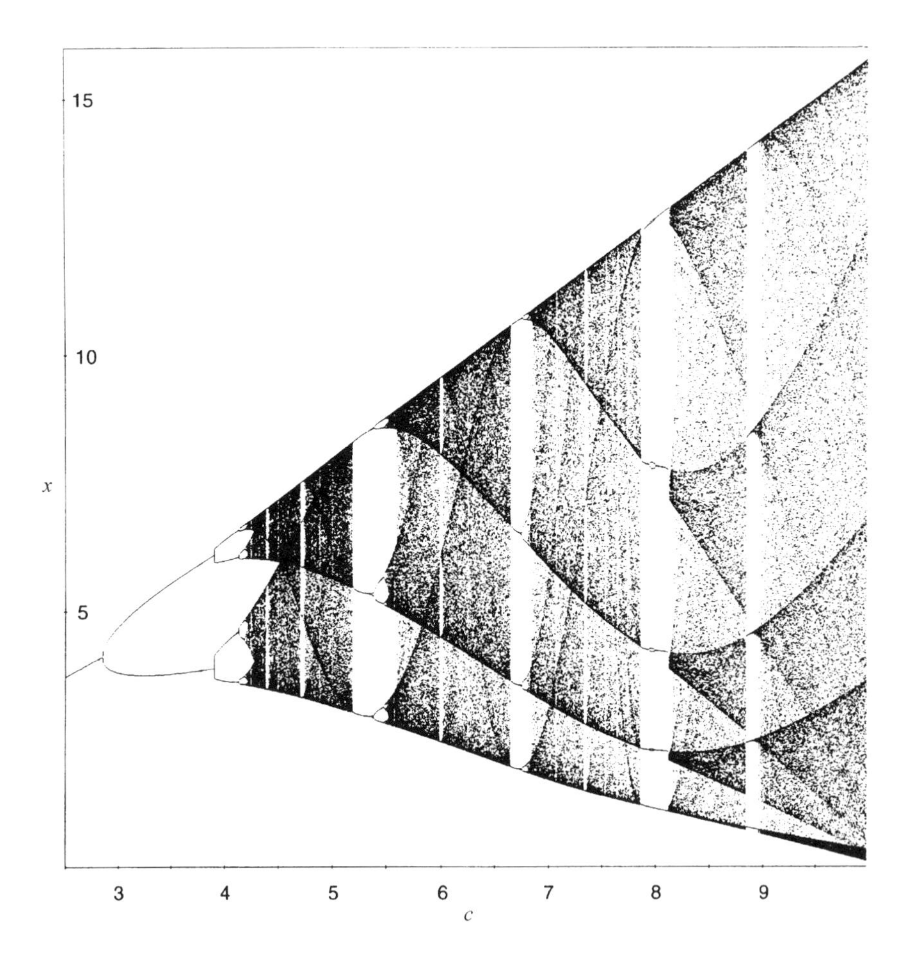

Fig. 6-7. Feigenbaum diagram (x plotted vs. c) for the Rössler system with the parameter values $a = b = 0.2$ in Eqs. (25), and $2.5 < c < 10$. At $c = 8$ we see the five $|x|$ values corresponding to the five spirals shown in Fig. 6-8. (From H. O. Peitgen et al., *Chaos and Fractals*, Springer-Verlag, Berlin, 1992, p. 692.)

the lowest values of c there is only one value of $|x|$ so the motion is a single spiral in the x, y plane, followed by an excursion along z, then another single spiral, etc. After one bifurcation of the Feigenbaum plot, as for example at $c = 3.5$, there are two x, y plane spirals per excursion along z. Figure 6-8 shows the five x, y plane spirals per excursion along z for the case $c = 8$, corresponding to five values of $|x|$ seen in Fig. 6-7 for this case.

Edward N. Lorenz proposed a similar set of equations, which we write without including a time dependence

$$\begin{aligned} x' &= \sigma(x + y) \\ y' &= Rx - y - xz \\ z' &= -Bz + xy \end{aligned} \tag{26}$$

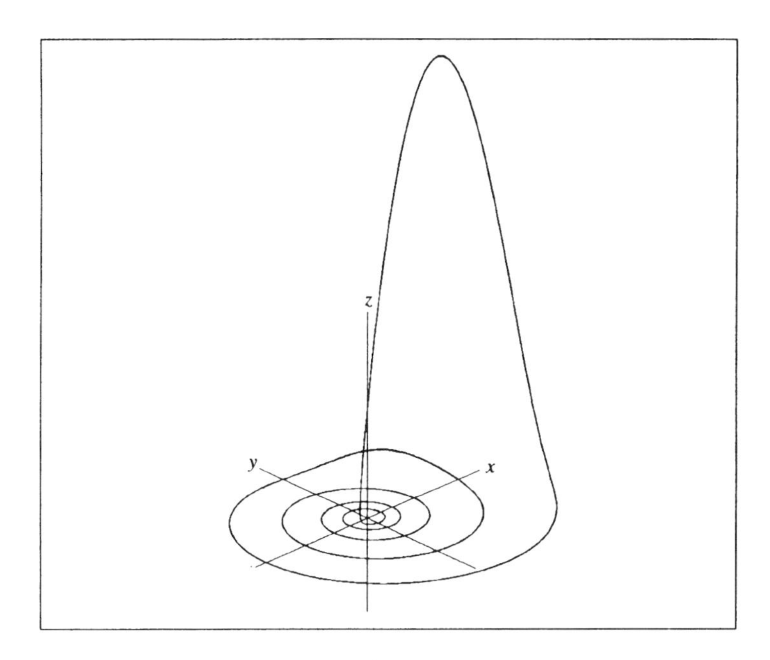

Fig. 6-8. Attracting periodic trajectory for the Rössler system with $a = b = 0.2$ and $c = 8$. Note that there are five cycles near the x, y plane for each excursion along z, corresponding to the five values of $|x|$ appearing at $c = 8$ on the Feigenbaum diagram of Fig. 6-7. (From H. O. Peitgen et al., *Chaos and Fractals*, Springer-Verlag, Berlin, 1992, p. 693.)

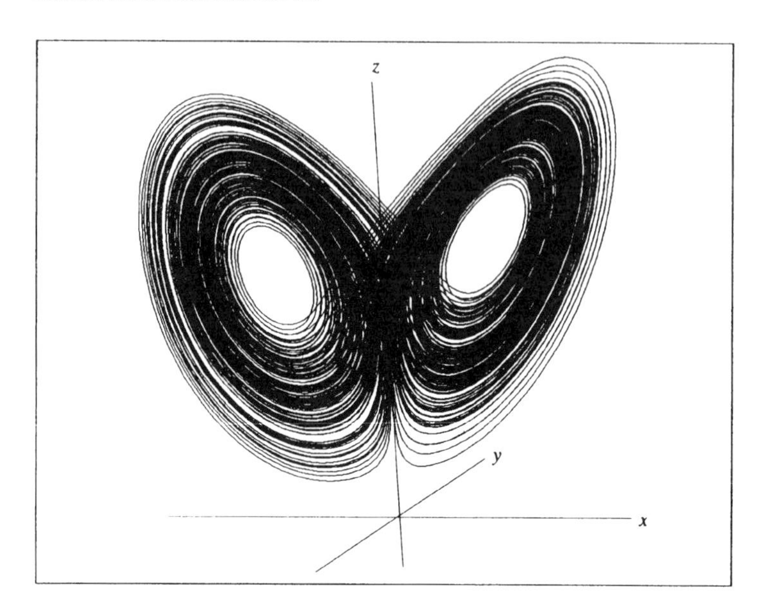

Fig. 6-9. Two spiraling trajectory sheets of Lorenz attractor. The system point continuously spirals outward in one sheet, then jumps from an outer orbit of that sheet to an inner orbit of the other where it resumes another outer spiraling motion followed by a jump, and so forth. (From H. O. Peitgen et al., *Chaos and Fractals*, Springer-Verlag, Berlin, 1992, p. 698.)

and Lorenz fixed the three-dimensionless parameters at the values

$$\sigma = 10, \quad B = 8/3, \quad R = 28 \tag{27}$$

We see from Fig. 6-9 that there are two sheets in which the trajectories spiral outwards, and when a threshold is reached in one sheet the position point is ejected from the outer part of its spiral to enter the inner part of the other sheet where it resumes a spiraling outward motion. The outward spiraling and intervening jumps repeat continuously with time.

CHAPTER 7

RELATIVITY

1 INTRODUCTION

This chapter surveys the basic concepts of special relativity and then applies them to four-vectors in mechanics and electrodynamics. The properties of four-vectors, their transformations, and their invariants are explained. The chapter ends with a discussion of translations in space and in time, as well as parity and time reversal.

2 NEWTONIAN MECHANICS

Newtonian mechanics was originally formulated with the implicit assumption that the laws of physics in a coordinate system x', y', z' moving at a constant velocity v relative to a stationary or laboratory system x, y, z are related to each other via a Galilean transformation. A vector $\mathbf{r}$ in the laboratory system is related to a vector $\mathbf{r}'$ in the moving system through the equation

$$\mathbf{r}' = \mathbf{r} - \mathbf{v}t \tag{1}$$

The first and second time derivatives of this expression give the velocities and accelerations, respectively

$$\dot{\mathbf{r}}' = \dot{\mathbf{r}} - \mathbf{v} \tag{2}$$

$$\ddot{\mathbf{r}}' = \ddot{\mathbf{r}} \tag{3}$$

or

$$\mathbf{a}' = \mathbf{a} \tag{4}$$

so the acceleration is the same in both the stationary and the moving system in Newtonian mechanics.

3 CONSTANCY OF SPEED OF LIGHT

A hundred years ago Michelson and Morely established that the experimentally measured velocity of light is the same in all uniformly moving coordinate systems, independent of the direction of propagation. We expect that measurements made within a particular coordinate system should not distinguish that system from others moving uniformly relative to it. More generally, the postulate of equivalence requires that the laws of physics be expressed the same way in all uniformly moving coordinate systems. It follows that transformations between such systems must preserve the form of these laws.

Since the velocity of light is found experimentally to be invariant and the Galilean transformation provides for changes in this velocity. Galilean transformations cannot be valid. The problem is resolved by taking into account time as a coordinate and using a space–time 'rotation' called a Lorentz transformation to convert between two systems moving uniformly relative to each other.

4 LORENTZ TRANSFORMATION

We work in the two four-dimensional space–time coordinate systems x, y, z, ct and x', y', z', z', ct' in Minkowski space where the time part is given by $x_4 = ct$. A Lorentz transformation between these systems with the primed system moving along the z direction at the speed v relative to the unprimed system as indicated in Fig. 7-1 is

$$\begin{pmatrix} 1 & 0 & 0 & 0 \\ 0 & 1 & 0 & 0 \\ 0 & 0 & \gamma & -\beta\gamma \\ 0 & 0 & -\beta\gamma & \gamma \end{pmatrix} \begin{pmatrix} x \\ y \\ z \\ ct \end{pmatrix} = \begin{pmatrix} x' \\ y' \\ z' \\ ct' \end{pmatrix} \tag{5}$$

and this gives

$$\begin{aligned} x' &= x \\ y' &= y \\ z' &= \gamma(z - \beta ct) \\ ct' &= \gamma(ct - \beta z) \end{aligned} \tag{6}$$

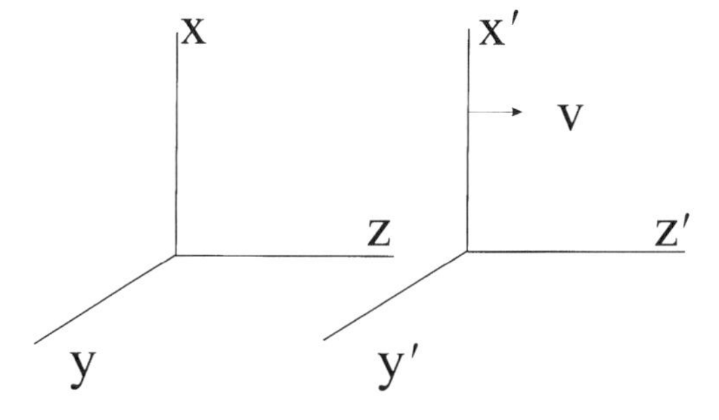

Fig. 7-1. Primed coordinate system x', y', z' shown moving to the right at the speed v relative to the unprimed system x, y, z.

where

$$\beta = v/c \tag{7}$$

$$\gamma = \frac{1}{(1-\beta^2)^{1/2}} \tag{8}$$

This is a Lorentz transformation along the z direction. In space–time notation it is a 'rotation' in the z, ct plane.

The Lorentz transformation preserves the 'length' or four-dimensional magnitude of the space–time four-vector

$$(x, y, z, -ct)\begin{pmatrix} x \\ y \\ z \\ ct \end{pmatrix} = (x^2 + y^2 + z^2 - c^2t^2) \tag{9}$$

where the contravariant column vector has a positive sign on its time component ct and the covariant row vector has a negative sign on ct. More generally, the magnitude of a four-vector with the components V_x, V_y, V_z, V_t

$$V_x^2 + V_y^2 + V_z^2 - V_t^2 \tag{10}$$

is preserved by a Lorentz transformation. The scalar product of two four-vectors is also preserved when each of them undergoes the same Lorentz transformation

$$V_xW_x + V_yW_y + V_zW_z - V_tW_t = V_x'W_x' + V_y'W_y' + V_z'W_z' - V_t'W_t' \tag{11}$$

or in a more compact notation

$$\mathbf{V} \cdot \mathbf{W} - V_tW_t = \mathbf{V}' \cdot \mathbf{W}' - V_t'W_t' \tag{12}$$

To gain some insight into the concept of a Lorentz transformation as a space–time rotation we can express such a rotation in terms of a boost parameter or rapidity ζ defined by

$$\tanh \zeta = \beta \tag{13a}$$

$$\sinh \zeta = \beta\gamma \tag{13b}$$

$$\cosh \zeta = \gamma \tag{13c}$$

which gives for Eqs. (6)

$$\begin{aligned} V_x' &= V_x \\ V_y' &= V_y \\ V_z' &= V_z \cosh \zeta - V_t \sinh \zeta \\ V_t' &= -V_z \sinh \zeta + V_t \cosh \zeta \end{aligned} \tag{14}$$

Consider two successive space–time rotations through the angles

$$\zeta'' = \zeta + \zeta' \tag{15}$$

in the same space–time plane, i.e., in the same velocity direction. The trigonometric identity

$$\tanh \zeta'' = \tanh(\zeta + \zeta') \tag{16}$$

$$= \frac{\tanh \zeta + \tanh \zeta'}{1 - \tanh \zeta \tanh \zeta'} \tag{17}$$

gives the following expression:

$$\beta'' = \frac{\beta + \beta'}{1 + \beta\beta'} \tag{18}$$

which is called the Einstein addition law for parallel velocities. It is easy to see that if one of the first two velocities is c, i.e., if either $\beta = 1$ or $\beta' = 1$, then β'' is also equal to 1. Thus, β'' can never exceed unity, and c is the highest attainable speed. For low velocities, $\beta, \beta' \ll 1$, the term $\beta\beta'$ can be neglected, and Eq. (18) reduces to the Galilean result, $v'' = v + v'$.

5 PAST AND FUTURE

The distance–time four-vector $\mathbf{r}$, ct may be said to represent an event in space–time relative to the origin $x = y = z = t = 0$ of the four-dimensional coordinate system, just as the coordinate space vector $\mathbf{r}$ designates a position in space relative to the origin $x = y = z = 0$ of the space coordinates.

There are kinds of four-vectors, depending on the sign of the magnitude, which in our case is $r^2 - c^2t^2$, and they are as follows:

$$\begin{aligned} r^2 - c^2t^2 &< 0 \quad \text{time-like} \\ r^2 - c^2t^2 &= 0 \quad \text{light-like} \\ r^2 - c^2t^2 &> 0 \quad \text{space-like} \end{aligned} \tag{19}$$

The time-like four-vector represents an event in the past for $t < 0$, and an event in the future for $t > 0$, as indicated in Fig. 7-2. The equals sign indicates a light-like four-vector corresponding to motion at the speed of light, $x = ct$. The space-like case designates an event that is neither in the past nor in the future, but somewhere else than at the origin. Points along the vertical time axis are events in the past ($t < 0$) or future ($t > 0$) that occur at the same point in space as the event at the origin, and points along the x axis are events elsewhere in space that take place at the same time as the event at the origin.

A Lorentz transformation can always be found to transform a future event at another location to a future event that takes place at the origin, $x = 0$, of the space system. In like manner, an event in the elsewhere region can always be transformed to an event that takes place at the same time as the event at the origin, i.e., simultaneously. This is the reason for the designation time-like and space-like for the cases $r^2 < c^2t^2$ and $r^2 > c^2t^2$, respectively.

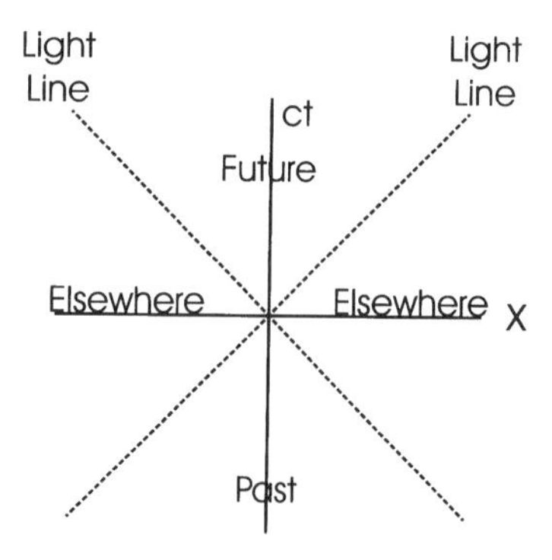

Fig. 7-2. Space–time diagram, ct versus x, showing the domains of the past, the future, and elsewhere separated by the light lines (- - - - -) corresponding to motion at the speed of light.

As an example, consider a flash of light emitted at a particular point on earth, and a second flash emitted τ seconds later at a position 1200 km away. We wish to know if the two events are space-like or time-like relative to each other. Since light goes at the speed of 300 km/ms, it takes light 4 ms to travel between the two points. If the second event occurs 8 m later, then it is theoretically possible for a spaceship to travel at 150 km/ms so it is passing over the starting point when the first flash goes off, and is passing over the final point when the second flash goes off. Both flashes are seen directly below the spaceship, i.e., at the same place in the spaceship coordinate system. They are space-like.

Now suppose that the second flash goes off only 2 ms later. The spaceship can no longer travel fast enough to make the two events occur at the same place in its coordinate system. For the two events to be simultaneous the light signals from the two events must arrive at the spaceship while it is passing over the midpoint between the two positions. To determine the velocity at which the spaceship must be traveling for this to occur we carry out a Lorentz transformation of the events

$$c\Delta t_{\text{sp-sh}} = 0 = \gamma(c\Delta t - \beta\Delta x) \tag{20}$$

where $\Delta x = 1200$ km and $\Delta t = 2$ ms. This gives a speed of $\beta = 0.5$. If the speed is less, then the signal from the first event will arrive first, and if the speed is greater, then the signal from the second event will arrive first.

Table 7-1 lists the characteristics of several four-vectors. Most of these can be either space-like or time-like depending on the values of their

TABLE 7-1. Space and time parts of several common four-vectors

Name	Space part	Time part	Type	Value in rest system
Position, time	$\mathbf{r}$	ct	–	–
Momentum, energy	$\mathbf{p} = \gamma m\mathbf{v}$	$E/c = \gamma mc$	time-like	mc
Force	$\gamma\mathbf{F}$	$\gamma\mathbf{F}\cdot\boldsymbol{\beta}$	space-like	$\mathbf{F}$
Electromagnetic potential	$\mathbf{A}$	ϕ/c	–	–
Gradient	∇	$-\frac{1}{c}\frac{\partial}{\partial t}$	–	–
Propagation	$\mathbf{k}$	ω/c	light-like	0
Current, charge	$\mathbf{J}$	$c\rho$	–	–

components. For example, the electromagnetic potential four-vector is space-like if $A > \phi/c$ and time-like if $A < \phi/c$. Other four-vectors, however, are intrinsically of a particular type. The four-vector for free space electromagnetic wave propagation is, by its nature, light-like because we always have the relation $k = \omega/c$. The energy–momentum four-vector is time-like since it can be transformed to a system in which the particle is at rest so $p = 0$ and $E \approx \mathrm{mc}^2$. If we equate the magnitude $p^2 - E^2/c^2$ with the magnitude $-m^2c^2$ in the rest system we obtain the well-known expression for the relativistic energy of a particle

$$E^2 = p^2c^2 + m^2c^4 \tag{21}$$

The related velocity four-vector U has space-part $\gamma\mathbf{v}$ and time-part γc.

6 LORENTZ CONTRACTION AND TIME DILATATION

Let us consider an observer looking at a spaceship moving at the velocity v relative to his earth-bound laboratory coordinate system. The spaceship in its rest system has the length ℓ_0 along the direction of the velocity and a clock with a hand that makes one revolution every Δt_0 seconds. We are interested in knowing what an observer in the laboratory system will report about the length ℓ of the sapceship and the time Δt for the clock hand to rotate.

The transformation equations for two events, x_1, t_1 and x_2, t_2, in the unprimed spaceship system to the primed laboratory system are

$$\begin{aligned} x_2' - x_1' &= \gamma[(x_2 - x_1) - \beta c(t_2 - t_1)] \\ t_2' - t_1' &= \gamma[(t_2 - t_1) - (\beta/c)(x_2 - -x_1)] \end{aligned} \tag{22}$$

When the clock is measured, $x_1 = x_2$, since it remains in the same place in the spaceship, so

$$t_2' - t_1' = \gamma(t_2 - t_1) \tag{23}$$

or

$$\Delta t' = \gamma \Delta t_0 \tag{24}$$

and time is dilated. For example, consider a pion with a lifetime of 0.026 μs at rest. If it is created moving at $\beta = 0.9999$, then $\gamma = 70.7$ and the lifetime observed in the laboratory is 1.84 μs. At this speed, an insect that normally lives for one week will be observed to live for well over a year!

Now consider the length $\ell_0 = x_2 - x_1$ of the spaceship. Both ends are measured at the same time on earth, i.e., $t_2' = t_1'$, so Eqs. (22) become

$$\begin{aligned} \ell' &= \gamma[\ell_0 - \beta c(t_2 - t_1)] \\ 0 &= \gamma[(t_2 - t_1) - (\beta \ell_0 / c)] \end{aligned} \tag{25}$$

and eliminating $t_2 - t_1$ from these expressions gives

$$\ell' = \ell_0 / \gamma \tag{26}$$

so the length appears shorter when viewed from earth. This is called the Lorentz contraction.

7 DOPPLER SHIFT

To obtain the relativistic Doppler shift formulae that apply to light we consider transforming the propagation four-vector **k**, ω/c to a primed coordinate system moving at the velocity v

$$\begin{aligned} k_{\parallel}' &= \gamma(k_{\parallel} - \beta\omega/c) \\ k_{\perp}' &= k_{\perp} \\ \omega' &= \gamma(\omega - \beta c k_{\parallel}) \end{aligned} \tag{27}$$

where for light

$$k^2 - \omega^2/c^2 = k'^2 - \omega'^2/c^2 = 0 \tag{28}$$

Setting $k_{\parallel} = k \cos\theta$ we obtain

$$\omega' = \omega\gamma(1 - \beta\cos\theta) \tag{29}$$

For the special cases of the longitudinal Doppler shift, $\theta = 0$, and the transverse Doppler shift, $\theta = \pi/2$, corresponding to $\boldsymbol{\beta}$ being parallel to and perpendicular to **k**, respectively, we obtain

$$\omega' = \omega\left(\frac{1-\beta}{1+\beta}\right)^{1/2} \qquad \begin{array}{c} \text{longitudinal} \\ \theta = 0 \\ \boldsymbol{\beta} \| \mathbf{k} \end{array} \tag{30}$$

$$\omega' = \gamma\omega \qquad \begin{matrix}\text{transverse}\\ \theta = \pi/2 \\ \boldsymbol{\beta} \perp \mathbf{k}\end{matrix} \tag{31}$$

We can also show that

$$\tan\theta' = \frac{k'_\perp}{k'_\parallel} = \frac{\sin\theta}{\gamma(\cos\theta - \beta)} \tag{32}$$

8 NEWTON'S LAW

The covariant form of Newton's first law is expressed in terms of the force four-vector K_μ and the momentum four-vector $p_\mu = mu_\mu$ where $u_\mu = (\gamma\mathbf{v}, \gamma c)$ is the velocity four-vector

$$K_\mu = \frac{d}{d\tau} p_\mu \tag{33}$$

and the proper time τ is related to the ordinary time t through the time-dilation expression

$$dt = \gamma d\tau \tag{34}$$

The force four-vector has the space and time parts

$$K_\mu = (\gamma\mathbf{F}, \gamma\mathbf{F}\cdot\boldsymbol{\beta}) \tag{35}$$

so the space and time parts of Newton's law are

$$\mathbf{F} = \frac{d}{dt} m\gamma\mathbf{v} \tag{36}$$

$$\mathbf{F}\cdot\mathbf{v} = c^2 \frac{d}{dt} m\gamma \tag{37}$$

It is easy to show that the scalar product of the force and velocity four-vectors is zero.

9 SCALAR PRODUCTS AND INVARIANTS

We mentioned above that four-vector scalar products are invariant. Consider the scalar product of the position, time four-vector with the propagation four-vector

$$(\mathbf{k}, \omega/c) \cdot (\mathbf{r}, ct) = \mathbf{k} \cdot \mathbf{r} - \omega t \tag{38}$$

This scalar product appears in the well-known expression $e^{ik \cdot r - i\omega t}$ for the propagation of a plane wave.

Some important invariant equations can be obtained as scalar products of the gradiant four-vector $\nabla, -(1/c)\partial/\partial t$. One should note that the time part of this four-vector has a minus sign that other four-vectors do not have. Two examples of such scalar product equations are the Lorentz condition satisfied by the electromagnetic potentials

$$\nabla \cdot \mathbf{A} + \frac{1}{c^2} \frac{\partial \phi}{\partial t} = 0 \tag{39}$$

and the charge-current continuity equation

$$\nabla \cdot \mathbf{J} + \frac{\partial \rho}{\partial t} = 0 \tag{40}$$

These expressions are invariant under Lorentz transformations.

The scalar product of the gradient four-vector with itself, called the d'Alembertian

$$\square^2 = \nabla^2 - \frac{1}{c^2} \frac{\partial^2}{\partial t^2} \tag{41}$$

is a scalar operator. When it operates on a scalar function we have the wave equation

$$\square^2 \Psi = 0 \tag{42}$$

The space and time parts of the electromagnetic potentials and the current four-vectors are related through the d'Alembertian

$$\square^2 \mathbf{A} = -\mu_0 \mathbf{J} \tag{43}$$

$$\square^2 \phi = -\rho/\varepsilon_0 \tag{44}$$

where we used the free-space expression $c^2 = 1/\mu_0 \varepsilon_0$.

10 MORE GENERAL TRANSFORMATIONS

Until now we have discussed space and space–time rotations without taking into account translations. A more general transformation in Minkowski space that preserves the velocity of light is the Poincaré transformation, also called the inhomogeneous Lorentz transformation. This operates on a four-vector V to convert it to V'.

$$V' = L_{\mathrm{G}}V + b \tag{45}$$

where L_{G} is a general Lorentz transformation, sometimes called a homogeneous Lorentz transformation, and b takes into account a translation of the origin of the coordinates in space–time. This Poincaré transformation is dependent on ten independent parameters, four for the translation along x, y, z, and ct, three for a boost velocity v_x, v_y, and v_z, and three Euler angles for a space rotation.

The general Lorentz transformation L_{G} can involve going from an unprimed system x, y, z, ct to a primed system x', y', z', ct' in which the new origin is moving at a velocity v and the new axes are rotated relative to the old ones. L_{G} may be decomposed into a product of a space rotation R and a pure Lorentz transformation or boost L involving no coordinate rotation

$$L_{\mathrm{G}} = RL = L'R' \tag{46}$$

where, in general, $R \neq R'$ and $L \neq L'$ Space rotations and Lorentz transformations do not commute unless the boost velocity is in the same direction as the axis of rotation. We saw in Chapter 3 that R and R' have the general form

$$R = \begin{pmatrix} \cos\theta_{xx} & \cos\theta_{xy} & \cos\theta_{xz} \\ \cos\theta_{yx} & \cos\theta_{yy} & \cos\theta_{yz} \\ \cos\theta_{zx} & \cos\theta_{zy} & \cos\theta_{zz} \end{pmatrix} \tag{47}$$

and the most general boost L can be written

$$\begin{pmatrix} 1+\dfrac{(\gamma-1)\beta_x^2}{\beta^2} & \dfrac{(\gamma-1)\beta_x\beta_y}{\beta^2} & \dfrac{(\gamma-1)\beta_x\beta_z}{\beta^2} & -\beta_x\gamma \\ \dfrac{(\gamma-1)\beta_y\beta_x}{\beta^2} & 1+\dfrac{(\gamma-1)\beta_y^2}{\beta^2} & \dfrac{(\gamma-1)\beta_y\beta_z}{\beta^2} & -\beta_y\gamma \\ \dfrac{(\gamma-1)\beta_z\beta_x}{\beta^2} & \dfrac{(\gamma-1)\beta_z\beta_y}{\beta^2} & 1+\dfrac{(\gamma-1)\beta_z^2}{\beta^2} & -\beta_z\gamma \\ -\beta_x\gamma & -\beta_y\gamma & -\beta_z\gamma & \gamma \end{pmatrix} \tag{48}$$

a matrix that is real and symmetric with a determinant of $+1$.

In the discussion of space rotations in Chapter 3 we mentioned that there are regular rotations R where the determinant of the matrix $|R| = +1$, as well as improper rotations that have a determinant $|R| = -1$. Proper and improper rotations form disjoint sets in the sense that all proper rotations can be transformed into each other by sequences of infinitesimal proper rotations, and an analogous statement applies to improper rotations, but there is no finite sequence of infinitesimal matrices that can convert a proper to an improper rotation, and vice versa. An inversion operation is needed for the transformation. Proper and improper rotations are never 'close' to each other. There are vectors in our real world, and vectors in a mirror image world that cannot be transformed into each other by space rotations.

When we consider the space–time manifold of Minkowski space there are four sets of disjoint transformations, and four sets of four-vectors. The general Lorentz transformations that we discussed until now (48) are proper in the sense that no inversion is included. In Minkowski space we can have space inversion associated with the matrix R of Eq. (46), time inversion associated with the boost L of this expression, and space–time inversion in which both R and L invert. The following matrices bring about these three classes of inversion

Matrix	Operation	Transformation
$\begin{pmatrix} -1 & & & \\ & -1 & & \\ & & -1 & \\ & & & +1 \end{pmatrix}$	space inversion (parity P)	$x \Rightarrow -x$ $\quad \lvert R \rvert = -1$ $y \Rightarrow -y$ $z \Rightarrow -z$ $\quad \lvert L \rvert = +1$ $t \Rightarrow t$
$\begin{pmatrix} +1 & & & \\ & +1 & & \\ & & +1 & \\ & & & -1 \end{pmatrix}$	time inversion (time reversal T)	$x \Rightarrow x$ $\quad \lvert R \rvert = +1$ $y \Rightarrow y$ $z \Rightarrow z$ $\quad \lvert L \rvert = -1$ $t \Rightarrow t$
$\begin{pmatrix} -1 & & & \\ & -1 & & \\ & & -1 & \\ & & & -1 \end{pmatrix}$	space–time inversion (PT)	$x \Rightarrow -x$ $\quad \lvert R \rvert = -1$ $y \Rightarrow -y$ $z \Rightarrow -z$ $\quad \lvert L \rvert = -1$ $t \Rightarrow -t$
$\begin{pmatrix} 1 & & & \\ & 1 & & \\ & & 1 & \\ & & & 1 \end{pmatrix}$	identity	$x \Rightarrow y$ $\quad \lvert R \rvert = +1$ $y \Rightarrow y$ $z \Rightarrow z$ $\quad \lvert L \rvert = +1$ $t \Rightarrow t$

We know that many laws of physics are invariant under the parity P and time-reversal T operations, associated with the first two matrices in this set, and these operations commute, $PT = TP$. There is also the third operation C called charge conjugation that brings particles to antiparticles, and invariance with respect to all three is called PCT invariance. This is discussed further in Chapter 26.

CHAPTER 8

THERMODYNAMICS

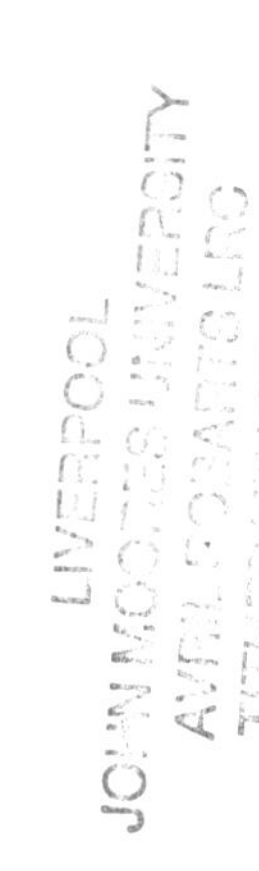

1 INTRODUCTION

Thermodynamics is a branch of physics that has not changed very much during the past century. In this chapter we cover classical thermodynamics, together with a number of applications, and end by saying a few words about deriving thermodynamics from statistical mechanics. For simplicity, some of the principles are illustrated in terms of the ideal gas law.

In thermodynamics we are particularly concerned with what are called reversible, quasistatic processes. Consider a system subject to constraints that keep it in an initial state, and assume that the constraints are removed so the system transforms to a final state. If the constraints are restored and the system returns to its original state, then the transformation is said to be brought about by a reversible process. If restoring the constraints does not restore the original state, then the process is irreversible. If the transformation is brought about by a gradual process in which the system always remains arbitrarily close to equilibrium, then the process is called quasistatic. Much of the present chapter deals with reversible quasistatic processes.

2 EXACT AND INEXACT DIFFERENTIALS

Some of the subtleties associated with thermodynamics involve the distinction between inexact differentials, such as the heat dQ, and exact differentials, such as the entropy $dS = dQ/T$, we we begin by clarifying this distinction. Consider the function $F(x, y)$ of two independent variables x and y. Its differential dF

$$dF = F(x + dx, y + dy) - F(x, y) \tag{1}$$

can often be written in the form

$$dF = A(x, y)dx + B(x, y)dy \tag{2}$$

This differential is exact if

$$\left(\frac{\partial A}{\partial y}\right)_x = \left(\frac{\partial B}{\partial x}\right)_y \tag{3}$$

An exact differential has the property that the line integral between points (x_1, y_1) and (x_2, y_2)

$$\int_1^2 dF = K \tag{4}$$

is independent of the path taken in the x, y plane. An inexact differential can also be written in the form (2), but Eq. (3) will not be satisfied. For an inexact differential the value of the integral K depends on the path taken to go from point 1 to point 2. For example, the differential of work dW is inexact if friction is present to make its line integral depend on the path.

3 LAWS OF THERMODYNAMICS

Much of thermodynamics is based on some very general laws, and we state and comment on each of these laws.

The Zeroth Law clarifies the meaning of equilibrium: Two systems in thermal equilibrium with a third system are in thermal equilibrium with each other.

The First Law of Thermodynamics is the conservation of energy. There are two alternative statements of the First Law that should be mentioned, and they are as follows:

First Statement of First Law: Work done in bringing an isolated system from an initial state to a final state is independent of the process.

Alternative Statement of First Law: When a system interacts and changes from one macrostate to another its change in internal energy U plus the work W done by the system is equal to the heat Q absorbed by the system. In differential form we write

$$dU = -dW + dQ \tag{5}$$

and for a gas the work W is of the pressure–volume type

$$dW = PdV \tag{6}$$

The First Law provides for energy balance without any restrictions on the relative changes of the various quantities involved in energy flow. The Second Law establishes a restriction on this flow associated with the change in entropy dS that accompanies the input of heat into a non-isolated system. For a quasistatic infinitesimal process this change in entropy is given by

$$dS = dQ/T \tag{7}$$

and dS of the system is positive when it absorbes a quantity of heat dQ.

The Second Law states that when a thermally isolated system changes from one macrostate to another the entropy tends to increase

$$\Delta S \geq 0 \tag{8}$$

For the special case of a reversible quasistatic process the entropy remains the same, i.e., $\Delta S = 0$. In addition to this general statement (8) we provide two specific formulations of the Second Law.

Kelvin–Planck Statement: It is not possible to construct an engine which, operating in a cycle, produces no other effect than to extract heat from a reservoir and perform an equivalent amount of work.

Clausius Statement: It is not possible to construct a perfect refrigerator, i.e., an engine that produces no other effect than to transfer heat from a colder to a hotter body.

In the next section, we discuss the closest to 'perfect' heat engine called the Carnot Cycle in which the equals sign applies in Eq. (8).

The Third Law concerns the unattainability of absolute zero temperature. The entropy of every system approaches zero as the temperature approaches zero.

The Zeroth Law seems somewhat trivial. The First Law of Energy Conservation is the foundation of much of physics. In relativity theory it is generalized to the conservation of the energy–momentum four-vector. The Second Law sets limits on the interchangeability of the different types of energy, especially as far as heat and work are concerned. It prevents, for example, ocean liners from extracting heat from the ocean as a source of energy to run the ship. The Second Law is not widely applied outside of thermodynamics, nor is the Third Law.

The differential of work dW takes various forms in different systems. For a gas we have pressure–volume work (6); namely, $dW = PdV$, for a surface film $dW = -SdA$, where S is the tension and A the area, in an electrical circuit with voltage V and charge Q we have $dW = -VdQ$, and for a magnetic system $dW = -MdB$, where M is the magnetization and B is the magnetic field. Some of the examples in this chapter involve gas laws with dW given by PdV.

For a reversible process we take the differential of heat to be $dQ = TdS$, where the entropy differential dS is exact, but dQ is not. The entropy change may be calculated by an integration

$$S_f - S_i = \int_i^f \frac{dQ}{T} \tag{9}$$

and this can be written in terms of the specific heat C where $dQ = CdT$

$$S_f - S_i = \int_i^f \frac{Cdt}{T} \tag{10}$$

This calculation can be carried out using the specific heat at constant volume C_V or at constant pressure C_P, and these can be expressed in terms of the internal energy U and enthalpy H, respectively

$$C_v = \left(\frac{\partial Q}{\partial T}\right)_v = \left(\frac{\partial U}{\partial T}\right)_v \tag{11}$$

$$C_p = \left(\frac{\partial Q}{\partial T}\right)_p = \left(\frac{\partial H}{\partial T}\right)_p \tag{12}$$

4 HEAT ENGINE AND CARNOT CYCLE

An example of the Second Law is the operation of a heat engine that removes the quantity of heat Q_H from a hot reservoir at the temperature T_H, does work W, and deposits the heat Q_C at a cold reservoir of temperature T_C, as shown in Fig. 8-1. The First Law ensures energy balance

$$Q_H = Q_C + W \tag{13}$$

and the Second Law which restricts the entropy change

$$\Delta S = \frac{Q_C}{T_C} - \frac{Q_H}{T_H} \geq 0 \tag{14}$$

can be written

$$\frac{Q_C}{Q_H} \geq \frac{T_C}{T_H} \tag{15}$$

The efficiency η of the heat engine is

$$\eta = \frac{W}{Q_H} \tag{16}$$

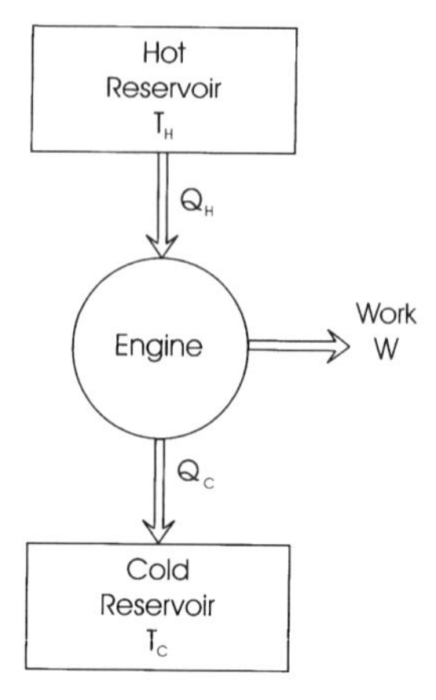

Fig. 8-1. Carnot cycle heat engine that takes the quantity of heat Q_H from the hot reservoir at temperature T_H performs work W, and expels heat Q_C to the cold reservoir at temperature T_C.

and with the aid of Eqs. (13) and (15) we obtain the inequality

$$\eta \leq 1 - \frac{T_C}{T_H} \tag{17}$$

A maximum output heat engine called a Carnot Cycle has the efficiency

$$\eta = 1 - \frac{T_C}{T_H} \tag{18}$$

Such a cycle using an ideal gas as a working substance takes the heat Q_H during an isothermal contraction in volume at the higher temperature T_H, undergoes an adiabatic contraction in volume, releases the heat W_C during an isothermal expansion at the temperature T_C, then expands adiabatically to its original state, after which it repeats the cycle. The work done per cycle is the area enclosed by these four curves on the P, V diagram shown in Fig. 8-2.

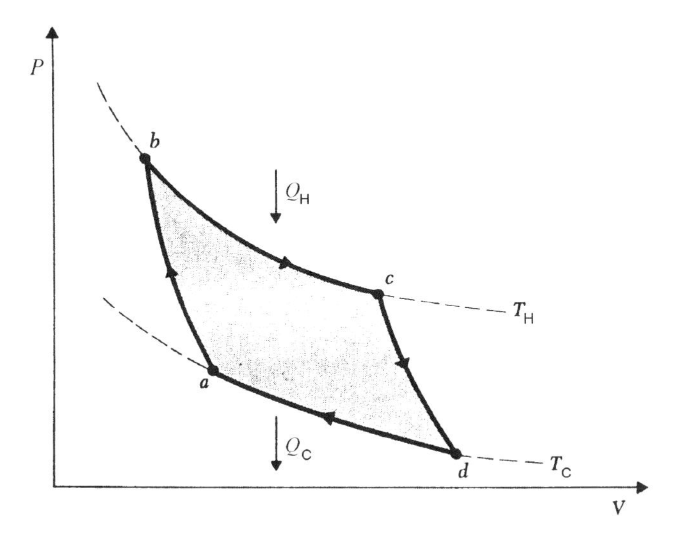

Fig. 8-2. *P*, *V* diagram for a Carnot Cycle showing the heat input at the upper isotherm T_H, the expelled heat at the lower isotherm T_C, and the two connecting adiabats. The enclosed area provides the work done in a cycle. (From F. Reif, *Statistical Thermal Physics*, McGraw-Hill, New York, 1965, p. 189.)

5 INTERNAL ENERGY, ENTHALPY, AND FREE ENERGIES

The discussion until now has been about internal energy, work, heat, and entropy. There are several additional thermodynamic functions that are of importance, and we introduce them in this section. In doing so we will assume that the system undergoes reversible quasistatic changes so that we can replace the heat differential dQ by the entropy expression TdS, and we use the gas law expression $dW = PdV$ for the work. We begin with the internal energy U, which is appropriate for the constant volume conditions that are popular among physicists, and then we define its counterpart the enthalpy H, which is more suitable for use when treating the constant pressure processes that are of greater interest to chemists and chemical engineers. The Helmholtz free energy F for constant volume and the Gibbs free energy for constant pressure are introduced since it is the state of lowest free energy that exists at a given temperature, and the free energy is the same before and after a change in phase. The section concludes with some partial derivative expressions involving the various thermodynamic functions that are called Maxwell relations.

The First Law given by

$$dU = TdS - PdV \tag{19}$$

can be expressed in terms of the enthalpy H

$$H = U + PV \tag{20}$$

as follows:

$$dH = TdS + VdP \tag{21}$$

The Helmholtz free energy F and the Gibbs free energy G defined by

$$F = U - TS \tag{22}$$

$$G = H - TS \tag{23}$$

have the respective differentials

$$dF = -SdT - PdV \tag{24}$$

$$dG = -SdT + VdP \tag{25}$$

We see below in Section 14 that the Helmholtz free energy plays a crucial role in the derivation of the laws of thermodynamics from those of statistical mechanics.

6 MAXWELL RELATIONS

The differentials of the internal energy dU, enthalpy dH and two free energies dF and dG that were written down in the previous section are all exact, and this property can be employed to derive differential relations between four thermodynamic variables P, V, S, and T on which they depend. Thus, the internal energy $U(S, V)$ considered as a function of entropy and volume has an exact differential dU, which is given by

$$dU = \left(\frac{\partial U}{\partial S}\right)_V dS + \left(\frac{\partial U}{\partial V}\right)_S dV \tag{26}$$

Comparing this with Eq. (19) we see that

$$T = \left(\frac{\partial U}{\partial S}\right)_V \quad \text{and} \quad P = -\left(\frac{\partial U}{\partial V}\right)_S \tag{27}$$

Carrying out another differentiation of each part of Eq. (27) and equating the resulting second derivatives gives

$$-\left(\frac{\partial T}{\partial V}\right)_S = \left(\frac{\partial^2 U}{\partial S \partial V}\right)_{VS} = \left(\frac{\partial^2 u}{\partial P \partial T}\right)_{SV} = \left(\frac{\partial P}{\partial S}\right)_V \tag{28}$$

and this provides us with the Maxwell relation

$$\left(\frac{\partial T}{\partial V}\right)_S = -\left(\frac{\partial P}{\partial S}\right)_V \tag{29}$$

which is also the condition for the exactness of the differential dU of Eq. (19). Similar reasoning applied to Eqs. (21), (24) and (25) for the enthalpy H, the Helmholtz free energy F and the Gibbs free energy G, provides the remaining three Maxwell relations

$$\left(\frac{\partial T}{\partial P}\right)_S = \left(\frac{\partial V}{\partial S}\right)_P \tag{30}$$

$$\left(\frac{\partial S}{\partial V}\right)_T = \left(\frac{\partial P}{\partial T}\right)_V \tag{31}$$

$$\left(\frac{\partial S}{\partial P}\right)_T = -\left(\frac{\partial V}{\partial T}\right)_P \tag{32}$$

The derivations of these relations depend on the fact that U, H, F, and G all have exact differentials.

7 GIBBS PHASE RULE

For a system with c components and ϕ phases in a state of equilibrium the number of degrees of freedom f is given by Gibbs phase rule

$$f = c - \phi + 2 \tag{33}$$

Our main interest is in one-component systems ($c = 1$) so we have $f = 3 - \phi$. If only one phase is present, $\phi = 1$, then $f = 2$, and two parameters are needed to specify uniquely the state of the system. One can choose P and T for a gas, and the volume is determined by the equation of state such as the ideal gas law or van der Waals equation. When two phases are in equilibrium as during the boiling process, then $\phi = 2$ and hence $f = 1$ so there is only one degree of freedom; namely, the position along the phase transition line. A transition from one phase to the other can occur anywhere along this line, and the Clausius–Clapeyron equation (40), which is given below, may be looked upon as the equation of state for this line. Finally, the triple point with three phases in equilibrium ($\phi = 3$) is unique for a substance since ther are no degrees of freedom ($f = 0$).

Sodium chloride dissolved in water constitutes a two-component system $c = 2$. The phases that can be present include NaCl in liquid solution, precipitated NaCl, frozen NaCl solution, etc. Binary alloys are examples of two-component systems.

8 CHANGES OF PHASE

A change in phase such as melting or boiling involves the transformation of a system from one state to another at a constant temperature, so $dT = 0$. We see from the Helmholtz free energy expression (24) that if the phase change occurs at constant volume (isochoric, $dV = 0$), then the Helmholtz free energy is constant

$$\left.\begin{aligned} dF &= 0 \\ F &= \text{const} \end{aligned}\right\} \quad \text{constant volume phase change} \tag{34}$$

If, on the other hand, the phase change occurs at constant pressure (isobaric, $dP = 0$), then the Gibbs free energy remains constant, and from Eq. (25) we have

$$\left.\begin{aligned} dG &= 0 \\ G &= \text{const} \end{aligned}\right\} \quad \text{constant pressure phase change} \tag{35}$$

In other words, the free energy of the system is the same before and after the transition. The Gibbs free energy of liquid water and ice are the same at 0°C and atmospheric pressure. Melting and vaporization depend on both T and P so there is a range of values for both along what is called the melting line and the vaporization line, respectively. Along, for example, the melting line, the Gibbs free energies of the liquid G_L and the solid G_S are equal, $G_L = G_S$, where the subscripts L and S denote liquid and solid phases, respectively. On the liquid side of the line, $G_L < G_S$, and on the solid side of the line $G_L > G_S$. At the triple point, the three phases, gas, liquid and solid, are in thermodynamic equilibrium so G is the same for all three of them: $G_G = G_L = G_S$. The triple point is unique for each substance.

Along a phase transition line the Clausius–Clapeyron equation

$$\frac{dP}{dT} = \frac{\Delta S}{\Delta V} \tag{36}$$

is satisfied, where ΔS is the entropy change that accompanies the volume change ΔV at the transition. Along the melting line, $\Delta V = V_L - V_S$, and ordinarily ΔV is positive, as noted below. When a first-order phase transition takes place at a temperature T_c there is a latent heat L, and this is accompanied by a change in entropy.

$$\Delta S = \frac{L}{T_c} \tag{37}$$

Using Eq. (37), the Clausius–Clapeyron equation becomes

$$\frac{dP}{dT} = \frac{L}{T(V_L - V_S)} \tag{38}$$

an expression that may be looked upon as the equation of state for the melting line. For most materials the solid phase is denser so $V_L > V_S$ and dP/dT is positive, meaning that the melting line has a positive slope. Water is an exception since ice floats.

If the specific heat is integrated over a broad range of temperature that includes one or more phase transitions, then the latent heat contributions L_i must be added to Eq. (10)

$$S_f - S_i = \int_i^f \frac{CdT}{T} + \sum \frac{L_j}{T_{ci}} \tag{39}$$

A second-order phase transition has no latent heat, but there can be a discontinuity in the specific heat to be taken into account in the integration. A discontinuity is present when mean field theory applies.

9 SPECIFIC HEAT OF IDEAL GAS

Each molecule in an ideal gas has the kinetic energy

$$KE = \tfrac{1}{2}mv_x^2 + \tfrac{1}{2}mv_y^2 + \tfrac{1}{2}mv_z^2 \tag{40}$$

and each of these terms contributes the amount $\frac{1}{2}k_B T$ to the thermal energy, for a total of $3k_B T/2$ per atom and $3RT/2$ per mole, where $R = N_A k_B$ is the gas constant, and N_A is Avogadro's number. Therefore, we have for the internal energy of an ideal monotomic gas

$$U = \frac{3}{2}RT \qquad \text{ideal gas} \tag{41}$$

The specific heat at constant volume from Eq. (11) is then

$$C_V = \frac{3}{2}R \qquad \text{monatomic gas} \tag{42}$$

A diatomic gas has additional energy E_{vib} arising from the kinetic and potential contributions of the harmonic stretching force $-kx$ along the bond axis

$$E_{\text{vib}} = \tfrac{1}{2}mv_x^2 + \tfrac{1}{2}m\omega^2 x^2 \tag{43}$$

where $k = m\omega^2$, and each term contributes $\frac{1}{2}RT$ to the thermal energy, to give

$$C_V = \frac{5}{2}R \qquad \text{diatomic gas} \tag{44}$$

The argument can be generalized to a non-linear polyatomic molecule of n atoms that has $3n - 6$ vibrational normal modes, each of which can contribute RT to the thermal energy, but in practice many of these modes are not yet excited at temperatures where the specific heat is measured.

To obtain an expression for the specific heat at constant pressure, we write the first law of Thermodynamics

$$dQ = C_V dT + PdV \tag{45}$$

and replace the term PdV by its value from the ideal gas law $PV = RT$ written in differential form

$$PdV + VdP = RdT \tag{46}$$

to obtain

$$dQ = (C_V + R)dT - VdP \tag{47}$$

and dQ/dT taken at constant pressure gives

$$C_P = C_V + R \tag{48}$$

Therefore, we have for an ideal gas

$$C_P = \frac{5}{2}R \qquad \text{monatomic gas} \tag{49}$$

$$C_P = \frac{7}{2}R \qquad \text{diatomic gas} \tag{50}$$

and these are close to experimental values for many gases.

10 ADIABATIC GAS LAW

The specific heat ratio, which we designate by γ, has the value

$$\gamma = C_P/C_V = 5/3 \tag{51}$$

for an ideal monatomic gas, where use was made of Eqs. (42) and (49). For a gas of diatomic molecules at a temperature sufficiently high so the vibrational modes are excited, Eqs. (44) and (50) give

$$\gamma = C_P/C_V = 7/5 \tag{52}$$

In an adiabatic process $dQ = 0$, and we have from Eqs. (47), (49), and (50)

$$\begin{aligned} VdP &= C_P dT \\ PdV &= -C_V dT \end{aligned} \tag{53}$$

The ratio of these two expressions

$$\frac{dP}{P} = -\gamma \frac{dV}{V} \tag{54}$$

can be integrated to give the adiabatic gas law

$$PV^{\gamma} = \text{const} \tag{55}$$

This law provides an experimental way to evaluate the specific heat ratio γ.

11 SPECIFIC HEAT OF SOLIDS

A solid has a Debye temperature Θ_{D} above which most vibrational normal modes are excited. In the high temperature region, $T \gg \Theta_{\mathrm{D}}$, each atom can vibrate along the x, y, or z direction, with each vibrational mode contributing RT to the specific heat

$$C_V \approx C_P = 3R \quad T \gg \Theta_{\mathrm{D}} \tag{56}$$

a result called the Law of Dulong and Petit. There is little difference between C_V and C_P for a solid because the pressure has so little effect on the volume. This specific heat arising from these vibrational or phonon modes is often called the phonon specific heat.

More generally, the Debye theory for the specific heat begins by writing down the total vibrational energy per unit volume U using the Debye approximation $\omega = kv$, the resultant phonon density of states $D_{\mathrm{ph}}(\omega) = \omega^2/2\pi^2 v^3$, and the Fermi-Dirac distribution function $f(\omega)$ of Eq. (56) in Chapter 9

$$U = \int \hbar\omega \, D_{\mathrm{ph}}(\omega) f(\omega) d\omega \tag{57}$$

$$= \int_0^{\omega_D} \hbar\omega \frac{\omega^2}{2\pi^2 v^3} \frac{1}{e^{\beta\hbar\omega} - 1} d\omega \tag{58}$$

Differentiating this equation with respect to T gives

$$C_V = 9R\left(\frac{T}{\Theta_{\mathrm{D}}}\right)^3 \int_0^{x_D} \frac{x^4 e^x dx}{(e^x - 1)^2} \tag{59}$$

where $x = \beta\hbar\omega$ and $x_{\mathrm{D}} = \hbar\omega/k_{\mathrm{B}}\Theta_{\mathrm{D}}$. The integration can be carried out in closed form for the low temperature limit $x_{\mathrm{D}} \Rightarrow \infty$ and for the high temperature limit $e^{x} \approx 1 + x$ to give

$$C_V \approx \frac{12\pi^4}{5} R\left(\frac{T}{\Theta_{\mathrm{D}}}\right)^3 \qquad \mathrm{T} \ll \Theta_{\mathrm{D}} \tag{60}$$

$$C_V \approx 3R \qquad \mathrm{T} \gg \Theta_{\mathrm{D}} \tag{61}$$

as observed experimentally, where the classical limit (61) is called the Law of Dulong and Petit.

If the solid is a conductor, then at low temperatures, $T \ll T_{\mathrm{F}}$, where the Fermi temperature T_{F} is defined in terms of the Fermi energy E_{F} through the expression

$$E_{\mathrm{F}} = k_{\mathrm{B}} T_{\mathrm{F}} \tag{62}$$

the conduction electrons contribute a specific heat term C_{e} that is linear in the temperature

$$C_{\mathrm{e}} = \gamma T \tag{63}$$

For many metals this is close to the free electron value

$$C_{\mathrm{e}} = \tfrac{1}{2}\pi^2 R(T/T_{\mathrm{F}}) \tag{64}$$

where the Fermi temperature T_{F} is typically several thousand degrees kelvin. In the low temperature region, $T \ll \Theta_{\mathrm{D}}, T_{\mathrm{F}}$, the overall specific heat C_{total} is the sum of contributions from the electrons (64) and the lattic vibrations (60)

$$C_{\mathrm{total}} = \gamma T + AT^3 \tag{65}$$

where the factor $A = 234R/\Theta_{\mathrm{D}}^3$ in the Debye theory. The coefficients γ and A can be evaluated experimentally from the intercept and the slope, respectively, of a plot of C_{total}/T versus T^2.

12 THERMOELECTRIC AND THERMOMAGNETIC EFFECTS

We begin this section with some comments on thermal conductivity, and then we discuss some thermal effects that depend on the presence of electric and magnetic fields.

The thermal current density **U**, the heat energy flow per unit time per unit cross-sectional area, represents the transport of entropy density S_ϕ at the velocity **v**

$$\mathbf{U} = TS_\phi \mathbf{v} = -K\nabla T \tag{66}$$

where use was made of Fourier's law

$$\mathbf{U} = -K\nabla T \tag{67}$$

and K is the coefficient of thermal conductivity. Normal metals are good conductors of heat via the law of Wiedermann and Franz (Chapter 24, Eq. (18)), with parallel electron and phonon heat conduction channels acting independently, so their thermal conductivities add: $K = K_e + K_{ph}$.

An open-circuited conductor with a temperature gradient can develop an electric field along the gradient direction (thermopower or Seebeck effect) or perpendicular to this gradient (Nernst effect). When an isothermal electric current flows, a thermal current can appear flowing parallel to (Peltier effect) or perpendicular to (Ettinghausen effect) the electric current direction. The two transverse effects require the presence of an applied magnetic field. The thermal analog of the Hall effect is called the Righi–Leduc effect. Figure 8-3 shows the measurement techniques for these various effects, and Table 8-1 summarizes their properties.

Seebeck Effect. A conductor with a temperature gradient ∇T but no electric current flow can develop a steady-state electric field in the gradient direction

$$E = S\nabla T \tag{68}$$

which gives rise to an electrostatic potential difference $V_2 - V_1$

$$V_2 - V_1 = S(T_2 - T_1) \tag{69}$$

shown in Fig. 8-3(a) where S is variously called the thermopower, thermoelectric power, or Seebeck coefficient. We should be careful not to confuse

TABLE 8-1. Thermoelectric and thermomagnetic effects. The applied quantity is in the y direction, an effect measured longitudinally is also in the y direction, an effect measured transversely is in the x direction, and in most cases there is an applied magnetic field in the z direction.

Effect	Electric field (E or ∇V)	Electric current (I)	Temperature gradient (∇T)	Heat current (dQ/dt)	Magnetic field B_{app}	Figure
Resistivity	Meas. y	Appl. y	0	—	0	—
Magnetores. longitudinal	Meas. y	Appl. y	0	—	Appl. y	—
Magnetores. transverse	Meas. y	Appl. y	0	—	Appl. z	—
Thermal conductivity	—	0	Meas. y	Appl. y	0	—
Hall	Meas. x	Appl. y	0	—	Appl. z	Chapter 24, Eq. (14)
Righi–Leduc	—	0	Meas. x	Appl. y	Appl. z	8–3(e)
Seebeck	Meas. y	0	Appl. y	—	0	8–3(a)
Magneto-Seebeck	Meas. y	0	Appl. y	—	Appl. x, y or z	—
Nernst	Meas. x	0	Appl. y	—	Appl. z	8–3(b)
Magneto-Nernst	Meas. x	0	Appl. y	—	Appl. x, y or z	—
Peltier	—	Appl. y	0	Meas. y	—	8–3(c)
Ettinghausen	—	Appl. y	Meas. x	—	Appl. z	8–3(d)

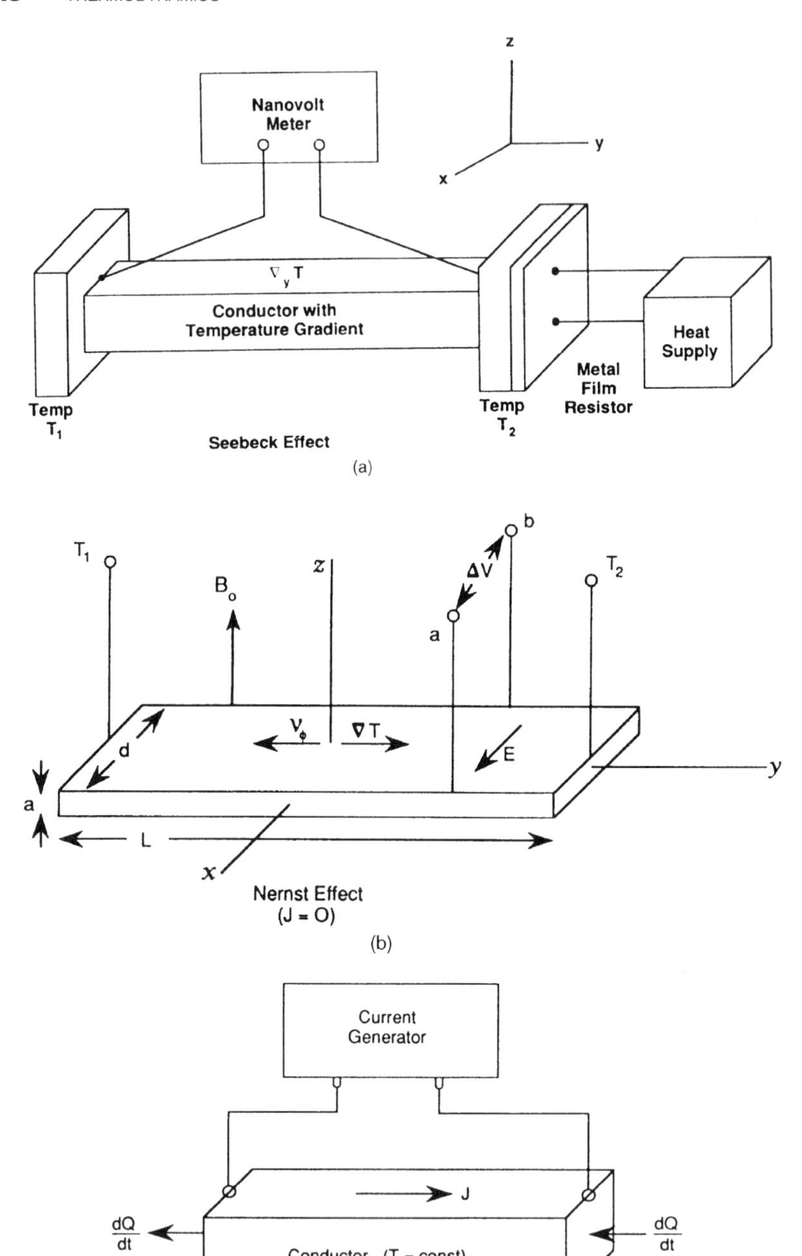
Nanovolt Meter
z
y
x
$\nabla_y T$
Conductor with Temperature Gradient
Heat Supply
Metal Film Resistor
Temp T_1
Temp T_2
Seebeck Effect
(a)
T_1
B_o
z
b
ΔV
a
T_2
v_ϕ
∇T
d
E
y
a
L
x
Nernst Effect (J = O)
(b)
Current Generator
J
dQ/dt
Conductor (T = const)
dQ/dt
Peltier Effect
(c)

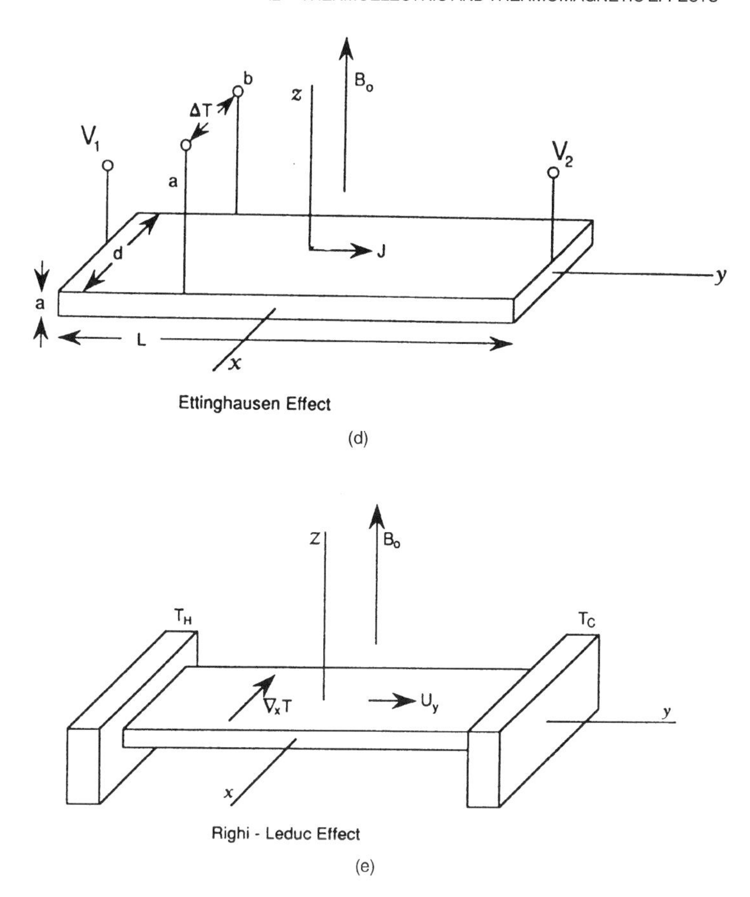

Fig. 8-3. Experimental arrangements for various magnetothermal effects: (a) Seebeck effect, (b) Nernst effect, (c) Peltier effect, (d) Ettinghausen effect, and (e) Righi–Leduc effect. (From C. P. Poole, Jr et al., *Superconductivity*, Academic Press, New York, 1995, Chapter 14.)

this symbol with S_ϕ for transport entropy. The free electron approximation gives

$$S = \left(\frac{\pi^2}{2}\right)\left(\frac{k_B}{e}\right)\left(\frac{T}{T_F}\right) \tag{70}$$

$$= 142(T/T_F) \quad \mu V/K \tag{71}$$

where the Fermi temperature T_F of Eq. (62) can be, typically, 10^4 to $10^5 K$.

Nernst Effect. A conductor with a temperature gradient ∇T in the presence of an applied magnetic field B and no electric current flow can develop a steady-state electric field $E_x = dV/dx$ transverse to the gradient direction, which gives rise to the voltage drop, ΔV across a sample of width d, as shown in Fig. 8-3(b).

$$\Delta V = -QBd[\nabla T - (\nabla T)_c] \tag{72}$$

where the factor Q is called the Nernst coefficient.

Peltier Effect. When a conductor is maintained at a constant temperature with a uniform electric current density J flowing through it, a thermal current density U arises to carry away the Joule heat generated by J, as shown in Fig. 8-3(c). These quantities are related by the Peltier coefficient π_P

$$U = \pi_P J \tag{73}$$

Lord Kelvin deduced what is called the Thomson relation

$$\pi_P = ST \tag{74}$$

between the temperature, the Peltier coefficient π_P and the thermopower S.

Ettinghausen Effect. When a conductor in an applied magnetic field is maintained at a constant temperature longitudinally ($\nabla_y T$) with a uniform electric current density J_y flowing through it, then heat energy can flow in the transverse direction to establish a transverse temperature gradient dT/dx given by

$$|dT/dx| = \epsilon J_y B_z \tag{75}$$

where ϵ is the Ettinghausen coefficient, as shown in Fig. 8-3(d).

Righi–Leduc Effect. In the Righi–Leduc effect, one end of the sample is heated and the resulting temperature gradient $\nabla_y T$ causes a thermal current of density U_y to flow from the hot to the cold end. A perpendicular magnetic field B_0 applied along z produces a transverse temperature gradient $\nabla_x T$ given by

$$\nabla_x T = R_L B_0 U_y \tag{76}$$

as indicated in Fig. 8-3(e), where R_L is the Righi–Leduc coefficient. The Law of Wiedermann and Franz (Chapter 24, Eq. (18)) relates R_L to the Hall coefficient R_H

$$R_H = R_L L_0 T \tag{77}$$

where $L_0 = (3/2)(k_B/e)^2$ is the Lorentz number. The thermal Hall angle Θ_{th}

$$\tan \Theta_{th} = \nabla_x T / \nabla_y T \tag{78}$$

is defined in analogy with its Hall-effect counterpart.

13 NEGATIVE TEMPERATURES

Infinite temperature corresponds to the equalization of the populations of all energy levels. This is because the ratio of the populations of two energy levels N_1 and N_2 is given by

$$N_1 / N_2 = e^{-\Delta E / K_B T} \tag{79}$$

where $\Delta E = E_2 - E_1$, and $N_2 \approx N_1$ for $k_B T \gg \Delta E$. When the temperature is zero, all of the atoms are in the lowest energy state.

The presence of a greater population in the upper energy state means that the temperature is negative, and when all of the atoms are in the upper level the temperature is −0, or negative zero. Systems at negative temperatures are thus hotter than those with positive temperatures, and the hottest possible case is $T = -0$.

Negative temperatures can occur in systems in which the number of energy levels is finite. Masers and lasers act on the principle of establishing negative temperatures, which results in the spontaneous emission of radiation from an over-populated upper energy level.

14 STATISTICAL MECHANICS

One of the important quantities determined for a system in statistical mechanics is the partition function Z

$$Z = \sum e^{-\beta E} \tag{80}$$

where $\beta = 1/k_{\mathrm{B}}T$, and the summation is over all energy states. The Helmholtz free energy is related to Z via

$$F = -k_{\mathrm{B}}T \ln Z \tag{81}$$

and the internal energy is

$$U = -\frac{\partial}{\partial \beta} \ln Z \tag{82}$$

These expressions for F and U provide us with the entropy S from Eq. (22), and, in like manner, H and G from Eqs. (20) and (23), respectively. Equations (81) and (82) are thus convenient starting points for the derivation of thermodynamic expressions for systems in which the partition function Z is known or can be calculated.

CHAPTER 9

STATISTICAL MECHANICS AND DISTRIBUTION FUNCTIONS

1 INTRODUCTION

Statistical physics, or statistical mechanics, is concerned with systems containing large numbers of particles. The number is so large that it is not feasible to follow the motions of individual particles, and statistical methods are applied to obtain averaged properties that characterize the system. Distribution functions can be found, such as how velocities are distributed among the molecules of a gas, or how electrons are distributed in energy levels, and these are used for calculating averages. It was mentioned at the end of the previous chapter that a particular distribution function called a partition function Z can be employed to calculate thermodynamic properties. These properties are, by nature, averages. For example, the pressure on the walls of a container of gas is due to fast-, medium-, and slow-moving molecules colliding with the walls, so the value of the pressure can be deduced from a knowledge of the velocity distribution.

Another approach to statistical mechanics is to consider a collection or ensemble of identical systems each prepared in the same way, and then carry out averages over the ensemble. For example, consider the probability that the number 4 will appear when a pair of dice is thrown. This can be evaluated statistically by gathering together a collection or ensemble of 1000 pairs of identical dice, by shaking each pair randomly (preparing each the same way), throwing them all, and counting the number of pairs in the ensemble that give the answer 4. More generally, we assume that similarly prepared systems containing many particles will behave the same.

The two dice in each individual system of the ensemble, i.e., in each pair, might be identical and hence indistinguishable from each other, or they might differ in color or size and hence be distinguishable. Systems in classical mechanics contain distinguishable particles since we can keep track of each one. In contrast to this, the particles treated by quantum mechanics are indistinguishable, and they come in two types; namely, those with integer spin $S = 0, 1, 2, \ldots$, and those with half-integer spin $S = 1/2, 3/2, 5/2, \ldots$. We see later that these three types of particles obey different rules concerning how they can populate the energy levels that are available to them, and this causes them to have different energy level distribution functions.

A fundamental assumption of statistical mechanics is that of equal, a priori probabilities. More precisely, we assert that all of the accessible states of an isolated system in equilibrium have the same probability of being occupied. The extent to which theoretically possible states are accessible can be limited by assumptions concerning, for example, the energy or the temperature. Another basic assumption is that a time average of a system can be replaced by an ensemble average.

2 MICROCANONICAL ENSEMBLE

We begin by limiting the number of accessible states of a system by restricting its energy. Consider an isolated system of N particles in a fixed volume V that is capable of being configured in a large number of states labeled by i, each of which has a characteristic energy E_i. If we restrict the energy E of the system to lie within the narrow range from E_0 to $E_0 + dE$

$$E_0 \leq E \leq E_0 + dE \tag{1}$$

then the probability P_i of finding the system in a particular state i is

$$P_i = \begin{cases} C & \text{for } E_0 \leq E_i \leq E_0 + dE \\ 0 & \text{otherwise} \end{cases} \tag{2}$$

where the constant C can be evaluated from the normalization condition

$$\sum P_i = 1 \tag{3}$$

of overall unit probability over all accessible states, i.e., over all states that satisfy the energy condition (2). The collection or ensemble of these accessible states is called a microcanonical ensemble. It is typically used for small energy ranges, and for isolated systems, without taking into account thermal effects.

The number of states per energy interval, called the density of states $D(E)$, is assumed to have a constant value $E(E_0)$ over the energy range (1) for the microcanonical ensemble, and to be zero outside this range. Degenerate states, or distinguishable states of the same energy, are counted as separate states.

3 CANONICAL ENSEMBLE

A canonical ensemble is characterized by a fixed temperature T. When an ensemble is in contact with a heat reservoir at the temperature T, then the population will redistribute itself so that the lower lying energy levels are more populated than the upper ones. If there are only two energy levels, then the ratio of the populations N_1 and N_2 of the two levels is related to their respective energies E_1 and E_2 by the Boltzmann factor

$$\frac{N_2}{N_1} = \exp\left[-\beta(E_2 - E_1)\right] \tag{4}$$

where

$$\beta = 1/k_B T \tag{5}$$

The population ratio N_2/N_1 is the same as the ratio of the probabilities of occupancy P_2/P_1

$$\frac{P_2}{P_1} = \frac{N_2}{N_1} \tag{6}$$

If the energy levels have the respective degeneracies g_1 and g_2 then Eq. (4) becomes

$$\frac{N_2}{N_1} = \frac{g_2}{g_1} \exp\left[-\beta(E_2 - E_1)\right] \tag{7}$$

but in the following discussion degeneracies will not be taken into account. Examples of such degeneracies are $g_s = 2$ for the two spin states $M_S = \pm\frac{1}{2}$ of an electron, and $g_L = 2L + 1$ for orbital motion. The presence of a magnetic field will, of course, remove these degeneracies.

If the relative populations P_i, P_j of every pair of states i, j of the ensemble are related to each other through the Boltzmann ratio (4), then the ensemble is called a canonical ensemble. This means that it has a uniform temperature throughout.

The probability of occupancy of a state in a canonical ensemble can be written as follows

$$P_i = \frac{e^{-\beta E_i}}{Z} \tag{8}$$

where the probabilities are normalized

$$\sum P_i = 1 \tag{9}$$

and the denominator Z of Eq. (8)

$$Z = \sum e^{-\beta E_i} \tag{10}$$

is called the partition function or sum over states, German Zustandsumme. Equation (8) may be considered as the defining equation for a canonical ensemble.

The average energy of a canonical ensemble is given by

$$\langle E\rangle = \frac{\sum e^{-\beta E_i} E_i}{Z} = -\frac{1}{Z}\frac{\partial Z}{\partial \beta} = -\frac{\partial \ln Z}{\partial \beta} \tag{11}$$

In addition, we have for the average squared energy

$$\langle E^2\rangle = \frac{1}{Z}\frac{\partial^2 Z}{\partial \beta^2} \tag{12}$$

and for what is called the dispersion

$$\langle (E - \langle E\rangle)^2\rangle = -\frac{\partial \langle E\rangle}{\partial \beta} = \frac{\partial^2 \ln Z}{\partial \beta^2} \tag{13}$$

which is a measure of the range of the energy distribution.

4 GRAND CANONICAL ENSEMBLE

We have discussed systems in isolation characterized by a fixed energy and a fixed number of particles (microcanonical ensembles) and systems that can exchange energy with a heat reservoir characterized by a fixed temperature and a fixed number of particles (canonical ensemble). In both cases it was implicitly assumed that the number of particles was a constant. Some physical systems are in contact with a reservoir with which they can exchange not only heat but also particles, so the number of particles is not constant. Such a system is described by a grand canonical ensemble, with the probability of occupancy P_i of a level with energy E_i and number of particles N_i given by

$$P_i = \frac{e^{-\beta E_i - \alpha N_i}}{\sum e^{-\beta E_i - \alpha N_i}} \tag{14}$$

The average energy $\langle E\rangle$ and average number of particles $\langle N\rangle$ are given by

$$\langle E\rangle = \frac{\sum e^{-\beta E_i - \alpha N_i} E_i}{\sum e^{-\beta E_i - \alpha N_i}} \tag{15}$$

$$\langle N \rangle = \frac{\sum e^{-\beta E_i - \alpha N_i} N_i}{\sum e^{-\beta E_i - \alpha N_i}} \tag{16}$$

The quantity α is sometimes written in terms of the chemical potential μ

$$\alpha = -\mu / k_B T \tag{17}$$

The chemical potential can be considered as a thermodynamic quantity related to the Helmholtz free energy F through the expression

$$dF = -S\,dT - P\,dV - \mu\,dN \tag{18}$$

which is a generalization of Eq. (24) in Chapter 8, and we have

$$S = -\left(\frac{\partial F}{\partial T}\right)_{V,N} \tag{19}$$

$$P = -\left(\frac{\partial F}{\partial V}\right)_{T,N} \tag{20}$$

$$\mu = -\left(\frac{\partial F}{\partial N}\right)_{V,T} \tag{21}$$

We can also define the grand potential Ω

$$\Omega = F - \mu N \tag{22}$$

The grand canonical ensemble is used to describe situations in which the number of particles does not remain constant, such as systems involving chemical reactions of the type

$$2H_2 + O_2 \Rightarrow 2H_2O \tag{23}$$

in which three molecules react to form two molecules so the number N is not conserved.

5 THERMODYNAMICS

Statistical mechanics is related to thermodynamics through the partition function (10). We have already given an expression for the average energy $\langle E \rangle$, and this can be identified with the internal energy U of thermodynamics (see Chapter 8, Eq. (82))

$$U = \langle E \rangle = -\frac{\partial \ln Z}{\partial \beta} \tag{24}$$

The Helmholtz free energy F is obtained from the expression

$$F = -k_B T \ln Z \tag{25}$$

and the entropy S

$$S = k_B \left(\ln Z - \beta \frac{\partial \ln Z}{\partial \beta} \right) \tag{26}$$

is provided by the defining relation

$$F = U - TS \tag{27}$$

of the Helmholtz free energy. In like manner, we easily obtain the enthalpy H and the Gibbs free energy G

$$H = U + PV \tag{28}$$

$$G = H - TS \tag{29}$$

to complete our catalog of fundamental thermodynamic functions (see Chapter 8, Section 4). Thus, once the partition function is known for a system all of the thermodynamic properties can, in principle, be calculated. In this section we do not take into account changes in the number of particles in the system as we did when we wrote down the Helmholtz free energy expressions such as Eq. (18) in the previous section.

As an example, the harmonic oscillator has the energies

$$E_n = (n + \tfrac{1}{2}) \hbar \omega \tag{30}$$

to give for the partition function

$$Z = \sum \exp[-\beta(n + \tfrac{1}{2})\hbar\omega] = e^{-\beta\hbar\omega/2}[1 + e^{-\beta\hbar\omega} + e^{-2\beta\hbar\omega} + \ldots] \tag{31}$$

$$= \frac{e^{-\beta\hbar\omega/2}}{1 - e^{+\beta\hbar\omega}} \tag{32}$$

Using Eq. (11) we obtain for the average energy

$$\langle E \rangle = \hbar\omega\left\{\frac{1}{2} + \frac{1}{e^{-\beta\hbar\omega} - 1}\right\} \tag{33}$$

$$\approx \begin{cases} k_B T & \hbar\omega \ll k_B T \quad (34a) \\ \frac{1}{2}\hbar\omega & \hbar\omega \gg k_B T \quad (34b) \end{cases}$$

This result is useful for deriving the Debye theory of specific heats. According to the equipartition theorem in the classical or high temperature limit each quadratic term in a Hamiltonian contributes the amount $\frac{1}{2}k_B T$ to the energy. The harmonic oscillator has two such terms, $\frac{1}{2}mv^2$ and $\frac{1}{2}kx^2$, so from Eq. (34(a)) it contributes the amount $k_B T$ at high temperatures.

6 PARAMAGNETISM

An atom or ion with angular momentum $\hbar\mathbf{J} = \hbar\mathbf{L} + \hbar\mathbf{S}$ in free space has a magnetic moment $\boldsymbol{\mu}$ given by

$$\boldsymbol{\mu} = \gamma\hbar\mathbf{J} = g\mu_B\mathbf{J} \tag{35}$$

where γ is the gyromagnetic ratio, sometimes called the magnetogryic ratio, g is the dimensionless g-factor, $\mu_B = e\hbar/2m$ is the Bohr magneton, and

$$g\mu_B = \gamma\hbar \tag{36}$$

In a magnetic field $\mathbf{B}$ the energy per spin is

$$U = -\boldsymbol{\mu} \cdot \mathbf{B} = g\mu_B m_J B \tag{37}$$

For the case of a free electron $L = 0, J = S = \frac{1}{2}$, and the energy per spin in the field is $\pm\frac{1}{2}g\mu_B B$.

Paramagnetism arises from an ensemble of spins which interact with an externally applied magnetic field B, but not with each other. There are two energy states $\pm\frac{1}{2}g\mu_B B$ for $S = \frac{1}{2}$, and this gives for the partition function Z

$$Z = e^{-\frac{1}{2}g\mu_B B\beta} + e^{\frac{1}{2}g\mu_B B\beta} \tag{38}$$

and the average magnetic moment per spin $\langle\mu\rangle$ is

$$\langle\mu\rangle = \frac{-\frac{1}{2}g\mu_B e^{-\frac{1}{2}g\mu_B B\beta} + \frac{1}{2}g\mu_B e^{\frac{1}{2}g\mu_B B\beta}}{Z} \tag{39}$$

$$\approx \frac{g^2\mu_B^2 S(S+1)B}{3k_B T} \qquad g\mu_B B \ll k_B T \tag{40}$$

where we include the factor $S(S+1)$ in the numerator to take into account higher spin values, $S \geq \frac{1}{2}$. If we insert the partition function from Eq. (38) into Eq. (39), then we can see that the magnetization M, or magnetic moment per unit volume is given by

$$M = \frac{1}{2}g\mu_B N \tan h(\tfrac{1}{2}g\mu_B B/k_B T) \tag{41}$$

with the high temperature limit for $S \geq \frac{1}{2}$

$$M \approx \frac{Ng^2\mu_B^2 S(S+1)B}{3k_B T} \qquad g\mu_B B \ll k_B T \tag{42}$$

where N is the number of spins per unit volume. It is easy to write down the susceptibility $\chi = \mu_0 M/B$.

The situation is more complicated when orbital motion is present and the quantum number J replaces S. Equation (41) now contains what is called a Brillouin function $B_J(x)$, as is explained in typical solid state physics texts

$$B_J(x) = \frac{2J+1}{2J}\coth\left(\frac{(2J+1)x}{2J}\right) - \frac{1}{2J}\coth\left(\frac{x}{2J}\right) \tag{43}$$

where

$$x = gJ\mu_B B/k_B T \tag{44}$$

and the magnetization M is given by

$$M = NgJ\mu_B B_J(x) \tag{45}$$

This reduces to Eq. (42) in the high temperature limit for $J = S$.

7 ENTROPY AND ADIABATIC DEMAGNETIZATION

Entropy is associated with the randomness of a system in occupying the states that are accessible to it. Let $\Omega(E)$ be the number of states accessible when the system is in the energy state E. Statistically, entropy S is defined as proportional to the natural logarithm of this number of states, or more specifically

$$S = k_B \ln \Omega \tag{46}$$

The absolute temperature of a system can be expressed in terms of Ω

$$\frac{1}{k_B T} = \frac{\partial \ln \Omega}{\partial E} \tag{47}$$

As an example of the usefulness of the above definition of entropy, consider a system of N non-interacting spins each of which has $2J + 1$ accessible states, since this is the spin degeneracy. From Eq. (46) the spin entropy is

$$S = k_B \ln(2J + 1)^N = Nk_B \ln(2J + 1) \tag{48}$$

If a magnetic field is turned on, the spin degeneracy will be lifted, the lower energy levels will become more populated than the upper ones, and the entropy will be reduced. If the temperature is now raised high enough so that the thermal energy $k_B T$ far exceeds the magnetic energy level separation, $k_B T \gg g\mu_B B$, then the entropy will approach its maximum value (48). If, on the other hand, the thermal energy is far less than the magnetic energy, $k_B T \ll g\mu_B B$, then all of the spins will begin to populate the lowest magnetic state. As a result, $k_B T/g\mu_B B \Rightarrow 0$, only one state is accessible, so we have $\Omega \Rightarrow 1$ and from Eq. (46) $S \Rightarrow 0$, in accordance with the third law of thermodynamics.

More quantitatively, in the presence of a magnetic field the entropy of a paramagnetic sample varies with the temperature in the manner shown in Fig. 9-1. We see from the figure that applying the field decreases the entropy as described above; the stronger the field the greater the decrease in entropy. This occurs because increasing the magnetic field increases the separation of the Zeeman sublevels. If the field is turned on isothermally at the temperature T_i the system goes from point *a* to point *b* on the figure, the atomic magnetic moments distribute themselves in the Zeeman sublevels in accordance with the Boltzmann factor for T_i, and the entropy decreases. If, now, the sample is isolated from the heat bath and the magnetic field is turned off under adiabatic conditions, the entropy cannot change so the system goes from point *b* to point *c* on the figure, thereby decreasing its temperature to T_f. This is the phenomenon of adiabatic demagnetization, a technique that can be used to reach very low temperatures.

8 MAXWELL DISTRIBUTION

An important application of statistical mechanics is to derive an expression for the speeds in a gas of non-interacting molecules assuming that the only

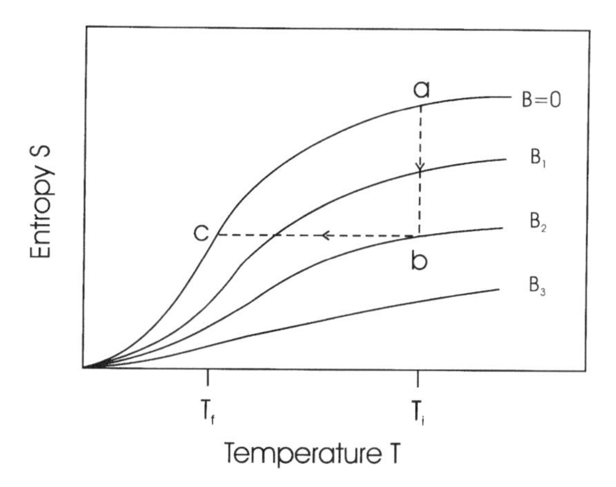

Fig. 9-1. Temperature dependence of the entropy of a paramagnetic spin system. The isothermal magnetization along path $a \Rightarrow b$ decreases the entropy and the adiabatic demagnetization along path $b \Rightarrow c$ lowers the temperature of the spin system.

energy that each molecule has is kinetic; namely, $\frac{1}{2}mv^2$. Every molecule has x, y and z components of kinetic energy, and by the equipartition theorem each component contributes the amount $\frac{1}{2}k_BT$ to the energy, so each molecule has the thermal energy $(3/2)k_BT$. The number of molecules in a unit volume with the speed between v and $v + dv$ is given by

$$F(v)dv = 4\pi\rho\left\{\frac{m}{2\pi k_B T}\right\}^{3/2} v^2 \exp[-mv^2/2k_BT]\,dv \tag{49}$$

where ρ is the number density of molecules per unit volume, and

$$\int_0^\infty F(v)\,dv = \rho \tag{50}$$

The function $F(v)$, called a Maxwellian distribution, is plotted in Fig. 9-2 for three different temperatures. The gas molecules that satisfy this distribution function are said to obey classical or Maxwell Boltzmann (MB) statistics.

The average velocity in a Maxwellian distribution is zero, $\langle \mathbf{v} \rangle = 0$, since positive and negative values appear equally often for each component. The speed $v_{\max}$ at the maximum of the curve, also called the most probable speed, is given by

$$v_{\max} = (2k_BT/m)^{1/2} \tag{51}$$

The mean square velocity v_{ms} is calculated from the integral

$$\langle v^2 \rangle = \frac{1}{\rho}\int F(v)\,v^2\,dv = 3k_BT/m \tag{52}$$

and has the value

$$v_{\text{ms}} = (\langle v^2 \rangle)^{1/2} = (3k_BT/m)^{1/2} \tag{53}$$

as may be seen by evaluating the integral (52) in terms of gamma functions.

9 CLASSICAL AND QUANTUM STATISTICS

The Maxwell Boltzmann statistics of the previous section apply to what we call distinguishable particles; we assume that we can tell the difference between each atom of a particular type, and that there is no limit to the

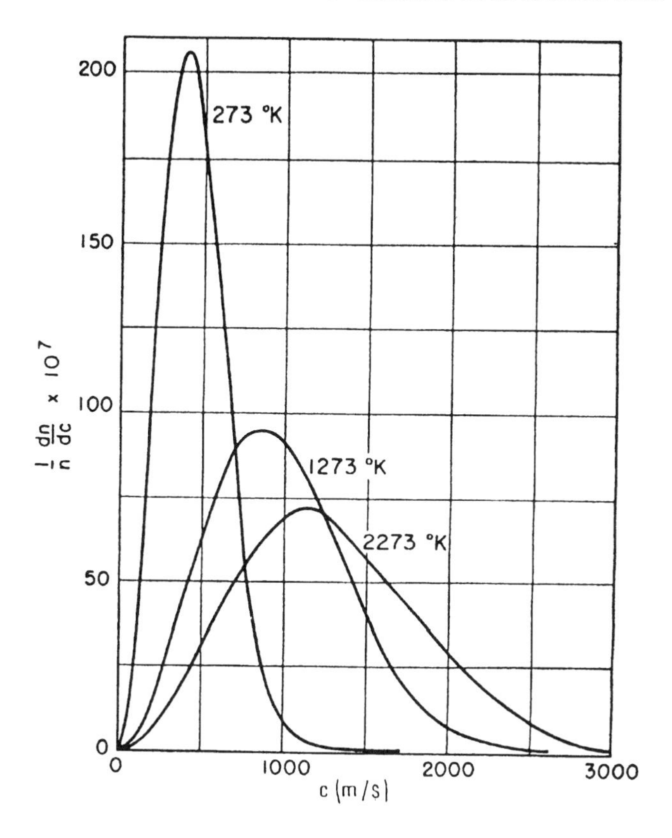

Fig. 9-2. Maxwellian velocity distribution at three temperatures. (From W. J. Moore, *Physical Chemistry*, Prentice-Hall, 1955, Fig. VII-12.)

number of particles that can be in an individual state, such as one with a particular value of energy. Interchanging two particles has no effect on the wave function of the problem since this function has no symmetry requirements.

In quantum statistics the particles are indistinguishable, and there are two cases to consider. Particles with integer spin ($S = 0, 1, 2, \ldots$) called bosons have wave functions that are symmetric, i.e., that do not change sign under the interchange of two particles

$$\Psi(r_1 r_2 r_3 \ldots) = \Psi(r_2 r_1 r_3 \ldots) \tag{54}$$

There are no restrictions on the number of particles that can occupy each state, and they are said to obey Bose–Einstein (BE) statistics. In contrast to this, particles with half integer spin ($S = \frac{1}{2}, \frac{3}{2}, \frac{5}{2}, \ldots$) called fermions have wave functions that are antisymmetric under the interchange of particles

$$\Psi(r_1 r_2 r_3 \ldots) = -\Psi(r_2 r_1 r_3 \ldots) \tag{55}$$

These particles obey Fermi–Dirac (FD) statistics according to which no two of them can be in the same quantum state. In other words, FD particles obey the Pauli exclusion principle. In wave functions (54) and (55) each particle is identified by its position vector r_i. In actuality, each particle has additional quantum numbers designating, for example, its orbital (ℓ, m_ℓ) and spin (s, m_s) states, but these are omitted for simplicity.

In Table 9-1 we compare the three types of statistics for the case of a system containing two particles and three individual states. We see that MB has the most and FD has the least number of overall states. In the MB case, one third of the states have two particles together, in the BE case half of them do, and in the FD case none of them do. Thus, statistically, bosons tend to be closer together and fermions further apart compared with the classical MB case.

One striking difference occurs in the limit of absolute zero where all of the bosons condense into the lowest energy state, while in a fermion (e.g.,

TABLE 9-1. Example of the number of overall states and the occupancy thereof for a two-particle (*A*,*B*) and three energy level (1,2,3) system obeying the MB, BE, and FD statistics

Maxwell–Boltzmann statistics 9 states			Bose–Einstein statistics 6 states			Fermi–Dirac statistics 3 states		
1	2	3	1	2	3	1	2	3
A	B		A	A		A	A	
B	A			A	A		A	A
	A	B	A		A	A		A
	B	A	AA					
A		B		AA				
B		A			AA			
AB								
	AB							
		AB						

electron) system all of the levels are doubly occupied up to a level called the Fermi energy E_F, and are empty above this level. The unique behavior of superfluid helium below the lambda point T_λ and of a superconductor below its transition temperature T_c have similarities to what might be expected from boson condensation since ^{4}He atoms and Cooper pairs formed from coupled electrons (or holes) are bosons.

10 DISTRIBUTION FUNCTIONS

It is important for us to know the distribution function $f(E)$, which gives the extent to which energy levels are occupied as a function of the temperature for these three types of statistics. In other words, $f(E)$ is proportional to the probability that a single particle will be in a state with the energy E. For the three types of statistics the distribution functions are as follows:

$$f(E) = \frac{1}{e^{\beta(E-\mu)} + 1} \qquad \text{Fermi–Dirac} \tag{56}$$

$$f(E) = \frac{1}{e^{\beta(E-\mu)} - 1} \qquad \text{Bose–Einstein} \tag{57}$$

$$(E_j) = \frac{Ne^{-\beta E_j}}{\sum e^{-\beta E_j}} \qquad \text{Maxwell–Boltzmann} \tag{58}$$

The so-called chemical potential μ is adjusted so that the probabilities sum to the total number of particles in the system

$$\sum f(E_i) = N \tag{59}$$

For the FD case we have

$$0 \le f(E) \le 1 \tag{60}$$

and for absolute zero temperature

$$f(E) = \begin{cases} 1 & \text{for } E < \mu \\ 0 & \text{for } E > \mu \end{cases} \tag{61}$$

where $\mu = E_F$ is the Fermi energy at absolute zero. Figure 9-3 plots $f(E)$ versus E for the FD case at $T = 0$ and for $T \ll T_F$ where the Fermi temperature T_F is defined by the expression

$$E_F = k_B T_F. \tag{62}$$

Many of the properties of conduction electrons in a metal are explained satisfactorily by the FD distribution function (57) for $T \ll T_F$, where for a typical metal $T_F \sim 10^4$K.

We can see from the form of Eq. (57) that for BE statistics $f(E)$ can become arbitrarily large for energies close to absolute zero, and at 0 K there can be a consolidation of all of the particles into the lowest energy state, a phenomenon called Bose–Einstein condensation. At very high temperatures, the three types of statistics become classical; in other words, $f(E) \sim e^{-\beta E}$ for all but the lowest energies, meaning $E \gg k_B T$.

11 PHOTON STATISTICS

Photons are bosons that are not limited in number since they can be created or destroyed in a system, so their statistics do not include a chemical poten-

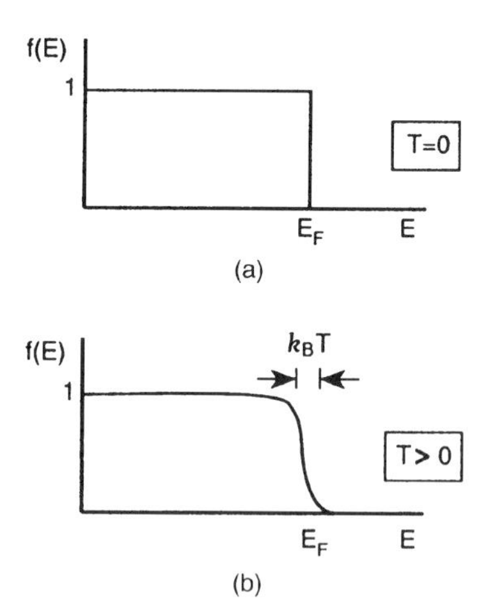

Fig. 9-3. Fermi–Dirac distribution function $f(E)$ plotted (a) at $T = 0$, and (b) for $0 < T \ll T_F$. (From C. P. Poole, Jr et al., *Superconductivity*, Academic Press, New York, 1995, p. 9.)

tial. The distribution function is given by Eq. (57) with $\mu = 0$. It is generally expressed in terms of frequency ω instead of energy E, where $\omega = E/\hbar$, and it is called the Planck distribution law

$$f(\omega) = \frac{1}{e^{\beta\hbar\omega} - 1} \tag{63}$$

The total energy density $u(T)$ obtained by taking into account the energy per photon $\hbar\omega$, the density of states $D(\omega)$, and carrying out an integration over all frequencies, is given by the Stefan–Boltzmann law

$$u(T) = \frac{\hbar}{\pi^2 c^3}\left(\frac{k_B T}{\hbar}\right)^4 \int_0^\infty \frac{x^3\,dx}{e^x - 1} \tag{64}$$

where $x = \beta\hbar\omega$. The plot of the integrand of Eq. (64) versus x that is presented in Fig. 9-4 provides the frequency distribution of the energy radiated by a black body. The integral of Eq. (64) has the value $\pi^4/15$, and the result of carrying out this integration is usually expressed in terms of the emitting power P of a black body

$$P = \sigma T^4 \tag{65}$$

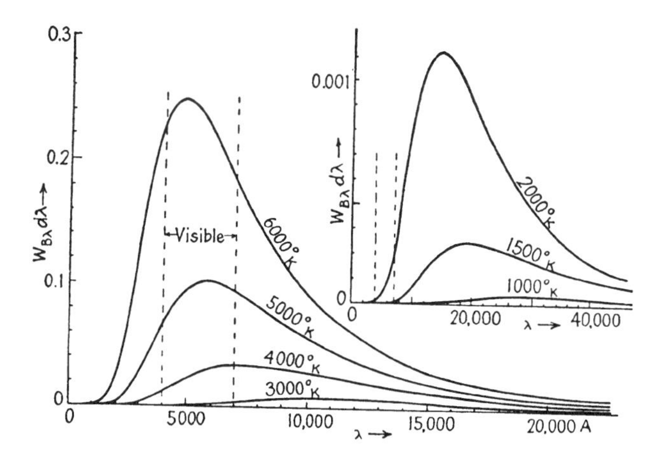

Fig. 9-4. Wavelength dependence of the photon energy density corresponding to the integrand of Eq. (64). This plot gives the distribution of the energy radiated by a black body for seven different temperatures. (From F. A. Jenkins and H. E. White, *Fundamentals of Optics*, McGraw-Hill, New York, 1950, p. 431.)

where the Stefan–Boltzmann constant σ is given by

$$\sigma = \frac{\pi^2 k_B^4}{60\hbar^3 c^2} \tag{66}$$

which has the value

$$\sigma = 5.67 \times 10^{-8} \quad \frac{\text{watts}}{\text{m}^2(\text{deg})^4} \tag{67}$$

The integrand of Eq. (64) has a maximum near the value $\beta\hbar\omega \approx 3$, so the frequency for the maximum output is proportional to the temperature. Thus, for two temperatures T_1 and T_2 we have two maximum frequencies, ω_1 and ω_2, given by

$$\frac{\omega_1}{T_1} = \frac{\omega_2}{T_2} \tag{68}$$

a result known as Wien's law. Figure 9-4 plots the integrand of Eq. (64) in wavelength units and shows how the maximum output point shifts with the temperature.

12 ORTHO-PARA STATISTICS

The proton that forms the nucleus of a hydrogen atom has a (nuclear) spin $I = \frac{1}{2}$, and in a hydrogen molecule the two individual nuclear spins I_i add vectorially

$$\mathbf{I}_1 + \mathbf{I}_2 = \mathbf{I} \tag{69}$$

to give a net nuclear spin I. The individual spins can align antiparallel to form a singlet ($I = 0$) or parallel to form a triplet ($I = 1$) state. The wave functions ψ of these states formed from the basis function $|m_1 m_2\rangle$ are antisymmetric for the singlet

$$\psi = \frac{1}{\sqrt{2}}[|+-\rangle - |-+\rangle] \qquad m = 0 \tag{70}$$

and symmetric for the triplet state

$$\psi = |--\rangle \qquad m = -1 \tag{71a}$$

$$\psi = \frac{1}{\sqrt{2}}[|+-\rangle + |-+\rangle \qquad m = 0 \tag{71b}$$

$$\psi = |++\rangle \qquad m = +1 \tag{71c}$$

Molecules in the singlet state are called parahydrogen and those in the triplet state are called orthohydrogen. The rotational energies of the molecule are labeled with the quantum number K, and the rotational wave function has the parity $(-1)^K$. The overall wave function must be antisymmetric since the nuclei are fermions, and therefore parahydrogen must have even K and orthohydrogen can only have odd K rotational states. The nuclear spin cannot change during rotational transitions, corresponding to the selection rule $\Delta m = 0$, and we also have the rotational transition selection rule $\Delta K = \pm 2$. All four spin states (70) and (71) are equally probable so there are three times as many orthohydrogen molecules. As a result, the rotational spectrum of hydrogen alternates between lines with the relative intensities 1:3:1:3:. . ., thereby confirming the effect of the nuclear statistics on the spectrum.

CHAPTER 10

ELECTROSTATICS AND MAGNETOSTATICS

1 INTRODUCTION

This chapter is concerned with electrostatics and magnetostatics; that is, electric and magnetic phenomena in the absence of any time dependence. General principles are covered here, then the next chapter treats a particular aspect of the subject; namely, electric and magnetic multipoles. SI units (formerly called MKS) are used throughout.

The discussion begins by clarifying some ideas about the nature of electric and magnetic fields. Then time is taken into account when we introduce Maxwell's equations, before again restricting the discussion to electrostatics and magnetostatics. Tesseral and spherical harmonics are introduced.

2 ELECTRIC AND MAGNETIC FIELDS

Consider a region of free space that contains an electric field **E** and a magnetic field **B**. A charge q moving at the velocity **v** in this region experiences the Lorentz force **F** given by

$$\mathbf{F} = q(\mathbf{E} + \mathbf{v} \times \mathbf{B}) \tag{1}$$

The electric field **E** can be considered as the force per unit charge, and the magnetic field **B** has been called the force per unit current. These are the fundamental electric and magnetic fields that are measured experimentally. Associated with **E** and **B** are the two derived fields **D** and **H**, respectively. A vacuum has a dielectric constant or permittivity ε_0 and a permeability μ_0, and in free space the derived fields are related to their fundamental counterparts by

$$\mathbf{D} = \varepsilon_0 \mathbf{E} \tag{2}$$

$$\mathbf{H} = \mathbf{B}/\mu_0 \tag{3}$$

A medium is characterized by a dielectric constant $\varepsilon \geq \varepsilon_0$ and a permeability $\mu \geq \mu_0$, so in a medium we have

$$\mathbf{D} = \varepsilon\ \mathbf{E} \tag{4}$$

$$\mathbf{B} = \mu \mathbf{B} \tag{5}$$

More generally, a medium can be anisotropic, with the components of **D** and **E** given by

$$\mathbf{D}_i = \sum \varepsilon_{ij} E_j \tag{6}$$

$$\mathbf{B}_i = \sum \mu_{ij} H_j \tag{7}$$

where the summation is over $j = z, y, z$. This means that **D** and **E** can be in different directions, and likewise for **B** and **H**. For the present, only isotropic media will be considered.

If a medium has an electric dipole moment per unit volume called the polarization **P**, and a magnetic dipole moment per unit volume called the magnetization **M**, then the fundamental and derived electric and magnetic fields are related by

$$\mathbf{D} = \varepsilon_0 \mathbf{E} + \mathbf{P} = \varepsilon_0 \mathbf{E}(1 + \chi_e) \tag{8}$$

$$\mathbf{B} = \mu_0(\mathbf{H} + \mathbf{M}) = \mu_0 \mathbf{H}(1 + \chi) \tag{9}$$

where $\chi_e = P/\varepsilon_0 E$ and $\chi = M/H$ are the dimensionless electric and magnetic susceptibilities, respectively.

3 MAXWELL'S EQUATIONS

The various fields introduced in the previous section are related to each other through Maxwell's equations. Two of these equations, the ones with the curl operation, depend explicitly on the time

$$\nabla \times \mathbf{H} - \frac{\partial \mathbf{D}}{\partial t} = \mathbf{J} \tag{10}$$

$$\nabla \times \mathbf{E} + \frac{\partial \mathbf{B}}{\partial t} = 0 \tag{11}$$

where **J** is the current density, and since for the present we are only interested in static phenomena we can rewrite these expressions without the time dependence

$$\nabla \times \mathbf{H} = \mathbf{J} \tag{12}$$

$$\nabla \times \mathbf{E} = 0 \tag{13}$$

Maxwell's two divergence equations

$$\nabla \cdot \mathbf{B} = 0 \tag{14}$$

$$\nabla \cdot \mathbf{D} = \rho \tag{15}$$

are already independent of the time, where ρ is the charge density. It will be convenient to use the expressions (4) and (5) to write static equations (12) and (15) in terms of the fundamental fields

$$\nabla \times \mathbf{B} = \mu \mathbf{J} \tag{16}$$

$$\nabla \cdot \mathbf{E} = \rho/\varepsilon \tag{17}$$

where we assume media that are uniform and isotropic so that the permeability μ and the dielectric constant ε are both independent of position. The energy densities U_{E} and U_{M} associated, respectively, with the presence of electric and magnetic fields in a region are given by

$$U_{\mathrm{E}} = \tfrac{1}{2}\mathbf{E} \cdot \mathbf{D} \qquad U_{\mathrm{M}} = \tfrac{1}{2}\mathbf{B} \cdot \mathbf{H} \tag{18}$$

Technically speaking, the quantities **E** and **B** are vector fields because they extend through regions of space with magnitudes and directions which depend on the position. Nevertheless, we shall usually refer to them as simply vectors.

4 ELECTROMAGNETIC POTENTIALS

It is often convenient to express the electric and magnetic fields in terms of potentials. We see from Maxwell's equation (14) that the magnetic field **B** has a vanishing divergence, $\nabla \cdot \mathbf{B} = 0$, which means that it is a solenoidal vector and hence can be derived from the curl of a vector potential **A**

$$\mathbf{B} = \nabla \times \mathbf{A} \tag{19}$$

We see from Maxwell's equation (13) for static phenomena that $\mathbf{E}$ has a vanishing curl, $\nabla \times \mathbf{E} = 0$, which means that it is an irrotational vector and hence can be derived from a scalar potential ϕ

$$\mathbf{E} = -\nabla\phi \tag{20}$$

For time dependent cases, Maxwell equation (11) must be used instead of (13), and E has a time dependent vector potential term

$$\mathbf{E} = -\nabla\phi - \frac{\partial \mathbf{A}}{\partial t} \tag{21}$$

The charge and current densities ρ and $\mathbf{J}$, respectively, may be considered as sources of the electric and magnetic fields because when they are present the fields are also.

5 DIELECTRIC CONSTANT AND PERMEABILITY

In free space the dielectric constant ε_0 and the permeability μ_0 provide the value of the velocity of light c in vacuo

$$(\mu_0\varepsilon_0)^{-1/2} = c = 2.9979 \times 10^8 \text{ m/s} \approx 3 \times 10^8 \text{ m/s} \tag{22}$$

Monochromatic light is an electromagnetic wave in which mutually perpendicular electric $\mathbf{E}$ and magnetic $\mathbf{H}$ fields oscillate at a frequency ω in a plane perpendicular to the direction of propagation. The power or energy flow is given by Poynting's vector $\mathbf{E} \times \mathbf{H}$. The energy densities $\frac{1}{2}\varepsilon E^2$ and $\frac{1}{2}\mu H^2$ stored in the electric and magnetic fields (18), respectively, are equal to each other

$$\tfrac{1}{2}\mu H^2 = \tfrac{1}{2}\varepsilon E^2 \tag{23}$$

and we obtain the impedance

$$E/H = (\mu/\varepsilon)^{1/2} \tag{24}$$

In a vacuum the characteristic impedance of free space has the value

$$(\mu_0/\varepsilon_0)^{1/2} = 120\pi \text{ ohms} \tag{25}$$

If we solve Eqs. (22) and (25) for ε_0 and μ_0 we obtain

$$\varepsilon_0 = \frac{1}{120\pi c} = \frac{1}{36\pi} 10^{-9} \text{ Farad/meter} = \frac{1}{36\pi} \times 10^{-9} \text{ C}^2/\text{Nm}^2 \tag{26}$$

$$\mu_0 = \frac{120\pi}{c} = 4\pi \times 10^{-7} \text{ Henry/meter} = 4\pi \times 10^{-7} \text{ N/A}^2 \tag{27}$$

where A denotes ampere, C = coulomb, N = newton, and m = meter. It is often convenient to use dimensionless relative dielectric constants $\varepsilon/\varepsilon_0$ and permeabilities μ/μ_0. The index of refraction n given by

$$n = (\varepsilon/\varepsilon_0)^{1/2} \tag{28}$$

is widely used in optics. Ordinarily in optics the assumption is made that $\mu = \mu_0$, and this makes the index of refraction the ratio of the velocity of light c in free space to that v in a medium

$$n = c/v \tag{29}$$

6 BOUNDARY CONDITIONS AT INTERFACES BETWEEN MEDIA

At an interface between two media, in the absence of electric charges and currents, the normal components of **D** and **B** and the tangential components of **E** and **H** are continuous across the interface, as shown in Fig. 10-1. If there are surface charge densities σ or surface current densities K they produce discontinuities in the normal components $D_\perp$ and the tangential components $H_\parallel$, respectively. The equations of these boundary conditions at the interface of medium 1 and medium 2 are

$$\mathbf{n} \cdot (\mathbf{B}_2 - \mathbf{B}_1) = 0 \tag{30a}$$

$$\mathbf{n} \times (\mathbf{E}_2 - \mathbf{E}_1) = 0 \tag{30b}$$

$$\mathbf{n} \cdot (\mathbf{D}_2 - \mathbf{D}_1) = \sigma \tag{30c}$$

$$\mathbf{n} \times (\mathbf{H}_2 - \mathbf{H}_1) = \mathbf{K} \tag{30d}$$

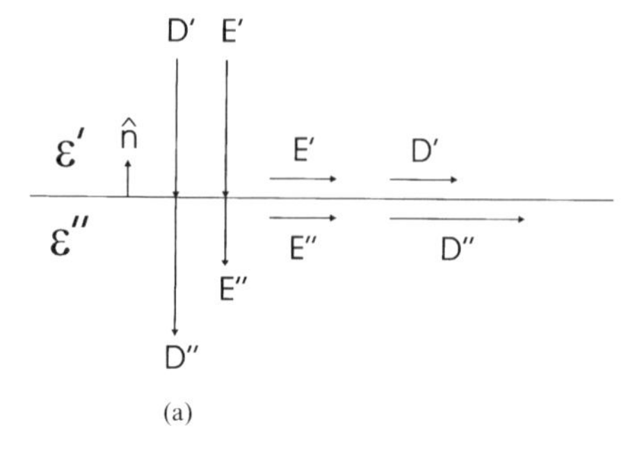

(a)

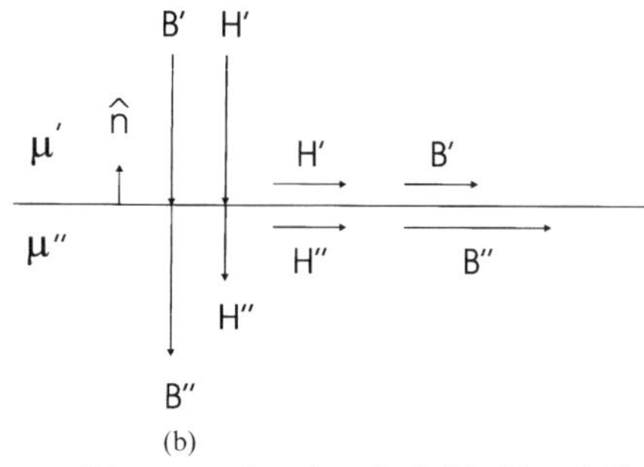

(b)

Fig. 10-1. Boundary conditions on the electric fields **E** and **D** (a) and the magnetic field **B** and **H** (b) at the interface between two media with the respective dielectric constants $\varepsilon'' = 2\varepsilon'$ and permeabilities $\mu'' = 2\mu'$. It is assumed that there are no charge or current densities at the interface.

Note that the surface current density **K** flows parallel to the interface, and is perpendicular to **H** at the surface. The units of the charge density σ are $\mathrm{C/m^2}$, and the units of the current density **K** are A/m.

7 SOURCES OF ELECTRIC FIELD

We proceed to examine the source terms ρ and **J** of the **E** and **B** fields (17) and (16), respectively. Consider first the **E**-field case. If we integrate the expression (17) over a volume that completely encloses a distribution of charge ρ we obtain

$$\int \nabla \cdot \mathbf{E}\, dV = \frac{1}{\varepsilon} \int \rho\, dv \tag{31}$$

We can convert the first integral to a surface integral, and note that the second is just the total charge Q enclosed by the surface

$$\int \mathbf{E} \cdot \mathbf{d\sigma} = Q/\varepsilon \tag{32}$$

a result called Gauss' law.

If the surface is a sphere that encloses a point charge q at the center, then by symmetry E at a particular distance r from the center must be the same in all directions, so we obtain from Gauss' law (32)

$$\mathbf{E} = \frac{q\hat{\mathbf{n}}}{4\pi\varepsilon r^2} \quad \text{(outside sphere)} \tag{33}$$

If q is positive, then $\mathbf{E}$ is a vector field directed outward from q, with $\hat{\mathbf{n}}$ a dimensionless unit vector in the $\mathbf{r}$ direction. The total electric flux, which may be evaluated by integrating $\mathbf{E} \cdot \mathbf{d\sigma}$ over any surface that completely encloses the charge, is always q/ε.

The charge density of a uniformly charged sphere is

$$\rho = \frac{3q}{4\pi R^3} \tag{34}$$

Within the sphere at a distance r from the center the charge $4\pi r^3 \rho/3$ lies inside that point and the effect of the remaining charge averages to zero, so the field inside is

$$\mathbf{E} = \frac{qr\hat{\mathbf{n}}}{4\pi\varepsilon R^3} \quad \text{(inside sphere)} \tag{35}$$

Thus, $\mathbf{E}$ inside increases linearly with distance r from the center, then falls off as $1/r^2$ outside.

Electric field lines begin on positive charges and end on negative charges. If there are equal amounts of positive and negative charge in a region, then the electric field lines can all begin and end in the region, although some of these field lines could extend outside before returning.

8 SOURCES OF MAGNETIC FIELD

Next, let us examine the case of currents as sources of magnetic fields. Consider a surface through which a current flows. The magnitude of the current can be determined by integrating the dot product $\mathbf{J} \cdot \mathbf{d}\boldsymbol{\sigma}$ over the surface, where $\mathbf{d}\boldsymbol{\sigma}$ is an element of the surface. From Maxwell's time-dependent equation (16) we have

$$\int (\nabla \times \mathbf{B}) \cdot \mathbf{d}\boldsymbol{\sigma} = \mu \int \mathbf{J} \cdot \mathbf{d}\boldsymbol{\sigma} \tag{36}$$

The right-hand side is the total current I through the surface, and the left-hand side can be converted to a line integral around it so for free space we have

$$\oint \mathbf{B} \cdot \mathbf{d}\boldsymbol{\ell} = \mu_0 I \tag{37}$$

If the current is along a straight line, then the $\mathbf{B}$ lines are circles of radius r centered on and perpendicular to the direction of the current line, as shown in Fig. 10-2 and the integral gives

$$\mathbf{B} = \mu_0 I/2\pi r \quad \text{(outside wire)} \tag{38}$$

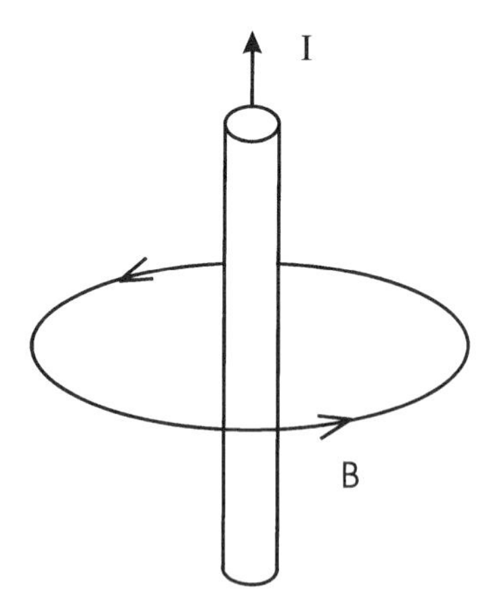

Fig. 10-2. Magnetic field line **B** encircling a wire carrying a current I.

The right-hand rule of pointing the thumb in the current direction gives the **B** field in the direction of the fingers encircling the thumb.

If the wire is carrying a current of uniform density

$$J = I/\pi R^2 \tag{39}$$

where R is the radius of the wire, then following the reasoning leading from Eq. (34) to Eq. (35) we find for the B field inside the wire

$$B = \frac{\mu_0 r I}{2\pi R^2} \quad \text{(inside wire)} \tag{40}$$

so B rises linearly with r inside, then falls off as $1/r$ outside, as shown in Fig. 10-3.

9 ELECTROSTATICS

Now that we have seen how charge and current act as sources for electric and magnetic fields, respectively, we examine the electrostatic case in more

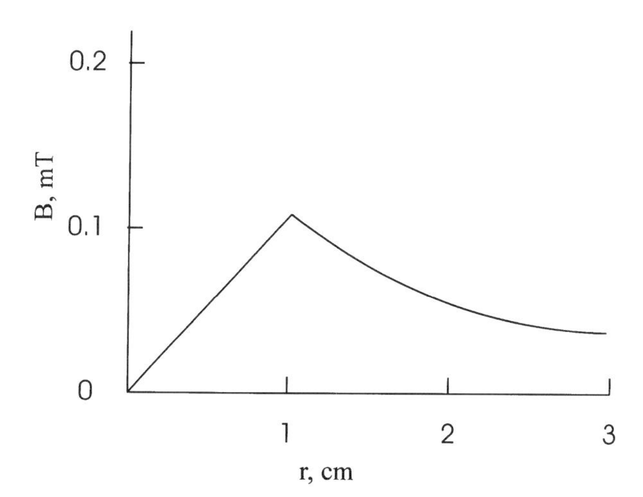

Fig. 10-3. Plot of the magnitude of the encircling magnetic field B_ϕ inside and outside a current carrying wire of radius $r = 1\,\text{cm}$. The figure is drawn for the case of uniform current density $\mathbf{J} = I/\pi r^2$ inside the wire.

detail. The force on an electric charge q placed in an electric field $\mathbf{E}$ is given by Eq. (1)

$$\mathbf{F} = q\mathbf{E} \tag{41}$$

Using Eq. (35) we obtain Coulomb's law for the force between two charges q and q'

$$\mathbf{F} = \frac{qq'}{4\pi\varepsilon_0 r^2} \tag{42}$$

where like charges repel and unlike charges attract.

The electric field at a point r arising from a charge distribution ρ in free space is given by the integral

$$\mathbf{E}(\mathbf{r}) = \frac{1}{4\pi\varepsilon_0} \int \frac{\rho(\mathbf{r}')(\mathbf{r} - \mathbf{r}')}{|\mathbf{r} - \mathbf{r}'|^3} d^3r' \tag{43}$$

which can also be written

$$\mathbf{E}(\mathbf{r}) = -\frac{1}{4\pi\varepsilon_0} \nabla_{\mathbf{r}} \int \frac{\rho(\mathbf{r}')}{|\mathbf{r} - \mathbf{r}'|} d^3r' \tag{44}$$

where the gradient $\nabla_{\mathbf{r}}$ is with respect to $\mathbf{r}$, and recalling Eq. (20)

$$\mathbf{E} = -\nabla \boldsymbol{\Phi} \tag{45}$$

we see that the scalar potential $\boldsymbol{\Phi}(\mathbf{r})$ is given by the expression

$$\boldsymbol{\Phi}(\mathbf{r}) = \frac{1}{4\pi\varepsilon_0} \int \frac{\rho(\mathbf{r}')}{|\mathbf{r} - \mathbf{r}'|} d^3r' \tag{46}$$

The work W done in moving a charge q from position A to position B is

$$W = -\int_A^B \mathbf{F} \cdot \mathbf{d}\ell = -q \int_A^B \mathbf{E} \cdot \mathbf{d}\ell = q(\Phi_{\mathrm{A}} - \Phi_B) \tag{47}$$

independent of the path from A to B.

Combining the expression $\mathbf{E} = -\nabla\boldsymbol{\Phi}$ and $\nabla \cdot \mathbf{E} = \rho/\varepsilon$ gives Poisson's inhomogeneous equation

$$\nabla^2\boldsymbol{\Phi} = -\rho/\varepsilon \tag{48}$$

for which we already have a solution (46) in principle. If no charge density is present we have Laplace's homogeneous equation

$$\nabla^2\boldsymbol{\Phi} = 0 \tag{49}$$

Many electrostatic problems involve solving one or the other of these two equations for specified boundary conditions.

10 BOUNDARY CONDITIONS AT SURFACES

Typical solutions of Laplace's equation provide the electric field distribution in a volume subject to the assignment of values of the electric field on the enclosing surfaces, or subject to the assignment of values of the potential on these surfaces. If the bounding surfaces are good conductors, then from Eq. (30b) and (30c) the electric field at the boundary will be normal to the surface with the value

$$E_\perp = \sigma/\varepsilon = \partial\boldsymbol{\Phi}/\partial n \tag{50}$$

where σ is the surface charge density, and from Eq. (45) $\partial\boldsymbol{\Phi}/\partial n$ is the normal derivative of the potential at the surface. The specification at the surface of $E_\perp$, or equivalently $\partial\boldsymbol{\Phi}/\partial n$, corresponds to Neumann boundary conditions. when the charges on the surface are fixed in position, then the potentials assume values consistent with the charges.

Instead of designating the charge distribution on the surface, one can apply what is called Dirichlet boundary conditions, which is the stipulation of the potential everywhere on the surface. When potentials are established on a conducting boundary, the charges move to conform to the potentials. Cauchy boundary conditions, whereby both $\boldsymbol{\Phi}$ and $\partial\boldsymbol{\Phi}/\partial n$ are specified, overdetermine the problem, and inconsistencies can result. It is, however, possible to use mixed boundary conditions, whereby $\boldsymbol{\Phi}$ is specified over part of the surface and $\partial\boldsymbol{\Phi}/\partial n$ over the remainder.

11 SOLVING POTENTIAL PROBLEMS

If a Green's function method is employed to solve Poisson's equation (48) we have, for the Neumann (G_N) and Dirichlet (G_D) boundary condition cases, respectively

$$\Phi(r) = \int_v \rho(\mathbf{r})G_N(\mathbf{r},\mathbf{r}')d^3r' + \frac{1}{4\pi}\int \frac{\partial\Phi}{\partial n} G_N(\mathbf{r},\mathbf{r}')da' \tag{51}$$

$$\Phi(r) = \int_v \rho(\mathbf{r})G_D(\mathbf{r},\mathbf{r}')d^3r' - \frac{1}{4\pi}\int \Phi(\mathbf{r}')\frac{\partial G_D}{\partial n'} da' \tag{52}$$

Chapter 28, Section 10 provides an explanation of the Green's function method.

Another way to work out potential problems is the method of images. This consists in simulating the boundary conditions by choosing one or several charges called image charges of appropriate signs and magnitudes and locating them outside the volume of interest. Symmetry can be of help in selecting the charges.

A third way to handle potential problems is to use Fourier series and evaluate the coefficients from the boundary conditions. Chapter 27, Section 9 provides some details on Fourier series.

One of the most common ways to solve potential problems is to make use of known solutions of the appropriate differential equations, such as those that are obtained by the separation of variables in Laplace's equation, as is explained in Chapter 28, Section 4. In cartesian coordinates there can be harmonic solutions $\sin kx$ and $\cos kx$, and growth, decay solutions $\sin h\kappa x$ and $\cosh \kappa x$. Alternative forms of these are $e^{\pm ikx}$ and $e^{\pm \kappa x}$, respectively. In cylindrical coordinates, the ϕ direction solutions are almost always of the harmonic type $\sin m\phi$ and $\cos m\phi$, in the z direction they can be growth, decay (e.g., $\sinh \kappa x$) or harmonic (e.g., $\sin kz$), and in the radial direction ρ one obtains Bessel functions $J_n(k\rho)$ or $N_n(k\rho)$. In spherical coordinates, the Tesseral harmonics $Z_{LM}^{S,C}(\theta,\phi)$, which are linear combinations of the two spherical harmonics with the same L and M, provide solutions for the angular part. They are real functions and hence more appropriate to use than spherical harmonics, which are complex. For Laplace's equation the solution for at last one coordinate must be harmonic, and at least one must be of the growth, decay type, as explained in Chapter 28, Section 4.

12 TESSERAL AND SPHERICAL HARMONICS

Tesseral harmonics $Z_{LM}^{S,C}(\theta, \phi)$ are real forms of the spherical harmonics

$$Y_{LM}(\theta, \phi) = (-1)^M \left\{ \frac{2L+1}{4\pi} \frac{(L-M)!}{(L+M)!} \right\}^{1/2} P_L^M(\cos\theta) e^{iM\phi} \tag{53}$$

obtained by taking the linear combinations $[Y_{LM}(\theta, \phi) \pm Y_{L-M}(\theta, \phi)]/\sqrt{2}$, as is mentioned in Chapter 28, Section 7. The potential in electrostatic problems is real, so the Tesseral harmonics provide the desired solutions to potential problems in cylindrical coordinates. They are defined as follows

$$Z_{LO}(\theta, \phi) = \left\{ \frac{2L+1}{4\pi} \right\}^{1/2} P_L(\cos\theta) \tag{54a}$$

$$Z_{LM}^C(\theta, \phi) = \left\{ \frac{2L+1}{2\pi} \frac{(L-M)!}{(L+M)!} \right\}^{1/2} P_L^M(\cos\theta) \cos M\phi \tag{54b}$$

$$Z_{LM}^S(\theta, \phi) = \left\{ \frac{2L+1}{2\pi} \frac{(L-M)!}{(L+M)!} \right\}^{1/2} P_L^M(\cos\theta) \sin M\phi \tag{54c}$$

where M must be positive, and $Z_{L0} = Y_{L0}$ for $M = 0$. In practical problems one can generally use symmetry to eliminate either Z_{LM}^C or Z_{LM}^S from consideration. If spherical harmonics are employed in electrostatic problems, as suggested by some texts, then they must be grouped together as Tesseral harmonics to provide the desired real solutions. The Tesseral harmonics share with the spherical harmonics various properties such as orthogonality, completeness, summation, and sum rule. We have, for example

$$\frac{1}{|\mathbf{r} - \mathbf{r}'|} = 4\pi \sum\sum\sum \frac{1}{2L+1} \frac{r_<^L}{r_>^{L+1}} Z_{LM}^j(\theta', \phi') Z_{LM}^j(\theta, \phi) \tag{55}$$

Some of the lower order Tesseral harmonics in unnormalized form are easily found from the corresponding spherical harmonic tabulation in Chapter 28, Section 7 and they are as follows:

$$Z_{00} = 1 \quad (\text{normalized } Z_{00} = 1/2\sqrt{\pi}) \tag{56}$$

$$
\begin{aligned}
Z_{10} &= z/r = \cos\theta \\
Z_{11}^{\mathrm{C}} &= x/r = \sin\theta\cos\phi \\
z_{11}^{\mathrm{S}} &= y/r = \sin\theta\sin\phi
\end{aligned}
\tag{57}
$$

$$
\begin{aligned}
Z_{20} &= (3z^2 - r^2)/r^2 = 3\cos\theta - 1 \\
Z_{21}^{\mathrm{C}} &= zx/r^2)/r^2 = \cos\theta\sin\theta\cos\phi \\
Z_{21}^{\mathrm{S}} &= yz/r^2)/r^2 = \cos\theta\sin\theta\sin\phi \\
Z_{22}^{\mathrm{C}} &= (x^2 - y^2)/r^2 = \sin^2\theta(\cos^2\phi - \sin^2\phi) \\
Z_{22}^{\mathrm{S}} &= xy/r^2 = \sin^2\theta\cos\phi\sin\phi
\end{aligned}
\tag{58}
$$

For convenience, both cartesian and spherical coordinate forms are listed here. Normalized expressions for the Tesseral harmonics for $L = 0, 1, 2$, and 3 are given in Chapter 28, Table 7.

13 MAGNETOSTATICS

Now that some of the essential facts of electrostatics have been covered, it is appropriate to say a few words about the related field of magnetostatics. Maxwell's equations (16) and (14) give

$$
\begin{aligned}
\nabla \times \mathbf{B} &= \mu_0 \mathbf{J} \\
\nabla \cdot \mathbf{B} &= 0
\end{aligned}
\tag{59}
$$

Since the divergence of a curl vanishes we have

$$\nabla \cdot \mathbf{J} = 0 \tag{60}$$

a relation that can also be obtained from the continuity equation

$$\nabla \cdot \mathbf{J} + \frac{\partial \rho}{\partial t} = 0 \tag{61}$$

by setting the time derivative term $\partial\rho/\partial t = 0$ for the static case.

We saw in Eq. (38) that a straight wire carrying a current I has an encircling B field with the value $\mu_0 I/2\pi r$. A more general expression is the

Biot–Savart law whereby a current element of length $d\ell$ of a wire carrying a current I produces the differential magnetic field $d\mathbf{B}$ given by the expression

$$d\mathbf{B} = \frac{\mu_0}{4\pi} \frac{I d\ell \times \mathbf{r}}{r^3} \tag{62}$$

To calculate the magnetic field $\mathbf{B}$ from this relation we evaluate the integral

$$\int \frac{d\ell \times \mathbf{r}}{r^3} \Rightarrow \int \frac{\sin\theta\, dx}{r^2} \Rightarrow \frac{1}{R}\int \sin\theta\, d\theta \Rightarrow \frac{2}{R} \tag{63}$$

where from Fig. 10-4 we have $x = R\cot\theta$, $dx = R\,d\theta/\sin^2\theta$, to give

$$\mathbf{B} = \frac{\mu_0 I}{2\pi R} \tag{64}$$

which is the result obtained above in Eq. (38).

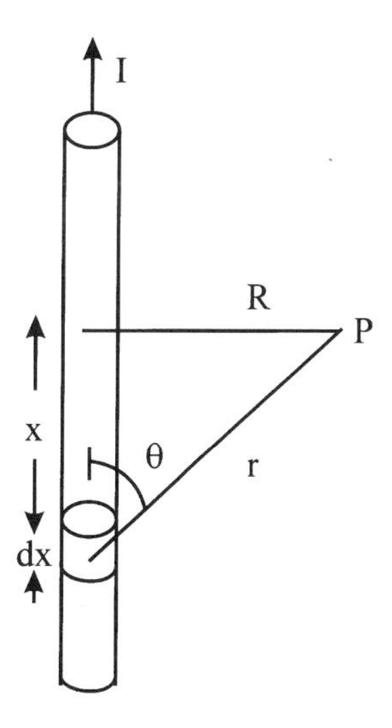

Fig. 10-4. Coordinate system for calculating the magnetic field at a point P a distance R from a wire carrying a current I. The integration is carried out over the range of angle $d\theta$ from $\theta = 0$ to $\theta = \pi$.

If the differential length $d\ell$ contains an amount of charge q in motion at the velocity v, then we can equate $Id\ell$ with $q\mathbf{v}$ in Eq. (62) and write

$$\mathbf{B} = \frac{\mu_0}{4\pi} \frac{q\mathbf{v} \times \mathbf{r}}{r^3} \tag{65}$$

for the $\mathbf{B}$ field of a moving charge.

If we replace $Id\ell$ by $\mathbf{J}d^3r'$ and integrate we obtain an expression for the magnetic field $\mathbf{B}(\mathbf{r})$ at the position $\mathbf{r}$ arising from the current density $\mathbf{J}(\mathbf{r})$

$$\mathbf{B}(\mathbf{r}) = \frac{\mu_0}{4\pi} \int \frac{\mathbf{J}(\mathbf{r}') \times (\mathbf{r} - \mathbf{r}')}{|\mathbf{r} - \mathbf{r}'|^3} d^3r' \tag{66}$$

which is similar to Eq. (43) for the electric field $\mathbf{E}$ of a charge distribution. This expression is equivalent to the analog of Eq. (44)

$$\mathbf{B}(\mathbf{r}) = \frac{\mu_0}{4\pi} \nabla_r \times \int \frac{\mathbf{J}(\mathbf{r}')}{|\mathbf{r} - \mathbf{r}'|} d^3r' \tag{67}$$

where the curl $\nabla_r\times$ operates on the components x, y, z of $\mathbf{r}$. Since we know that $\mathbf{B}$ can be derived from the curl of a vector potential $\mathbf{A}$ we can identify the integral of Eq. (67) as the vector potential

$$\mathbf{A}(\mathbf{r}) = \frac{\mu_0}{4\pi} \int \frac{\mathbf{J}(\mathbf{r}')}{|\mathbf{r} - \mathbf{r}'|} d^3r' \tag{68}$$

a result similar to Eq. (46) for the scalar potential $\Phi(\mathbf{r})$.

We conclude by writing down integrals for the force $\mathbf{F}$ and torque $\mathbf{N}$ exerted by an applied magnetic field $\mathbf{B}$ on a current density $\mathbf{J}$

$$\mathbf{F}(\mathbf{r}) = \int \mathbf{J}(\mathbf{r}') \times \mathbf{B}(\mathbf{r}') d^3r' \tag{69}$$

$$\mathbf{N}(\mathbf{r}) = \int \mathbf{r}' \times (\mathbf{J} \times \mathbf{B}) d^3r' \tag{70}$$

14 FARADAY'S LAW

The discussion until now has involved situations in which there is no time dependence. Before concluding the chapter we say a few words about Faraday's law of induction, which involves a time change of magnetic flux.

Consider a loop of wire bounded by a surface S with a unit vector $\mathbf{n}$ defined everywhere normal to the surface. If there is a magnetic field $\mathbf{B}$ in the region passing through the loop, then the magnetic flux $\boldsymbol{\Phi}$ passing through it is given by the integral of $\mathbf{B}$ over the loop surface

$$\boldsymbol{\Phi} = \int_{\text{surf}} \mathbf{B} \cdot \hat{\mathbf{n}}\, dA \tag{71}$$

Faraday's law states that a change in this flux induces an electromotive force (emf) in the loop given by

$$\text{emf} = -\frac{\partial \boldsymbol{\Phi}}{\partial t} \tag{72}$$

The negative sign arises from Lenz's law whereby the current induced in the loop by the emf is in a direction that opposes the change in flux. The emf equals the line integral of the electric field $\mathbf{E}$ around the loop of wire

$$\text{emf} = \oint \mathbf{E} \cdot d\ell \tag{73}$$

An electric generator is a device that transforms mechanical energy into electrical energy. Mechanical power can be used to move an electric circuit across a magnetic field so the induced emf produces a current that does work.

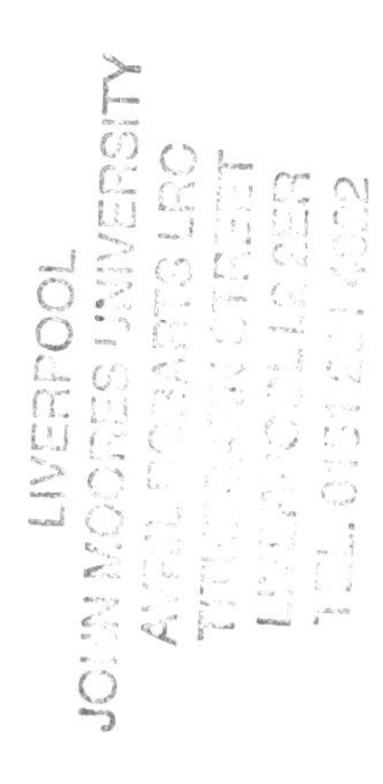

CHAPTER 11

MULTIPOLES AND MEDIA

1 INTRODUCTION

The two topics for this chapter are individual multipoles, i.e., monopoles, dipoles, quadrupoles, etc., of the electric and magnetic types, and continuous media that have polarizations and magnetizations arising from the presence of volume distributions of microscopic electric or magnetic dipoles, respectively. We begin by discussing electric multipoles, then magnetic multipoles, media, and finally materials with ellipsoidal shapes. We conclude by saying something about multipoles in quantum mechanics, atoms, and nuclei.

2 ELECTRIC MULTIPOLES

We found in chapter 10 that the electric scalar potential is given by Eq. (46)

$$\Phi(\mathbf{r}) = \frac{1}{4\pi\epsilon_0}\int \frac{\rho(\mathbf{r}')d^3r'}{|\mathbf{r}-\mathbf{r}'|} \tag{1}$$

and if we expand the denominator $(r^2 + r'^2 - 2\mathbf{r}\cdot\mathbf{r}')^{-1/2}$ in spherical and tessorial harmonics we obtain, respectively

$$\Phi(\mathbf{r}) = \frac{1}{4\pi\epsilon_0}\sum \frac{4\pi}{2\ell+1}\left\{\int Y^*_{\ell m}(\theta'\phi')r'^{\ell}\rho(\mathbf{r}')d^3r'\right\}\frac{Y_{\ell m}(\theta,\phi)}{r^{\ell+1}} \tag{2}$$

$$= \frac{1}{4\pi\epsilon_0}\sum \frac{4\pi}{2\ell+1}\left\{\int Z^j_{\ell m}(\phi',\phi')r'^{\ell}\rho(\mathbf{r}')d^3r'\right\}\frac{Z^j_{\ell m}(\theta,\phi)}{r^{\ell+1}} \tag{3}$$

Properties of these harmonics are given in Chapters 10 and 28. The coefficients $M^j_{\lambda m}$ in brackets

$$M^j_{\ell m} = \int Z^j_{\ell m}(\theta',\phi')r'^{\ell}\rho(\mathbf{r}')d^3r' \tag{4}$$

are characteristic of the charge distribution, and are called multipole moments as follows:

$$\begin{aligned} &\ell = 0 \quad \text{monopole} \\ &\ell = 1 \quad \text{dipole} \\ &\ell = 2 \quad \text{quadrupole} \\ &\ell = 3 \quad \text{octapole} \\ &\ell = 4 \quad \text{hexadecapole} \end{aligned} \tag{5}$$

The monopole, dipole, and quadrupole moment expressions for $m = 0$ are

$$M_{00} = \int \rho(\mathbf{r}')d^3r' = q \qquad \text{monopole} \tag{6}$$

$$M_{10} = \int z'\rho(\mathbf{r}')d^3r' = p_z \qquad \text{dipole} \tag{7}$$

$$M_{20} = \int (3z'^2 - r'^2)\rho(\mathbf{r}')d^3r' = Q_{zz} \qquad \text{quadrupole} \tag{8}$$

where normalization factors from the definitions of the Tesseral harmonics are omitted. More generally, the quadrupole moment components have the form

$$Q_{ij} = \int (3r_i'r_j' - r'^2\delta_{ij})\rho(\mathbf{r}')d^3r' \tag{9}$$

and Eqs. (6) to (9) are factors in the following alternative form of the potential expansion

$$\Phi(\mathbf{r}) = \frac{1}{4\pi\epsilon_0}\left\{\frac{q}{r} + \frac{\mathbf{p}\cdot\mathbf{r}}{r^3} + \tfrac{1}{2}\sum Q_{ij}\frac{r_i r_j}{r^5} + \cdots\right\} \tag{10}$$

which can be useful in applications.

A dipole is a vector with three components p_x, p_y, p_z, and these may be given as its magnitude p and the polar angles θ, ϕ of its direction. A quadrupole is a traceless, symmetric, second-rank tensor

$$Q_{ij} = Q_{ji} \tag{11a}$$

$$Q_{xx} + Q_{yy} + Q_{zz} = 0 \tag{11b}$$

with five independent components. It is most convenient to use three Euler angles to transform it to its principle axis system where it is diagonal, i.e., $Q_{ij} = 0$ for $i \neq j$, with two independent components. Using the convention

$$|Q_{xx}| \leq |Q_{yy}| \leq |Q_{zz}| \tag{12}$$

we customarily call Q_{zz} the quadrupole moment, and if it is not axially symmetric (i.e., if $Q_{xx} \neq Q_{yy}$), then we define an asymmetry parameter η

$$\eta = -\frac{Q_{yy} - Q_{xx}}{Q_{zz}} \tag{13}$$

$$0 \leq \eta \leq 1 \tag{14}$$

which measures the extent to which the quadrupole deviates from axial symmetry. Thus, Q_{zz} and η uniquely specify the quadrupole moment. The transverse components of the tensor are given by

$$Q_{xx} = -Q_{zz}\frac{1-\eta}{2} \qquad Q_{yy} = -Q_{zz}\frac{1+\eta}{2} \tag{15}$$

When $\eta = 0$ the quadrupole is axially symmetric ($Q_{xx} = Q_{yy}$), and when $\eta = 1$, a less likely possibility occurs; namely, $Q_{xx} = 0$ and $Q_{yy} = -Q_{zz}$.

From another viewpoint, multipole moment generation starts with a monopole moment or point charge. A pure dipole moment consists of two point charges or monopoles q, equal in magnitude and opposite in sign, located a short distance d apart, and its moment is given by

$$p = qd \tag{16}$$

The charges balance so the monopole moment of this pair is zero.

Following this same procedure, a pure quadrupole moment can be generated by bringing two oppositely directed dipole moments close together, and Fig. 11-1 shows examples of the two cases of a longitudinal and a transverse quadrupole moment generated in this manner. These pure quadrupole moments have no monopole or dipole contributions. In like manner, an octapole moment, $\ell = 3$, can be generated by bringing together two oppositely directed quadrupole moments, and the lower order $\ell = 0, 1, 2$ moments, of course, cancel for this charge distribution.

A general charge distribution can have many multipole moments. The values $M^{j}_{\ell m}$ of the lowest order non-vanishing one are independent of the choice of origin of the coordinate system in the region of the charge, but higher order moments can depend on this choice of origin.

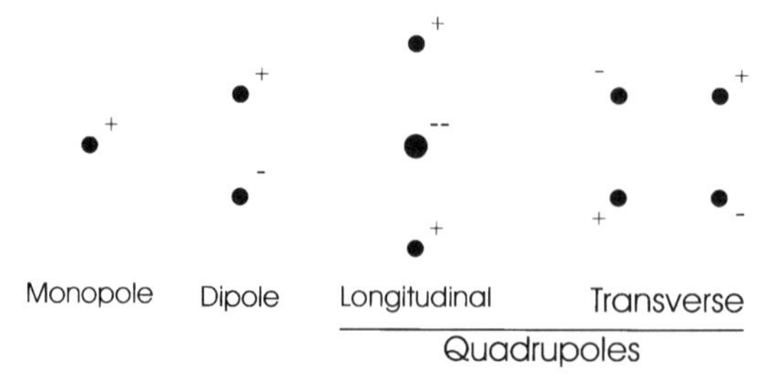

Fig. 11-1. Examples, from left to right, of an electric monopole, an electric dipole, a longitudinal electric quadrupole ($\eta = 0$) and a transverse electric quadrupole ($\eta = 1$).

3 ELECTRIC DIPOLE FIELDS

The electric field **E** produced by a charge distribution can be obtained from the gradient of the potential $\boldsymbol{\Phi}(\mathbf{r})$

$$\mathbf{E} = -\nabla\boldsymbol{\Phi} \tag{17}$$

and for a monopole we have Coulomb's law

$$\mathbf{E} = \frac{q\mathbf{r}}{4\pi\epsilon_0 r^2} \tag{18}$$

with a radially directed field.

The electric field components of an electric dipole $\mathbf{p} = p\mathbf{k}$ oriented along the z direction, expressed in spherical coordinates, are as follows

$$E_r = \frac{2p\cos\theta}{4\pi\epsilon_0 r^3} \tag{19a}$$

$$E_\theta = \frac{p\sin\theta}{4\pi\epsilon_0 r^3} \tag{19b}$$

$$E_\phi = 0 \tag{19c}$$

4 ENERGY AND TORQUE

The energy U of a charge distribution is given by the integral

$$U = \int \rho(\mathbf{r}')\boldsymbol{\Phi}(\mathbf{r}')d^3r' \tag{20}$$

and with the aid of Eqs. (10) and (18) this becomes

$$U = q\Phi(\mathbf{r}) - \mathbf{p} \cdot \mathbf{E} - \frac{1}{6}\sum Q_{ij}\frac{\partial E_j(0)}{\partial r_i} + \cdots \tag{21}$$

The interaction energy U_{dd} between two electric dipoles $\mathbf{p}_1$ and $\mathbf{p}_2$ separated by the distance $\mathbf{r}_1 - \mathbf{r}_2$ is given by the interaction of each with the electric field arising from the other, i.e., $U_{dd} = -\mathbf{p}_1 \cdot \mathbf{E}_2 = -\mathbf{p}_2 \cdot \mathbf{E}_1$, where

$$U_{dd} = \frac{\mathbf{p}_1 \cdot \mathbf{p}_2 - 3(\hat{\mathbf{n}} \cdot \mathbf{p}_1)(\hat{\mathbf{n}} \cdot \mathbf{p}_2)}{4\pi\epsilon_0|\mathbf{r}_1 - \mathbf{r}_2|^3} \tag{22}$$

and $\hat{\mathbf{n}}$ is a unit vector along the $(\mathbf{r}_1 - \mathbf{r}_2)$ direction.

A charge q placed in an electric field $\mathbf{E}$ experiences the force $\mathbf{F} = q\mathbf{E}$ that accelerates it along the field direction. A dipole $\mathbf{p}$ placed in a uniform field $\mathbf{E}$ experiences a torque $\mathbf{N}$

$$\mathbf{N} = \mathbf{p} \times \mathbf{E} \tag{23}$$

that orients it in the field, and provides it with the potential energy of orientation

$$U = -\mathbf{p} \cdot \mathbf{E} = -pE\cos\theta \tag{24}$$

When the dipole becomes aligned with the field so that $\theta = 0$ it is in its minimum energy state $U = -pE$ so it remains stationary in position. A pure quadrupole and all higher moments are unaffected by the presence of a uniform electric field, i.e., they experience no forces or torques.

Now consider an electric field in the x direction that has a uniform longitudinal gradient, i.e., a gradient along x with a constant value

$$\frac{\partial E_x}{\partial x} = \text{const} \tag{25}$$

as shown in Fig. 11-2. A monopole will, of course, accelerate, and the acceleration itself will increase if it moves toward stronger fields, and decrease if it moves toward weaker fields. The force will be different on the two ends of a dipole, as indicated in Fig. 11-2, so the dipole will begin by rotating and accelerating, then after it becomes oriented it will continue to accelerate. A quadrupole in such a field will experience a torque and orient to assume a position of minimum energy, after which it will remain

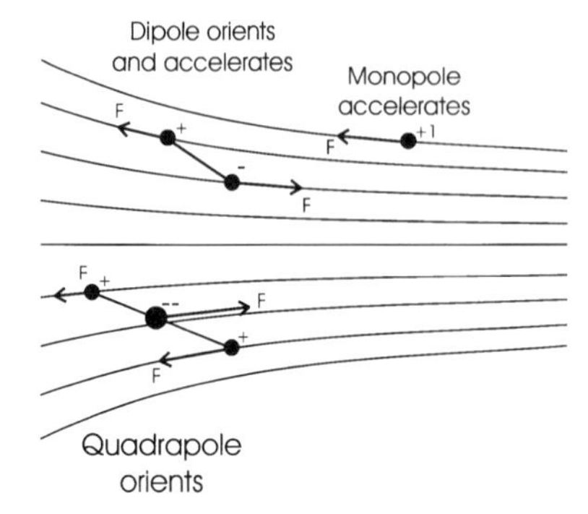

Fig. 11-2. Forces acting on an electric monopole, an electric dipole, and a longitudinal electric quadrupole in a magnetic field with a uniform gradient that increases in magnitude as one moves to the right.

in place. Octapoles and higher order moments will not be affected by the presence of uniform gradient electric fields. These results are easily generalized to higher degrees of gradients and to higher order multipoles.

5 MAGNETIC MULTIPOLES

Now that we have discussed the various properties of electric multipole moments it is appropriate to examine their magnetic analogs. There are some essential differences between the two cases because magnetic monopoles do not exist, and magnetic multipoles arise from moving charges or currents. There are also many similarities since the relationships between magnetic moments and magnetic fields are similar to the relationships between electric moments and electric fields. We discuss the latter similarities, and then return to examine the differences in origin.

The magnetic field components of a magnetic dipole $\boldsymbol{\mu} = \mu\mathbf{k}$ oriented along the z direction, expressed in spherical coordinates, are in the same form as their electric analogs (19)

$$B_r = \frac{\mu_0}{4\pi}\frac{2\mu\cos\theta}{r^3} \tag{26a}$$

$$B_\Theta = \frac{\mu_0}{4\pi}\frac{\mu\sin\theta}{r^3} \tag{26b}$$

$$B_\phi = 0 \tag{26c}$$

where one should not confuse the uses of the symbol of μ for magnetic moment and μ_0 for the permeability of free space.

The interaction energy between two magnetic dipoles $\boldsymbol{\mu}_1$ and $\boldsymbol{\mu}_2$ arises from the interaction of each with the magnetic field of the other, i.e., $U_{\mathrm{dd}} = -\boldsymbol{\mu}_1 \cdot \mathbf{B}_2 = -\boldsymbol{\mu}_2 \cdot \mathbf{B}_1$, corresponding to

$$U_{\mathrm{dd}} = \frac{\mu_0}{4\pi} \frac{\boldsymbol{\mu}_1 \cdot \boldsymbol{\mu}_2 - 3(\hat{\mathbf{n}} \cdot \boldsymbol{\mu}_1)(\hat{\mathbf{n}} \cdot \boldsymbol{\mu}_2)}{|\mathbf{r}_1 - \mathbf{r}_2|^3} \tag{27}$$

analogous to the electric dipole case (22).

The expressions for the torque $\mathbf{N}$ and the interaction energy U of a magnetic dipole moment $\boldsymbol{\mu}$ in a uniform magnetic field $\mathbf{B}$

$$\mathbf{N} = \boldsymbol{\mu} \times \mathbf{B} \tag{28}$$

$$U = -\boldsymbol{\mu} \cdot \mathbf{B} = -\mu B \cos\theta \tag{29}$$

are in the same form as their electric analogs (23) and (24), respectively. In atomic physics this magnetic energy U of an atom with total angular momentum $\mathbf{J}$ is called the Zeeman effect. There are $2J + 1$ quantized energies $U = g\mu_{\mathrm{B}} M_J$, where $-J \leq M_J \leq +J$.

6 MAGNETIC MOMENTS

When an electric current density $J(\mathbf{r})$ is present in a region there will be a magnetic moment density or magnetization $\mathbf{M}$ given by

$$\mathbf{M} = \tfrac{1}{2}\mathbf{r} \times \mathbf{J}(\mathbf{r}) \tag{30}$$

and a localized current density has an associated magnetic dipole moment $\boldsymbol{\mu}$

$$\boldsymbol{\mu} = \tfrac{1}{2}\int \mathbf{r}' \times \mathbf{J}(\mathbf{r}')d^3r' \tag{31}$$

This expression is easily integrated for a circular current loop of radius a carrying a current I, to give for the magnitude of the magnetic moment

$$\mu = \pi a^2 I \tag{32}$$

where the direction of $\boldsymbol{\mu}$ is along the axis, perpendicular to the plane of the loop. More generally, for a planar current loop of arbitrary shape

$$\mu = I \times (\text{area}) \tag{33}$$

where the area is enclosed by the loop. A spinning charge distribution, such as a rotating charged sphere, is equivalent to a collection of current loops, so it has a magnetic moment along the axis of the spin.

We have discussed classical magnetic moments. In quantum mechanics magnetic moments are proportional to quantized angular momenta. An electron with mass m and charge e in orbital motion with the angular momentum $\hbar\mathbf{L}$ has a magnetic moment $\boldsymbol{\mu}$ given by

$$\boldsymbol{\mu} = \mu_{\mathrm{B}}\mathbf{L} \tag{34}$$

where $\mu_{\mathrm{B}} = e\hbar/2m$ is the unit magnetic moment called the Bohr magneton. For the case of intrinsic spin with angular momentum $\hbar\mathbf{S}$ we have

$$\boldsymbol{\mu} = g\mu_{\mathrm{B}}\mathbf{S} \tag{35}$$

where the dimensionless g-factor has the value $g = 1$ classically, as in the orbital case (34), but $g = 2$ quantum mechanically. More precisely, to five significant figures, $g = 2.0023$ for a free electron.

Early direct experimental evidence for the spin of an electron came from the Stern–Gerlach experiment. Beams of neutral silver atoms in their ${}^2S_{1/2}$ electronic ground state with $L = 0$ and $S = \frac{1}{2}$ were sent through strong magnetic fields with field gradients. The gradient produces the force

$$F = \pm\mu\frac{\partial B}{\partial x} \tag{36}$$

which is opposite in direction for the up and down, i.e., $M_{\mathrm{S}} = \pm\frac{1}{2}$, spin states. The observed deflection showed that the magnetic moment of Ag is one Bohr magneton μ_{B}.

7 DIELECTRIC CONSTANT AND PERMEABILITY

Media other than free space can have atomic scale or close to atomic scale distributions of electric dipole moments, and this produces an electric dipole moment per unit volume called a polarization P. In like manner, media can have molecular scale distributions of electric currents and atoms with intrinsic spins that give the medium a magnetic dipole moment density called a magnetization M. These factors are taken into account by the dielectric constant ϵ and by the permeability μ, or by the dimensionless susceptibilities χ_e and χ, respectively, as was explained in Section 5 of the previous chapter.

The susceptibilities are measures of the extent to which the medium differs from free space. Some familiar materials have electric susceptibilities between 1 and 10, and many common non-magnetic materials are diamagnetic, meaning that their magnetic susceptibilities are negative. Paramagnetic materials have positive magnetic susceptibilities, and ferromagnets have positive values in the thousands. Both ϵ and μ can have a very pronounced frequency dependence, and sometimes they are complex: $\epsilon = \epsilon' + i\epsilon''$, $\mu = \mu' + i\mu''$. For the present we ignore the imaginary parts ϵ'' and μ''.

There are many ways in which ϵ and μ are important in physics. For example, the expressions for the velocity of light $c = (1/\mu_0\epsilon_0)^{1/2}$ and the characteristic impedance $\eta = (\mu_0/\epsilon_0)^{1/2}$ of free space were used in Section 5 of the previous chapter to provide the numerical values for ϵ_0 and μ_0. The capacitance C of a parallel plate capacitor of area A and separation d is proportional to the dielectric constant ϵ of the medium between the plates

$$C = \epsilon A/d \tag{37}$$

and the inductance L of a solenoid of length ℓ, area A with n turns is proportional to the permeability μ of the medium filling the coil

$$L = \mu n^2 \ell A \tag{38}$$

These expressions provide ways to determine the values of ϵ and μ for materials.

Coulomb's law for the force between two charges q and q' a distance r apart is inversely proportional to dielectric constant ϵ the medium

$$F = \frac{qq'}{4\pi\epsilon r^2} \tag{39}$$

An interesting consequence of the presence of the dielectric constant in the denominator of Coulomb's law is the phenomenon of weakly bound excitons in solids. These are electron-hole pairs interacting through the Coulomb potential and they have effective masses m^* that are less than the electron mass m_e. They have hydrogen-like energies

$$E_n = -\frac{m^* e^4}{2\hbar^2(4\pi\epsilon)^2 n^2} \tag{40}$$

which are rather low in magnitude because the dielectric constant ϵ is large, and the radius r of their orbits

$$r = n^2(\epsilon/\epsilon_0)(m_0/m^*)a_0 \tag{41}$$

can be very large, where $a_0 = 4\pi\epsilon_0\hbar^2/m_e e^2 = 0.53$ Å is the Bohr radius, and we assume the ground electronic state $n = 1$.

8 ELLIPSOIDS IN ELECTRIC AND MAGNETIC FIELDS

Until now we have assumed implicitly that a material of dielectric constant ϵ placed in an externally applied electric field E_0 will acquire an internal field given by $E_{in} = (\epsilon/\epsilon_0)E_0$. We have also assumed implicitly that a material of permeability μ placed in an applied magnetic field B_0 acquires an internal field with the value $B_{in} = (\mu/\mu_0)B_0$. Both of these statements are far from reality since the strengths of the internal fields have a pronounced dependence on the shape of the sample.

Before proceeding to analyze this shape dependence it is appropriate to point out that in the empty space outside a material the fundamental (E_0, B_0) and derived (D_0, H_0) fields are related in the usual manner

$$E_0 = D_0/\epsilon_0 \qquad B_0 = \mu_0 H_0 \tag{42}$$

and that inside the material the fields satisfy analogous relationships

$$E_{in} = D_{in}/\epsilon \qquad B_{in} = \mu H_{in} \tag{43}$$

The shape dependence arises from the boundary conditions given in Section 6 of the previous chapter, whereby in the absence of surface charges and currents ($\sigma = 0, \mathbf{K} = 0$) the normal components of **D** and **B** are continuous across the interface, and the tangential components of **E** and **H** are continuous there, as shown in Fig. 10-1. This means that for the circumstance of a long rod oriented parallel to the applied field the boundary conditions give for the electric case

$$\mathbf{E}_{in} = \mathbf{E}_0 \quad \text{and} \quad \mathbf{D}_{in} = \epsilon\mathbf{E}_0 \quad \text{longitudinal rod} \tag{44}$$

and for the magnetic case

$$\mathbf{H}_{in} = \mathbf{H}_0 \quad \text{and} \quad \mathbf{B}_{in} = (\mu/\mu_0)\mathbf{B}_0 \quad \text{longitudinal rod} \tag{45}$$

For the condition of a flat slab oriented perpendicular to the applied field we have for the electric case

$$\mathbf{D}_{\text{in}} = \mathbf{D}_0 \quad \text{and} \quad \mathbf{E}_{\text{in}} = (\epsilon_0/\epsilon)\mathbf{E}_0 \quad \text{flat slab} \tag{46}$$

and for the magnetic case

$$\mathbf{B}_{\text{in}} = \mathbf{B}_0 \quad \text{and} \quad \mathbf{H}_{\text{in}} = \mathbf{B}_0/\mu \quad \text{flat slab} \tag{47}$$

These expressions can be checked by comparing them with Fig. 10-1 of the previous chapter.

A long rod may be considered as the limit of an elongated prolate ellipsoid with principal axes $a = b \ll c$, and a flat slab constitutes the limit of a compressed oblate ellipsoid with $a = b \gg c$. It is of interest to determine the internal fields of a general ellipsoid placed in an external field because when the external field is applied along a principal direction, then the fields inside are parallel to the applied field, and, in addition, they are uniform everywhere inside. Another advantage is that closed-form solutions can be found for the field configurations inside and outside ellipsoids. For other cases the fields inside a material in an applied field can differ in direction from the applied field, and can vary from place to place inside the material.

An ellipsoid in an electric field is characterized by depolarization factors N_i, and for the three principal directions these satisfy the normalization condition

$$N_x + N_y + N_z = 1 \tag{48}$$

For the present case of axial symmetry, $a = b \neq c$, we have

$$N_x = N_y = N_\perp \quad \text{and} \quad N_z = N_\parallel \tag{49}$$

with the normalization condition

$$2N_\perp + N_\parallel = 1 \tag{50}$$

Table 11-1 provides expressions for $N_\perp$ and $N_\parallel$ for several cases and Fig. 11-3 shows how $N_\perp$ and $N_\parallel$ depend on the c/a ratio of the principal axes. For the rod and slab cases treated above we have

$$N_\perp = \tfrac{1}{2} - \tfrac{1}{2}\delta_4 \qquad N_\parallel = \delta_4 \quad \text{rod} \tag{51}$$

$$N_\perp = \tfrac{1}{2}\delta_1 \qquad N_\parallel = 1 - \delta_1 \quad \text{slab} \tag{52}$$

TABLE 11-1. Depolarization and equivalently demagnetization factors N_i for prolate and oblate ellipsoids of revolution with semiaxes $a = b$ and c. Correction factors $\delta_i \ll 1$ are given for the disk ($c \ll a$), sphere($c \sim a$), and rod ($c \gg a$) limits.

Ellipsoid	Condition	$N_{\parallel}$	$N_{\perp}$
Disk limit	$c \Rightarrow 0$	1	0
Flat disk	$c \ll a$	$1 - \delta_1$	$\frac{1}{2}\delta_1$
Oblate	$c \leq a$	$\frac{1}{3} + \delta_2$	$\frac{1}{3} - \frac{1}{2}\delta_2$
Sphere	$c = a$	$1/3$	$1/3$
Prolate	$c \geq a$	$\frac{1}{3} - \delta_3$	$\frac{1}{3} + \frac{1}{2}\delta_3$
Long rod	$c \gg a$	δ_4	$\frac{1}{2}(1 - \delta_4)$
Rod limit	$c \Rightarrow \infty$	0	$\frac{1}{2}$

Values for the correction factors δ_i are:

$$\delta_1 = \tfrac{1}{2}\pi \frac{c}{a}$$

$$\delta_2 = \tfrac{4}{15}\left(1 - \frac{c}{a}\right)$$

$$\delta_3 = \tfrac{4}{15}\left(1 - \frac{a}{c}\right)$$

$$\delta_4 = \left[\tfrac{1}{2}\ln\left(2\frac{c^2}{a^2}\right) - 1\right]\frac{a^2}{c^2}$$

where expressions for δ_1 to δ_4 are provided in Table 11-1 for the limit $\delta_i \ll 1$ in each case.

If an ellipsoid is placed in an external electric field E_0 oriented along a principal direction, then the internal fields D_{in}, E_{in} and P are all parallel to E_0, and the first two are related by the expression

$$ND_{\text{in}} + (1 - N)\epsilon_0 E_{\text{in}} = \epsilon_0 E_0 \tag{53}$$

where $D_0 = \epsilon_0 E_0$ and $D_{\text{in}} = \epsilon E_{\text{in}}$. When E_0 is not along a principal direction, then the internal fields are, in general, not parallel to E_0. Equations (44) to (47) were written down for the applied fields directed along the parallel direction ($N_{\parallel}$) of the ellipsoids. Equation (44) for the longitudinal rod corresponds to the limit $N_{\parallel} = 0$ and $\delta_4 = 0$. For the flat slab, we set $\delta_1 = 0$ and $N_{\parallel} = 1$ to obtain Eq. (46). The magnetic expressions (45) and (47) are treated analogously using Eq. (57) below. For a sphere $a = b = c$ and all directions are equivalent so from Eq. (48) $N = 1/3$, and we have the internal fields

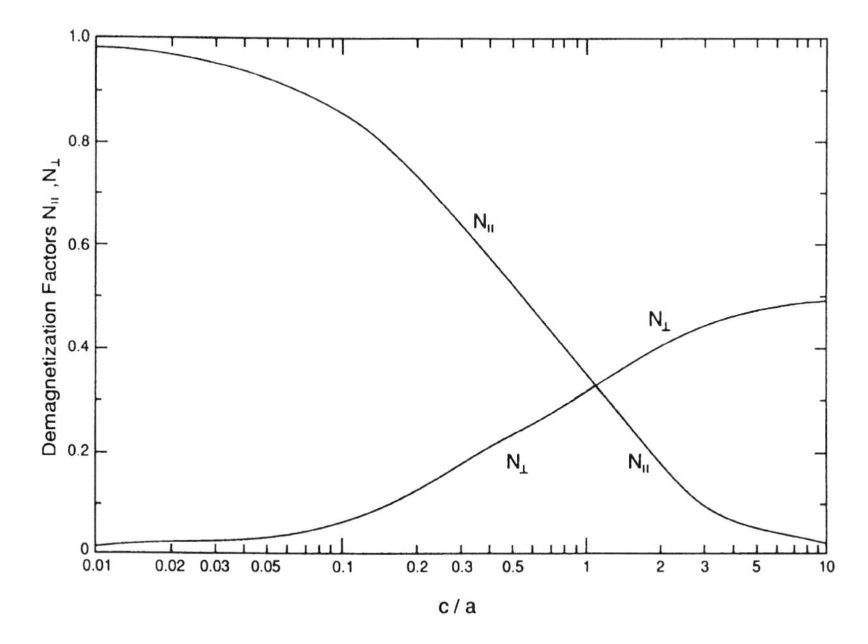

Fig. 11-3. Dependence of the perpendicular and parallel components $N_\perp$ and $N_\parallel$ of the depolarization and demagnetization factors on the principal axis ratio c/a of an ellipsoid. (From C. P. Poole, Jr et al., *Superconductivity*, Academic Press, New York, 1995, p. 327.)

$$E_{\text{in}} = \frac{3\epsilon_0}{\epsilon + 2\epsilon_0} E_0 \tag{54}$$

$$D_{\text{in}} = \frac{3\epsilon}{\epsilon + 2\epsilon_0} \epsilon_0 E_0 \tag{55}$$

and the polarization P

$$P = \frac{3(\epsilon - \epsilon_0)}{\epsilon + 2\epsilon_0} \epsilon_0 E_0 \tag{56}$$

all three of which are constant throughout the volume of the sphere. Figure 11-4 shows how electric field lines are concentrated within the dielectric sphere.

For a magnetic material in an applied magnetic field B_0 we have the magnetic analog of Eq. (53)

$$NB_{\text{in}} + (1 - N)\mu_0 H_{\text{in}} = B_0 \tag{57}$$

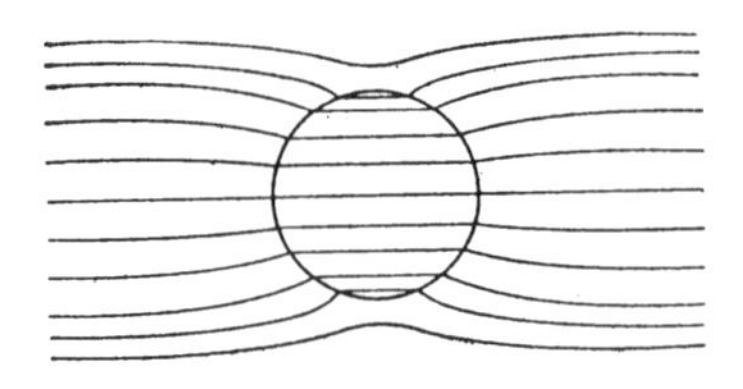

Fig. 11-4. Concentration of electric field lines in a dielectric sphere placed in an applied electric field. (From J. A. Stratton, *Electromagnetic Theory*, McGraw Hill, New York, 1941, p. 207.)

where N is now called a demagnetization factor. If we use this expression and Eqs. (51) and (52) with $\delta_1 = 0$ and $\delta_2 = 0$, then Eqs. (45) and (47) follow immediately. In addition, the analogs to Eqs. (54)–(56) are easily written down for the magnetic case of the sphere. Table 11-1 and Fig. 11-3 apply to the magnetic case also.

Magnetization can also occur in the absence of an applied field, so setting $B_0 = 0$ in Eq. (57) we find

$$\mu_0 H_{\text{in}} = -\frac{N}{1-N} B_{\text{in}} \tag{58}$$

and making use of the expression $B_{\text{in}} = \mu_0(H_{\text{in}} + M)$ gives the magnetization

$$\mu_0 M = \frac{B_{\text{in}}}{1-N} \tag{59}$$

Thus, the internal H_{in} field is opposite in direction to B_{in} and M. For a sphere $N = 1/3$ and its magnetic moment μ is equal to its magnetization times its volume

$$\boldsymbol{\mu} = \frac{4}{3}\pi a^3 M \tag{60}$$

9 NUCLEAR AND ATOMIC MULTIPOLES

A nucleus of atomic number Z has a monopole electric moment equal to its total charge Ze. It cannot have an electric dipole moment, but if the charge is distorted from a spherical shape it has a quadrupole moment Q_{zz}, which is a measure of the distortion. It cannot have a hexadecapole moment; in fact, it can only have even-order electric moments, with even ℓ, since all odd ℓ electric moments vanish identically.

The opposite is the case for the magnetic moments of a nucleus: only odd-order magnetic moments are possible, i.e., those with odd ℓ. A nucleus with an odd number of neutrons or odd number of protons has a magnetic dipole moment and a non-zero nuclear spin I. A deuteron, for example, has nuclear spin $I = 1$ and a magnetic dipole moment arising from oppositely directed proton and neutron moments. Nitrogen-14 which is 99.63% abundant is a similar case since it has seven protons and seven neutrons, and its nuclear spin $I = 1$. All isotopes with $I = 0$, such as ^{12}C and ^{16}O, are of the even–even type.

When a photon of the proper frequency to induce a transition between atomic energy levels strikes an atom its electric field E polarizes the atom, and the transition probability for inducing a transition from level n to level n' is calculated from the square of the dipole moment matrix element $\langle n|\mathbf{p}|n'\rangle$, where $\mathbf{p}$ is the electric dipole moment induced by the incoming radiation field. The electric dipole selection rules are $\Delta\ell = \pm 1$ and $\Delta j = 0, \pm 1$. Magnetic dipole transitions $\langle n|\boldsymbol{\mu}|n'\rangle$ and electric quadrupole transitions $\langle n|Q_{zz}|n'\rangle$ can also occur, each with characteristic selection rules, but the electric dipole ones are ordinarily much stronger. A similar situation occurs with nuclei, since nuclear transitions can arise from matrix elements of various types of nuclear electric and magnetic moments.

CHAPTER 12

RELATIVISTIC ELECTRODYNAMICS

1 INTRODUCTION

In Chapter 7 we discussed the main features of special relativity. The emphasis was on four-vectors and applications to mechanics, with very little mention made of higher rank tensors or electric and magnetic fields. This chapter extends the treatment to higher rank tensors, and discusses aspects of electrodynamics that are strongly dependent on relativity. Maxwell's equations, for example, are Lorentz invariant.

2 ELECTROMAGNETIC FIELD TENSOR

The electromagnetic fields form an antisymmetric tensor called the field strength tensor, which has the contravariant form $F^{\mu\nu}$

$$F^{\mu\nu} = \begin{pmatrix} 0 & -cB_z & cB_y & E_x \\ cB_z & 0 & -cB_x & E_y \\ -cB_y & cB_x & 0 & E_z \\ -E_x & -E_y & -E_z & 0 \end{pmatrix} \tag{1a}$$

and the covariant form $F_{\mu\nu} = g_{\mu\mu} F^{\mu\nu} g_{\nu\nu}$

$$F_{\mu\nu} = \begin{pmatrix} 0 & -cB_z & cB_y & -E_x \\ cB_z & 0 & -cB_x & -E_y \\ -cB_y & cB_x & 0 & -E_z \\ E_x & E_y & E_z & 0 \end{pmatrix} \tag{1b}$$

where the contravariant and covariant metric tensors, $g^{\mu\nu}$ and $g_{\mu\nu}$, respectively, have the same form

$$g^{\mu\nu} = g_{\mu\nu} = \begin{pmatrix} 1 & 0 & 0 & 0 \\ 0 & 1 & 0 & 0 \\ 0 & 0 & 1 & 0 \\ 0 & 0 & 0 & -1 \end{pmatrix} \tag{2}$$

A tensor such as F is transformed to its form F' in another coordinate system by means of similarity transformation

$$F' = S^{-1} F S \tag{3}$$

and if the second coordinate system is rotated and moving uniformly relative to the first one, then S is a general Lorentz transformation matrix L_{G}

$$F' = L_{\mathrm{G}}^{-1} F L_{\mathrm{G}} \tag{4}$$

Analogous expressions may be written for a space rotation R and for a special Lorentz transformation or boost L that involves no rotation

$$F' = R^{-1}FR \tag{5a}$$

$$F' = L^{-1}FL \tag{5b}$$

If both R and L are z-direction transformations, then we have for the matrices of the rotation R

$$R = \begin{pmatrix} \cos\theta & -\sin\theta & 0 & 0 \\ \sin\theta & \cos\theta & 0 & 0 \\ 0 & 0 & 1 & 0 \\ 0 & 0 & 0 & 1 \end{pmatrix} \tag{6}$$

and of the boost L

$$L = \begin{pmatrix} 1 & 0 & 0 & 0 \\ 0 & 1 & 0 & 0 \\ 0 & 0 & \gamma & -\beta\gamma \\ 0 & 0 & -\beta\gamma & \gamma \end{pmatrix} \tag{7}$$

As an example we carry out the pure Lorentz transformation (5b) using the particular boost of Eq. (7) with the field strength tensor simplified by setting two field components equal to zero, namely $E_x = 0$ and $B_x = 0$. The result is the transformed field strength tensor F'

$$F^{\mu\nu} = \begin{pmatrix} 0 & -cB_z & \gamma cB_y & -\beta\gamma cB_y \\ cB_z & 0 & -\beta\gamma E_y & \gamma E_y \\ -\gamma cB_y & \beta\gamma E_y & 0 & E_z \\ \beta\gamma cB_y & -\gamma E_y & -E_z & 0 \end{pmatrix} \tag{8}$$

This result can be written more generally by components as

$$\begin{aligned} E'_{\parallel} &= E_{\parallel} & B'_{\parallel} &= B_{\parallel} \\ \mathbf{E}'_{\perp} &= \gamma(\mathbf{E}_{\perp} + c\boldsymbol{\beta} \times \mathbf{B}) & c\mathbf{B}'_{\perp} &= \gamma(c\mathbf{B}_{\perp} - \boldsymbol{\beta} \times \mathbf{E}) \end{aligned} \tag{9}$$

Thus, we see that, in contrast to the four-vector case, the field components parallel to **v** do not change, but the ones perpendicular to **v** do change.

Before concluding this section we should mention that there is a dual field strength tensor $\mathcal{F}$ that interchanges the E and B fields, and its contravariant form is

$$\mathcal{F}^{\mu\nu} = \begin{pmatrix} 0 & E_z & -E_y & cB_x \\ -E_z & 0 & E_x & cB_y \\ E_y & -E_x & 0 & cB_x \\ -cB_x & -cB_y & -cB_z & 0 \end{pmatrix} \tag{10}$$

This tensor transforms like the regular field strength tensor. It can be generated by the tensor contraction

$$\mathcal{F}^{\alpha\beta} = \tfrac{1}{2}\varepsilon^{\alpha\beta\gamma\delta}F_{\gamma\delta} \tag{11}$$

where an element $\varepsilon^{\alpha\beta\gamma\delta} = 1$ for an even permutation of its four indices, equals -1 for an odd permutation, and is zero if any two indices are the same.

3 INVARIANTS

We recall learning that a four-vector has an invariant magnitude under space rotations and Lorentz transformations. A second-rank antisymmetric tensor has two invariants. One is obtained by a contraction of the $F_{\alpha\beta}$ tensor with its counterpart $F^{\alpha\beta}$, and the other by contracting $F_{\alpha\beta}$ with its dual $\mathcal{F}^{\alpha\beta}$

$$\sum F_{\alpha\beta}F^{\alpha\beta} = 2(c^2B^2 - E^2) \tag{12}$$

$$\sum F_{\alpha\beta}\mathcal{F}^{\alpha\beta} = 4c\mathbf{B}\cdot\mathbf{E} \tag{13}$$

The magnitude of the electromagnetic potential four-vector $\mathbf{A}, \boldsymbol{\Phi}/c$ is also invariant

$$\sum A_\alpha^2 = A^2 - \phi^2/c^2 \tag{14}$$

The values of these three invariants remain the same when the electromagnetic fields are transformed to new coordinate frames by space rotations or Lorentz transformations.

In the fields **E** and **B** are perpendicular to each other so $\mathbf{B} \cdot \mathbf{E} = 0$, then they will be perpendicular in all Lorentz frames, and the smaller of the two, E or cB, can be transformed away so only the other remains. If they are not perpendicular to each other, then $c\mathbf{B} \cdot \mathbf{E}$ as well as $c^2B^2 - E^2$ must remain the same, and neither field can be transformed away.

As an example, consider the case of fields with the values E_0 and B_0 along the y direction. A Lorentz transformation with a velocity along the z direction gives

$$\begin{aligned} E'_x &= -\beta\gamma c B_0 \qquad & cB'_x &= \beta\gamma E_0 \\ E'_y &= \gamma E_0 & cB'_y &= \gamma B_0 \\ E'_z &= 0 & cB'_z &= 0 \end{aligned} \tag{15}$$

This result is shown in Fig. 12-1. We see that for large γ the fields cB' and E' become large in magnitude and almost perpendicular to each other. It is easy to see that the invariants are satisfied

$$c^2B'^2 - E'^2 = c^2B^2 - E^2 \tag{16}$$

$$c\mathbf{B}' \cdot \mathbf{E}' = c\mathbf{B} \cdot \mathbf{E} \tag{17}$$

4 FIELD TENSOR AND POTENTIALS

The electromagnetic field tensor $F_{\alpha\beta}$ (1, 2) is related to the potential four-vector $\mathbf{A}, \Phi/c$ through the expression

$$F^{\alpha\beta} = c\left(\frac{\partial A^\beta}{\partial x_\alpha} - \frac{\partial A^\alpha}{\partial x_\beta}\right) \tag{18}$$

Derivatives of this tensor provide Maxwell's equations. The four-gradient

$$\sum_\alpha \frac{\partial}{\partial x_\alpha} F^{\alpha\beta} = (\mu/\varepsilon)^{1/2} J^\beta \tag{19}$$

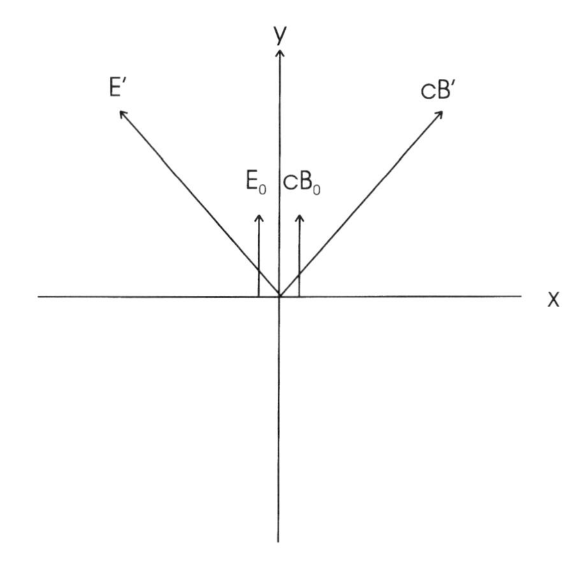

Fig. 12-1. Parallel electric and magnetic fields E_0 and B_0 oriented initially along the y direction are shown after a Lorentz transformation to a coordinate system moving along the z direction at close to the speed of light. The figure shows the transformed fields E' and B' greatly increased in magnitude and rotated in opposite directions so that the angle between them is almost $\pi/2$.

in conjunction with the **J**, $c\rho$ four-vector gives the inhomogeneous relations

$$\nabla \times \mathbf{H} - \frac{\partial \mathbf{D}}{\partial t} = \mathbf{J} \qquad \beta = x, y, z \tag{20}$$

$$\nabla \cdot \mathbf{D} = \rho \qquad \beta = t \tag{21}$$

and the expression

$$\frac{\partial}{\partial x_\alpha} F_{\beta\gamma} + \frac{\partial}{\partial x_\beta} F_{\gamma\alpha} + \frac{\partial}{\partial x_\gamma} F_{\alpha\beta} = 0 \tag{22}$$

provides the homogeneous Maxwell equations

$$\nabla \times \mathbf{E} + \frac{\partial \mathbf{B}}{\partial t} = 0 \quad t \text{ included in } \alpha, \beta, \gamma \tag{23}$$

$$\nabla \cdot \mathbf{B} = 0 \quad \alpha, \beta, \gamma \neq t \tag{24}$$

The inhomogeneous Maxwell equations can also be obtained from the dual field tensor via the expression

$$\sum \frac{\partial}{\partial x_\alpha} \mathcal{F}^{\alpha\beta} = 0 \tag{25}$$

as may be checked easily by working out some terms.

5 GAUGE TRANSFORMATIONS

For a given field strength tensor (1), meaning a given set of electromagnetic fields with components E_i and B_i, the electromagnetic four-vector is not unique. Since the magnetic field $\mathbf{B}$ is derived from the curl of a vector potential $\mathbf{A}$

$$\mathbf{B} = \nabla \times \mathbf{A} \tag{26}$$

we can add to $\mathbf{A}$ the gradient of a scalar function $\Lambda(x, y, z, t)$ without changing $\mathbf{B}$

$$\mathbf{A}' = \mathbf{A} + \nabla\Lambda \tag{27}$$

If this done, then to satisfy the electric field equation

$$\mathbf{E} = -\nabla\boldsymbol{\Phi} - \frac{\partial \mathbf{A}}{\partial t} \tag{28}$$

a time derivative of Λ must be added to $\boldsymbol{\Phi}$

$$\boldsymbol{\Phi}' = \boldsymbol{\Phi} - \frac{1}{c}\frac{\partial \Lambda}{\partial t} \tag{29}$$

Equations (27) and (29) constitute a gauge transformation from $\mathbf{A}, \boldsymbol{\Phi}$ to $\mathbf{A}', \boldsymbol{\Phi}'$. Thus, there is some arbitrariness in the choice of the potentials, and Maxwell's equations are invariant under this type of transformation.

To remove this arbitrariness, it is a common practice to require the electromagnetic potential four-vector to have a vanishing four-divergence

$$\Box \cdot A = \nabla \cdot \mathbf{A} + \frac{1}{c^2}\frac{\partial \boldsymbol{\Phi}}{\partial t} = 0 \tag{30}$$

and this expression (30) is called the Lorentz condition. There is still some arbitrariness, and to remove this one can require that the scalar function Λ satisfy the wave equation

$$\nabla^2 \Lambda - \frac{1}{c^2}\frac{\partial^2 \Lambda}{\partial t^2} = 0 \tag{31}$$

and potentials (28) and (29) with a scalar function Λ that satisfies Eq. (31) are said to belong to the Lorentz gauge. the Lorentz gauge has the advantage that it makes the space and time components of the electromagnetic potential four-vector satisfy uncoupled wave equations

$$\nabla^2 \boldsymbol{\Phi} - \frac{1}{c^2}\frac{\partial^2 \boldsymbol{\Phi}}{\partial t^2} = -\rho/\varepsilon_0 \tag{32}$$

$$\nabla^2 \mathbf{A} - \frac{1}{c^2}\frac{\partial^2 \mathbf{A}}{\partial t^2} = -\mu_0 \mathbf{J} \tag{33}$$

For other choices of gauge the wave equations are coupled together.

Another way to remove the arbitrariness, called the Coulomb or London gauge, involves choosing a vanishing three-divergence for the vector potential

$$\nabla \cdot \mathbf{A} = 0 \tag{34}$$

and as a result the scalar potential wave equation (32) becomes independent of time

$$\nabla^2 \boldsymbol{\Phi} = -\rho/\varepsilon_0 \tag{35}$$

This gauge is used in superconductivity, as is appropriate for its name.

6 MOTION OF CHARGE IN ELECTROMAGNETIC FIELDS

Consider a charge q of rest mass m moving in a region containing electromagnetic fields. The Lorentz force K_α on this charge can be written

$$K_\alpha = \frac{dp_\alpha}{d\tau} = \frac{q}{c}\sum F_{\alpha\beta} u_\beta \tag{36}$$

where the momentum four-vector p_α has the components $m\gamma\mathbf{v}$, $m\gamma c$, the velocity four-vector u is $\gamma\mathbf{v}$, γc, and the proper time $d\tau = dt/\gamma$. This gives, respectively, for the three-space and the one-time components of Eq. 936)

$$
\begin{aligned}
m\frac{d}{dt}\gamma\mathbf{v} &= \frac{q}{\gamma}(\mathbf{E} + \mathbf{v} \times \mathbf{B}) \quad && \alpha = x, y, z \\
mc\frac{d\gamma}{dt} &= \frac{q}{\gamma}\mathbf{E} \cdot \mathbf{v} && \alpha = t
\end{aligned}
\tag{37}
$$

The Lagrangian is

$$
L = -\frac{mc^2}{\gamma} + \frac{q}{c}\mathbf{v} \cdot \mathbf{A} - q\Phi \tag{38}
$$

and the Hamiltonian may be written in terms of the conjugate momentum **P**

$$
\mathbf{P} = \gamma m\mathbf{v} + q\mathbf{A} \tag{39}
$$

as follows

$$
\mathcal{H} = [(c\mathbf{P} - qc\mathbf{A})^2 + m^2c^4]^{1/2} + q\Phi \tag{40}
$$

For low velocities, $v \ll c$, the rest energy term m^2c^4 is dominant, and this expression is well approximated by

$$
\mathcal{H} = \frac{(\mathbf{P} - q\mathbf{A})^2}{2m} + q\Phi + mc^2 \tag{41}
$$

an operator that provides the Zeeman effect in quantum mechanics.

7 PERPENDICULAR FIELDS

Let us examine the perpendicular field case, $\mathbf{E}\perp\mathbf{B}$. Consider the fields (8) after transforming with a velocity $\boldsymbol{\beta} = \mathbf{v}/c$ perpendicular to **E** and **B**

$$
\begin{aligned}
E'_{\parallel} &= 0 \qquad & \mathbf{E}'_{\perp} &= \gamma(\mathbf{E}_{\perp} + \boldsymbol{\beta} \times c\mathbf{B}) \\
B'_{\parallel} &= 0 & c\mathbf{B}'_{\perp} &= \gamma(c\mathbf{B}_{\perp} - \boldsymbol{\beta} \times \mathbf{E})
\end{aligned}
\tag{42}
$$

where the parallel fields remain zero.

There are two cases to consider, depending on whether **E** is larger than $c\mathbf{B}$, or vice versa. For $|\mathbf{E}| < c|\mathbf{B}|$ one can select the velocity

$$\mathbf{v} = \frac{\mathbf{E} \times \mathbf{B}}{B^2} \tag{43}$$

which makes $\mathbf{E}'_\perp = 0$, and we obtain

$$\mathbf{B}'_\perp = \frac{\mathbf{B}_\perp}{\gamma} = \left\{ \frac{c^2 B^2 - E^2}{c^2 B^2} \right\} \mathbf{B}_\perp \tag{44}$$

Since there is no **E** field in this transformed system the motion of a charged particle is a circle around the **B** field superimposed on a translation in the direction perpendicular to both **B** an **E** as shown in Fig. 12-2.

For the other case, with $|\mathbf{E}| > c|\mathbf{B}|$, one can select the velocity

$$\mathbf{v} = c^2 \frac{\mathbf{E} \times \mathbf{B}}{E^2} \tag{45}$$

which makes $\mathbf{B}'_\perp = 0$, and we have for the transformed $\mathbf{E}'_\perp$ field

$$\mathbf{E}'_\perp = \frac{\mathbf{E}_\perp}{\gamma} = \left\{ \frac{E^2 - c^2 B^2}{E^2} \right\} \mathbf{E}_\perp \tag{46}$$

This motion is hyperbolic with an ever-increasing velocity.

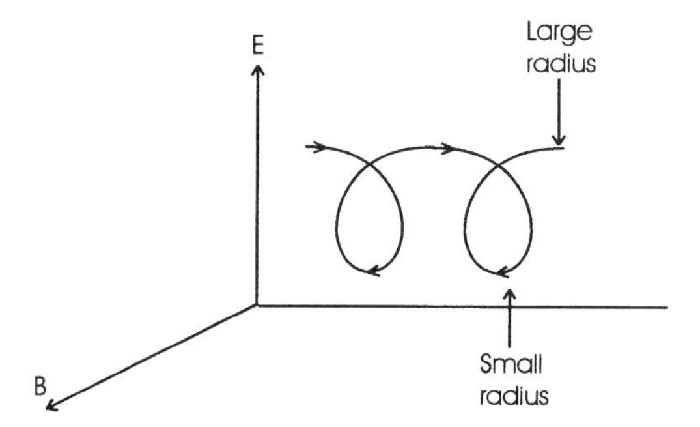

Fig. 12-2. Motion of a charged particle in crossed electric and magnetic fields for the case $|\mathbf{E}| < c|\mathbf{B}|$. The trajectory is a circle superimposed on a translation along a direction perpendicular to both **E** and **B**.

8 ADIABATIC INVARIANTS

If a charge is circulating in an orbit perpendicular to a magnetic field and the field is stronger on one side of the orbit and weaker on the other side, as indicated in Fig. 12-3, the charge will move in a spiral as shown, and eventually it will reflect when the strength of the field exceeds a critical value.

This motion can be analyzed in terms of the invariance of the action integral J

$$J = \oint \mathbf{P}_{\perp} \cdot \mathbf{d\ell} \tag{47}$$

where $\mathbf{P}$ is the conjugate momentum defined by Eq. (39). Since $\mathbf{v}_{\perp}$ is parallel to $\mathbf{d\ell}$ and

$$\omega_B = eB/\gamma m_0 c = v_{\perp}/a \tag{48}$$

where a is the orbit radius, we can show, after some careful analysis, that

$$J = qB\pi a^2 \tag{49}$$

so a decreases as B increases. The kinetic energy is also conserved in accordance with the expression

$$v_{\|}^2 + v_{\perp}^2 = v_0^2 \tag{50}$$

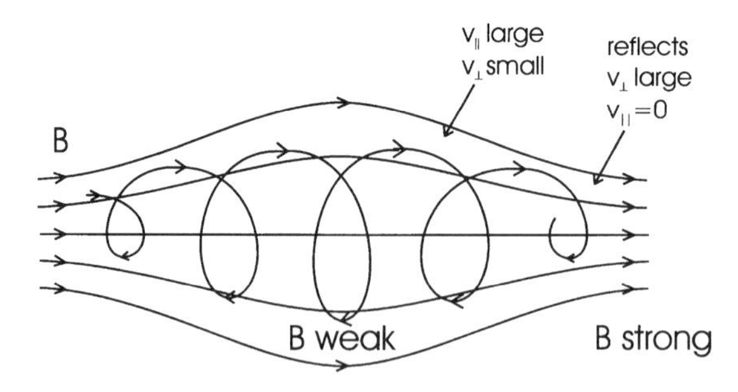

Fig. 12-3. Spiraling motion of a charge confined in a localized longitudinal magnetic field B that is weak in the center and strong at the ends.

where v_0 is the initial velocity of injection into the field. As the charge spirals toward higher fields the orbit radius a decreases via Eq. (49), and the velocity component $v_\perp$ increases via Eq. (48). The spiraling continues until $v_\perp$ reaches the initial velocity value v_0, at which point $v_\parallel$ vanishes through Eq. (50), and the charge stops, changes direction, and spirals out toward weaker fields. If the field is weak in the middle and stronger at both ends, then the charge can be made to spiral back and forth between the two ends, in the state of confinement indicated in Fig. 12-3.

CHAPTER 13

WAVE PROPAGATION

1 INTRODUCTION

This chapter is concerned with the propagation of plane waves in free space and in media, including especially dielectric media. It examines various phenomena commonly covered in optics texts. Conducting media and waveguide propagation is treated. The chapter concludes with a discussion of modulated waves and solitons.

2 PLANE WAVES

A plane wave traveling through space has the position and time dependence given by

$$I(\mathbf{r}, t) = I_0 e^{ik\cdot\mathbf{r} - i\omega t} \tag{1}$$

and in one dimension this becomes

$$I = I_0 e^{ikx - i\omega t} \tag{2}$$

This is a wave traveling to the right, in the direction of positive x. A wave traveling to the left has the form

$$I = I_0 e^{-ikx - i\omega t} \tag{3}$$

The sum and difference of traveling waves (2) and (3) is a standing wave

$$I = 2I_0 \cos(kx - \omega t) \quad \text{(sum)} \tag{4}$$

$$I = 2iI_0 \sin(kx - \omega t) \quad \text{(difference)} \tag{5}$$

We deal with electromagnetic waves for which the velocity in free space is x. The factor $\mathbf{k} \cdot \mathbf{r} - \omega t$ is the scalar product of the space–time and the propagation four-vectors, $(\mathbf{r}, \omega t)$ and $(\mathbf{k}, \omega/c)$, respectively, as was explained in Chapter 7.

In free space the velocity of light c is given by

$$c = \frac{1}{(\mu_0 \epsilon_0)^{1/2}} \tag{6}$$

and in a medium it is

$$v = \frac{c}{[(\mu/\mu_0)(\epsilon/\epsilon_0)]^{1/2}} \tag{7}$$

$$= c/n \tag{8}$$

where we make use of the index of refraction n, which is the square root of the relative dielectric constant

$$n = (\epsilon/\epsilon_0)^{1/2} \tag{9}$$

for a non-magnetic medium for which $\mu = \mu_0$. We also have the expressions

$$\lambda\nu = v \tag{10a}$$

$$\omega = vk \tag{10b}$$

where $\omega = 2\pi\nu$ and $k = 2\pi/\lambda$. When the wave enters a dielectric medium the frequency stays the same, the velocity of propagation decreases by Eq. (8), and the wavelength decreases by the amount

$$\lambda = \lambda_0/n \tag{11}$$

where λ_0 is the wavelength in free space.

A plane electromagnetic wave traveling in free space is associated with an energy flow density given by Poynting's vector $\boldsymbol{\Pi}$.

$$\boldsymbol{\Pi} = \mathbf{E} \times \mathbf{H} \qquad \text{J/m}^2\text{s} \tag{12}$$

and the transport of electromagnetic momentum $\mathbf{g}$ is given by a similar expression

$$\mathbf{g} = \mathbf{D} \times \mathbf{B} = \boldsymbol{\Pi}/c^2 \qquad \text{Js/m}^4 \tag{13}$$

The ratio of the electric to the magnetic field vectors equals $(\mu_0/\epsilon_0)^{1/2}$, the characteristic impedance of free space

$$E/H = (\mu_0/\epsilon_0)^{1/2} = 120\pi \qquad \text{ohms} \tag{14}$$

For propagation in a non-magnetic medium ($\mu = \mu_0$) we have

$$E/H = 120\pi/n \qquad \text{ohms} \tag{15}$$

where n is the index of refraction (9).

3 REFLECTION AND REFRACTION

We are all familiar with the basic law of reflection, that the angle of incidence equals the angle of reflection, as shown in Fig. 13-1. For refraction at an interface between two regions with the respective indices of refraction n_1 and n_2 and permeabilities $\mu_1 = \mu_2 = \mu_0$ we have Snell's law

$$n_1 \sin\theta_1 = n_2 \sin\theta_2 \tag{16}$$

which is illustrated in Fig. 13-2. This is easily proved by assuming that light takes the path of least time to go from a point P_1 in medium 1 to a point P_2 in medium 2. In other words, the sum of the time t_1 spend in medium 1 plus the time t_2 spent in medium 2 is a minimum

$$t_1 + t_2 = \text{minimum} \tag{17}$$

When the light passes through several successive media the optical path length OPL defined by

$$OPL = \sum_{n_i s_i} \tag{18}$$

is a minimum, where s_i is the path length in medium i with index of refraction n_i.

When the refraction is from an optically more dense (larger n) medium into a less dense medium, i.e., $n_2 < n_1$, there is a critical angle of incidence $\theta_1 = i_c$ which makes $\theta_2 = \frac{1}{2}\pi$, so we have

$$\sin i_c = \frac{n_2}{n_1} \tag{19}$$

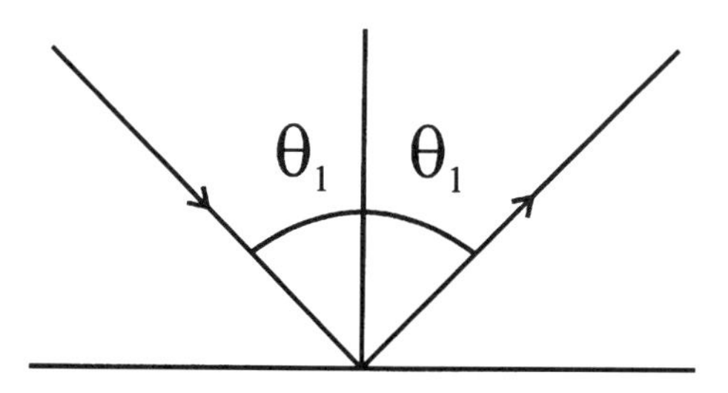

Fig. 13-1. Reflection of a light beam at a surface.

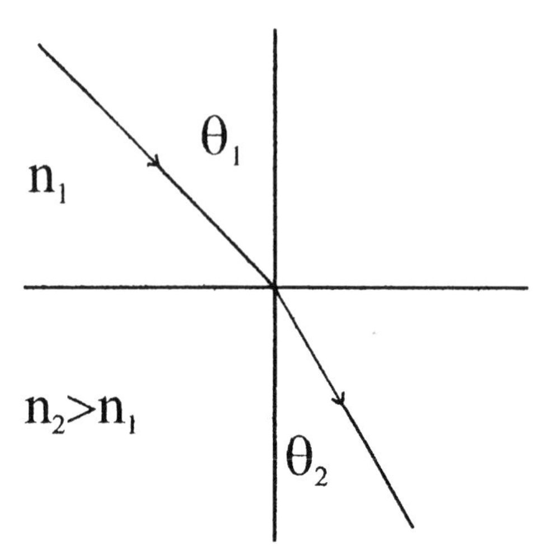

Fig. 13-2. Refraction of a light beam coming from a medium of lower index of refraction n_1 and entering a medium of higher index of refraction n_2.

and beams with larger angles of incidence, $\theta > i_c$, cannot refract into the medium n_2. Therefore, such a beam with $i_c < \theta_1 < \frac{1}{2}\pi$ is totally reflected.

Consider a plane wave with Poynting's vector $\boldsymbol{\Pi} = \mathbf{E}_0 \times \mathbf{H}_0$ traveling in a medium with index of refraction n_0 normally incident at an interface with another medium with index of refraction n_t. There can be a transmitted wave $\mathbf{E}_t \times \mathbf{H}_t$ and a reflected wave $\mathbf{E}_r \times \mathbf{H}_r$. The E and H fields are all parallel at the interface, so we have from the boundary conditions

$$E_0 \pm E_r = E_t \tag{20}$$

$$H_0 \mp H_r = H_t \tag{21}$$

The $\pm$ and $\mp$ signs appear because the reflected wave is going in a direction opposite to that of the incident wave, and either H_0 and H_r are parallel and E_0 and E_r are antiparallel, or vice versa. The upper choice of sign applies when $n_0 > n_t$, and for both cases we obtain for the reflected E_r and transmitted E_t (i.e., refracted) waves

$$E_r = \frac{n_0 - n_t}{n_0 + n_t} E_0 \tag{22}$$

$$E_{\mathrm{t}} = \frac{2n_0}{n_0 + n_{\mathrm{t}}} E_0 \tag{23}$$

We have examined the wave amplitudes at the interface, and it is also of interest to examine the energy flow across the interface. The intensity I is proportional to the square of the electric field amplitude, and for energy balance we must have the energy carried off by the reflected and transmitted waves equal to the incoming energy flux

$$n_0 I_0 = n_0 I_{\mathrm{r}} + n_{\mathrm{t}} I_{\mathrm{t}} \tag{24}$$

$$n_0 E_0^2 = n_0 E_{\mathrm{r}}^2 + n_{\mathrm{t}} E_{\mathrm{t}}^2 \tag{25}$$

Reflection and transmission coefficients R and T, respectively can be defined in terms of intensity with the property

$$R + T = 1 \tag{26}$$

where

$$R = \frac{E_{\mathrm{r}}^2}{E_0^2} \tag{27}$$

$$T = \frac{n_{\mathrm{t}} E_{\mathrm{t}}^2}{n_0 E_0^2} \tag{28}$$

These expressions may be checked with Eqs. (22) and (23).

4 REFLECTION, REFRACTION, AND TRANSMISSION AT OBLIQUE INCIDENCE

For oblique incidence with an angle of incidence i and an angle of refraction r there are two cases to consider, the E vector perpendicular to the plane of incidence

$$\left.\begin{aligned} E_{\mathrm{r}} &= \frac{n_0 \cos i - n_t \cos r}{n_0 \cos i + n_t \cos r} E_0 \qquad (29a)\\ E_{\mathrm{t}} &= \frac{2n_0 \cos i}{n_0 \cos i + n_t \cos r} E_0 \qquad (29b)\end{aligned}\right\} \text{perpendicular incidence}$$

and the E vector parallel to the plane of incidence.

$$\left.\begin{aligned} E_{\mathrm{r}} &= \frac{n_t \cos i - n_0 \cos r}{n_t \cos i + n_0 \cos r} E_0 \qquad (30a)\\ E_{\mathrm{t}} &= \frac{2n_0 \cos i}{n_t \cos i + n_0 \cos r} E_0 \qquad (30b)\end{aligned}\right\} \text{parallel incidence}$$

These expressions all reduce to Eqs. (22) and (23) for normal incidence where $i = r = \frac{1}{2}\pi$.

For the particular case of reflection from air ($n_0 \approx 1$) off another medium ($n_{\mathrm{t}} > 1$) we know from Snell's law (16) that $n_0 \sin i = n_{\mathrm{t}} \sin r$, meaning that

$$\sin i > \sin r \qquad (31)$$

$$\cos i \ < \cos r \qquad (32)$$

For this condition the reflected amplitude E_r for the parallel incidence case can be zero corresponding to

$$n_t \cos i = n_0 \cos r \qquad (33)$$

and this occurs at the Brewster angle $i = i_{\mathrm{B}}$ where the incident and the refracted rays make an angle of $\frac{1}{2}\pi$ with respect to each other (i.e., $\theta_1 + \theta_2 = \frac{1}{2}\pi$ on Fig. 13-2). Thus, $i_{\mathrm{B}} + r = \frac{1}{2}\pi$, so $\cos r = \sin i_{\mathrm{B}}$, and

$$\tan i_{\mathrm{B}} = n_t / n_0 \qquad (34)$$

When unpolarized light is incident at this angle the reflected beam is 100% polarized perpendicular to the plane of incidence. It is not possible to make $E_r = 0$ for the perpendicular polarization case since from Eq. (29a) this requires $\cos i = n_t \cos r$, and we already saw from Eq. (32) that $\cos i$ is less than $\cos r$.

5 POLARIZATION

When an electromagnetic wave enters a region of space it interacts with the charges and currents present there. The force $\mathbf{F} = q\mathbf{E}$ from the wave can displace the outer negative electronic charge cloud $-q$ of an atom by the distance Δx relative to the positive nucleus $+q$, thereby inducing an electric dipole moment $p = q\Delta x$ in the atom given by

$$p = \alpha E \tag{35}$$

where α is the polarizability. If we consider the model of an atom as an harmonically bound charge with the restoring force $\mathbf{F} = -m\omega_0^2\mathbf{r}$, then we can write

$$e\mathbf{E} = m\omega_0^2\mathbf{r} \tag{36}$$

and for the dipole moment $\mathbf{p} = e\mathbf{r}$ a comparison of Eqs. (35) and (36) gives for the polarizability

$$\alpha = e^2/m\omega_0^2 \tag{37}$$

The distribution of dipoles throughout the medium produces a dipole moment per unit volume called the polarization P, and we have

$$D = \epsilon E = \epsilon_0 E + P \tag{38}$$

If there are molecules in the medium with permanent electric dipoles, then they will orient in the applied field E and contribute to the polarization and dielectric constant.

The charges only respond to an incoming wave if its electric field oscillates slowly enough. At very low frequencies the dielectric constant ϵ is high because all of the permanent dipoles and polarizable charge can respond to the incoming field. As the frequency of the wave increases some charges can no longer follow the rapid oscillations and hence no longer contribute to ϵ. This causes various absorption mechanisms to become inoperative, such as rotations of molecules and molecular groups in the microwave region ($\approx 10^{11}$ Hz), vibrations in the infrared region ($\approx 10^{13}$ Hz) and electronic transitions in the visible and near ultraviolet ($\approx 10^{15}$ Hz). At the highest frequencies, in and beyond the X-ray region, none of the charges is able to respond, and ϵ approaches the free space value ϵ_0.

6 ABSORPTION AND DISPERSION

The dielectric constant ϵ has real and imaginary parts, ϵ' and ϵ'', respectively. The real part ϵ' is what we usually think of as the dielectric constant; it determines the velocity and the wavelength of an electromagnetic wave traveling through a medium. The imaginary part ϵ'', on the other hand, is responsible for absorption in the medium, for losses in a capacitor, for heating in a microwave oven, etc. In the absence of absorption $\epsilon'' = 0$, and if the medium is non-magnetic with $\mu = \mu_0$, then we have for the propagation constant k.

$$k = (\epsilon'/\epsilon_0)^{1/2}\omega/c \tag{39}$$

When the imaginary part ϵ'' of ϵ is taken into account, the k is complex and may be written as the sum of a propagation constant β and an absorption coefficient α

$$k = \beta + i\alpha = [\epsilon' + i\epsilon'')/\epsilon_0]^{1/2}\omega/c \tag{40}$$

corresponding to a propagating electric field

$$E = E_0 \mathrm{e}^{i\beta x - \alpha x - i\omega t} \tag{41}$$

For weak absorption, $\epsilon'' \ll \epsilon'$, we have from a power series expansion

$$\alpha \approx \tfrac{1}{2}\beta(\epsilon''/\epsilon') \tag{42}$$

The real and imaginary parts of the dielectric constant, ϵ' and ϵ'', respectively, are not independent but are related to each other through the Kramers–Kronig relations

$$\epsilon'(\omega) = \epsilon_0 + \frac{2}{\pi}\mathcal{P}\int \frac{\omega'\epsilon''(\omega')d\omega'}{\omega'^2 - \omega^2} \tag{43}$$

$$\epsilon''(\omega) = -\frac{2\omega}{\pi}\mathcal{P}\int \frac{[\epsilon'(\omega') - \epsilon_0]d\omega'}{\omega'^2 - \omega^2} \tag{44}$$

where $\mathcal{P}$ denotes the principal part of the complex integration.

A simple model for a dielectric constant assumes that the electronic charge that interacts with an incoming E field has an harmonic restoring force with the fundamental frequency ω_0, as in Eq. (36) above, and in addition we add a damping force γ, to give for the equation of motion

$$m(\ddot{x} + \gamma\dot{x} + \omega_0^2 x) = -eE_x e^{-i\omega t} \tag{45}$$

This has the solution $x(t) = x_0 e^{-i\omega t}$, and the induced dipole p is given by

$$p_x = -ex = \frac{e^2/m}{\omega_0^2 - \omega^2 - i\omega\gamma} E_x \tag{46}$$

If the medium has N molecules per unit volume and Z electrons per molecule with individual binding frequencies ω_j and damping constants γ_j we obtain Lorentzian line shapes for the case of narrow resonances, $\gamma_j \ll \omega_j$,

$$\frac{\epsilon(\omega)}{\epsilon_0} = 1 + \frac{Ne^2}{m} \sum \frac{f_j}{2\omega_j} \left\{ \frac{(\omega_j - \omega) + \frac{1}{2} i\gamma}{(\omega_j - \omega)^2 + (\frac{1}{2}\gamma)^2} \right\} \tag{47}$$

where the oscillator strengths f_j for the electrons obey the sum rule

$$\sum f_j = Z \tag{48}$$

We see from Fig. 13-3(a) that for this model over a broad range of frequencies the real part of the dielectric constant ϵ' exhibits a gradual decrease, with an up and down fluctuation at each resonant frequency ω_j. The imaginary part ϵ'' only becomes appreciable in the neighborhood of a resonance, in the manner illustrated in Fig. 13-3(b). Figure 13-4 provides details of the behavior of the real part ϵ' in this model. Between resonances it gradually increases with frequency until it reaches a maximum value at $\omega = \omega_j - \frac{1}{2}\gamma_j$ for the resonance centered at ω_j. Beyond this maximum, ϵ' precipitously drops to a minimum at $\omega = \omega_j + \frac{1}{2}\gamma_j$, after which it again gradually increases toward the next resonance at ω_{j+1}. The background of ϵ' between resonances is labeled ${}^{\mathrm{BK}}\epsilon$ on the figure.

7 CONDUCTING MEDIA

If the medium is conducting, then Maxwell's equation

$$\nabla \times \mathbf{H} = \mathbf{J} + \frac{\mathbf{D}}{\partial t} \tag{49}$$

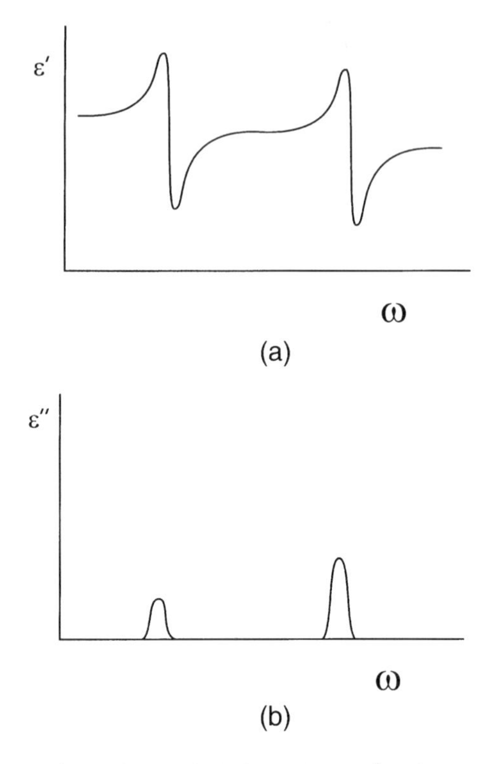

Fig. 13-3. Frequency dependence of (a) the real part ϵ' and (b) the imaginary part ϵ'' of the complex dielectric constant $\epsilon = \epsilon' + i\epsilon''$ near two resonances.

for a harmonic time dependence $e^{-i\omega t}$ gives

$$\nabla \times \mathbf{H} = [\sigma - i\omega\epsilon]\mathbf{E} \tag{50}$$

$$= -i\omega\epsilon E\left\{1 + \frac{i\sigma}{\omega\epsilon}\right\} \tag{51}$$

where use was made of Ohm's law $\mathbf{J} = \sigma\mathbf{E}$ and the expression $\mathbf{D} = \epsilon\mathbf{E}$. This means that the conductivity σ provides the losses, and the absorption coefficient α of a conductor originates from the presence of this finite conductivity.

The electrons that carry current in a conductor move about among positively charged nuclei, and hence they constitute an electrically neutral

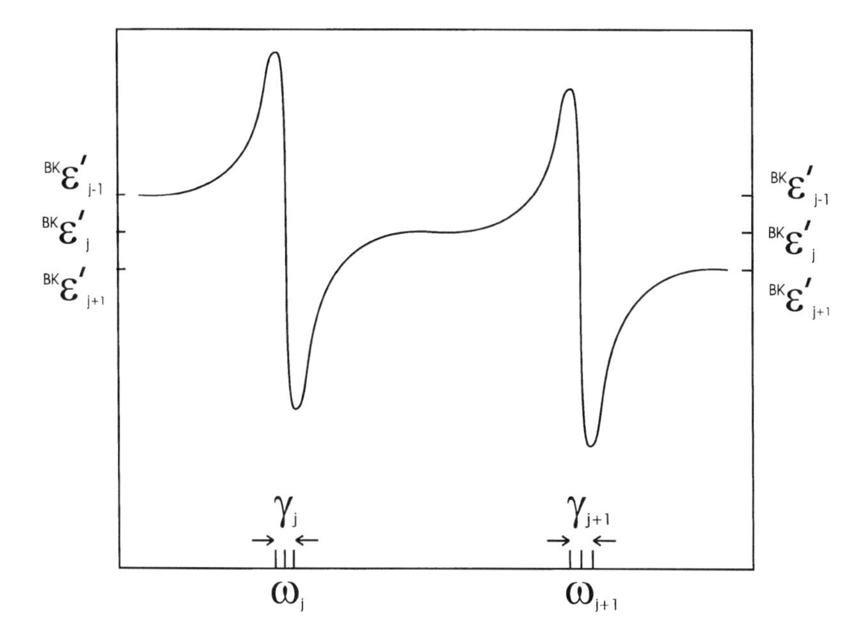

Fig. 13-4. Details of the frequency dependence of the real part of the dielectric constant ϵ' between two resonances centered at ω_j and ω_{j+1} for the narrow line Lorentzian shape model of Eq. (47).

mobile charge system called a plasma. At high frequencies this medium has the dielectric constant

$$\epsilon(\omega) = \epsilon_0 \left\{ 1 - \frac{\omega_p^2}{\omega^2} \right\} \tag{52}$$

where the plasma frequency is given by

$$\omega_p = (Ne^2/\epsilon_0 m)^{1/2} \tag{53}$$

Below the plasma frequency, $\omega < \omega_p$, the propagation constant is purely imaginary and electromagnetic waves are attenuated, i.e., the conductor is opaque. Above ω_p, in the ultraviolet region, the material becomes transparent. The experimental plasma wavelengths $\lambda_p = 2\pi c/\omega_p$ of alkali metals are in the range from 2000 Å for Li to 4400 Å for Cs.

8 GUIDED WAVES

An electromagnetic wave travels through free space at the velocity of light c with the product of its wavelength λ and frequency ν given by

$$\lambda\nu = c = 1/(\mu_0\epsilon_0)^{1/2} \tag{54}$$

This wave is transverse electromagnetic (TEM) in nature, which means that the electric vector E is perpendicular to the magnetic vector H, both are perpendicular to the direction of propagation z and the ratio of their amplitudes equals the impedance of free space

$$E_0/H_0 = (\mu_0/\epsilon_0)^{1/2} = 120\pi \text{ ohms} \tag{55}$$

When the same wave is constrained to travel through a hollow pipe called a waveguide it moves at a slower velocity of propagation v_g, and its wavelength λ_g, called the guide wavelength, is longer than in unbounded space. For a given waveguide, which we will assume to have a rectangular cross-section, there is a maximum wavelength called the cutoff wavelength λ_c that permits propagation along the guide. These three wavelengths are related to each other through the expression

$$\frac{1}{\lambda^2} = \frac{1}{\lambda_g^2} + \frac{1}{\lambda_c^2} \tag{56}$$

The waves that can propagate in waveguides are either transverse electric (TE), which means that they have a component of the H field along the direction of propagation, or they are transverse magnetic (TM) with a component of E along the propagation direction. When the waveguide has a circular cross-section, the variations of the wave amplitudes in the radial direction are determined by Bessel functions. In a rectangular waveguide of dimensions $a < b$ the dominant mode, meaning the mode with the smallest cutoff wavelength λ_c, is of the TE type with the following configurations for its fields

$$H_x = H_0 \sin(\pi x/a) e^{ikz - i\omega t} \tag{57}$$

$$H_z = H_0(\lambda_g/2a) \cos(\pi x/a) e^{ikz - i\omega t} \tag{58}$$

$$E_y = -Z_{\mathrm{TE}} H_0 \sin(\pi x/a) e^{ikz - i\omega t} \tag{59}$$

where z is the direction of propagation, $\lambda_c = 2a$ and Z_{TE} is the characteristic impedance for this mode

$$Z_{\mathrm{TE}} = (\mu/\epsilon)^{1/2} \frac{\lambda_g}{\lambda} \tag{60}$$

which gives the magnitude of the ratio E_y/H_x of the transverse electric field to the transverse magnetic field. This characteristic impedance is greater than the characteristic impedance Z_0 of the unbounded medium for transverse electromagnetic waves,

$$Z_0 = (\mu/\epsilon)^{1/2} \tag{61}$$

and more generally we have

$$Z_{\mathrm{TM}} < Z_0 < Z_{\mathrm{TE}} \tag{62}$$

The wave travels at the velocity

$$v_g = (\mu\epsilon)^{-1/2} \frac{\lambda}{\lambda_g} \tag{63}$$

which is slower than the velocity $v = (\mu\epsilon)^{-1/2}$ of a TEM wave in the unbounded medium characterized by μ and ϵ that fills the waveguide. Thus, both TE and TM waves travel more slowly than TEM waves.

In a typical case, a waveguide has a diameter of about 2 cm and operates at the microwave frequency of 10^{10} Hz. Light pipes function like waveguides but their diameters are ordinarily much greater than the wavelength of the light ($\lambda_g \ll 2a$) so the propagation is essentially of the TEM type.

9 MODULATED WAVES

We have been considering simple sinusoidal waves. In amplitude modulated (AM) radio transmission, use is made of waves that oscillate at a high carrier frequency denoted by ω, and are amplitude modulated at a lower frequency $\omega_{\mathrm{m}} \ll \omega$ called the modulation frequency. Consider a high frequency carrier wave of amplitude E_0 that is AM modulated by a wave of amplitude $E_{\rm m}$ superimposed on the carrier. The amplitude $E(t)$ of the modulated wave varies with time in accordance with the expression

$$E(t) = (E_0 + E_{\rm m} \sin \omega_{\rm m} t) \sin \omega t \tag{64}$$

and if we make use of the trigonometric identity

$$\cos(\omega t \mp \omega_{\rm m} t) = \cos \omega t \cos \omega_{\rm m} t \pm \sin \omega t \sin \omega_{\rm m} t \tag{65}$$

then Eq. (64) assumes the form

$$E(t) = E_0 \sin \omega t + \tfrac{1}{2} E_m \cos(\omega - \omega_m)t - \tfrac{1}{2} E_m \cos(\omega + \omega_m)t \tag{66}$$

Thus, the overall signal consists of three frequency components, the carrier frequency ω, and the sideband sum and difference frequencies $\omega \pm \omega_m$. The overall signal has the following minimum and maximum amplitudes

$$\begin{aligned} E_{\rm min} &= E_0 - E_m \\ E_{\rm max} &= E_0 + E_m \end{aligned} \tag{67}$$

as shown in Fig. 13-5, with the percentage modulation defined by

$$\%\ \text{modulation} = 100 E_m / E_0 \tag{68}$$

The respective wavelengths $\lambda = c/\omega$ and $\lambda_m = c/\omega_m$ of the carrier and the modulation signals are indicated in the figure.

Another type of modulation called frequency modulation is used for FM radio and television transmission, but space limitations prevent analyzing this.

10 SOLITONS

A solitary wave or soliton is one that has a localized shape and that propagates without change in form. It was first observed in 1834 by John Scott Russell as a heap or rounded elevation of water of peak height h that moved forward in a barge channel of depth d at a speed

$$v = [g(h + d)]^{1/2} \tag{69}$$

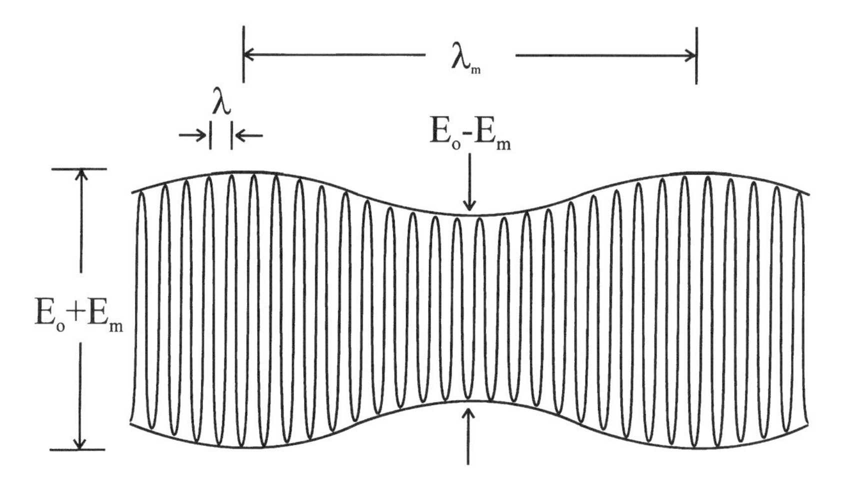

Fig. 13-5. Amplitude modulated wave showing the rapidly oscillating carrier frequency ω within the envelope of the slowly oscillating modulation frequency ω_m. The minimum and maximum amplitudes as well as the carrier and modulation wavelengths $\lambda = 2\pi c/\omega$ and $\lambda_m = 2\pi c/\omega_m$, respectively, are shown.

which averaged about 14 km/h. Russell's original observations were made by following the wave on horseback. Russell also found that a taller, and hence faster moving (from Eq. (69)) solitary wave can catch up with and pass through a slower moving wave.

The profile of the solitary wave $\psi(x, t)$ for $h \ll d$ is given by

$$\psi(x, t) = h\,\mathrm{sech}^2[(x - vt)/w] \tag{70}$$

and the width parameter w

$$w = 2d[(h + d)/3h]^{1/2} \tag{71}$$

has the approximate value

$$w \sim 2(d^3/3h)^{1/2} \tag{72}$$

for $h \ll d$. The waveforms for three amplitudes and widths are shown in Fig. 13-6. It was not until 1895 that Korteweg and deVries showed that Russell's solitary waves can be accounted for by what became known as the KdeV equation

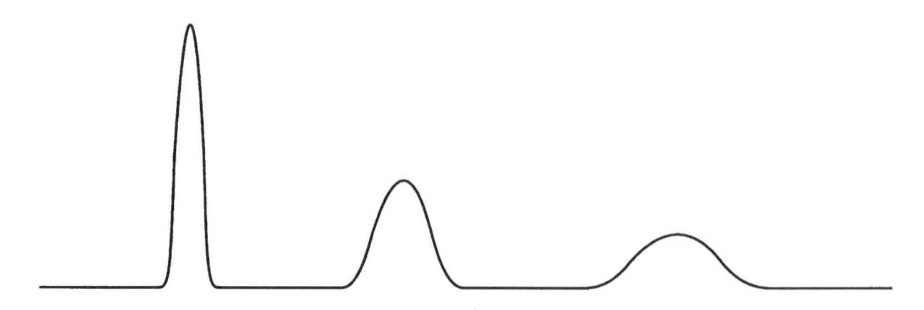

Fig. 13-6. Three solitary waves of differing amplitudes and widths. (From R. K. Dodd et al., *Solitons and Nonlinear Wave Equations*, Academic Press, New York, 1982, p. 5.)

$$\frac{\partial\psi}{\partial t} = (g/h)^{1/2}\left[\epsilon\frac{\partial\psi}{\partial x} + (3/2)\psi\frac{\partial\psi}{\partial x} + \tfrac{1}{2}\sigma\frac{\partial^3\psi}{\partial x^3}\right] \tag{73}$$

where

$$\sigma = \tfrac{1}{3}h^3 - hT/g\rho \tag{74}$$

T is the surface tension, g is the acceleration of gravity, ρ is the density of the liquid, and ϵ is an arbitrary parameter. The surface tension term can often be neglected.

One of the early wave equations of quantum mechanics was the Klein–Gordon equation, which has the following form in one space dimension

$$\frac{\partial^2}{\partial x^2}\psi - \frac{1}{c^2}\frac{\partial^2}{\partial t^2}\psi = (mc/\hbar)^2\psi \tag{75}$$

where $\hbar/m_e c = \lambda_c/2\pi$, and $\lambda_c = 2.426 \times 10^{-12}$ m is the Compton wavelength of an electron. This equation has harmonic wave solutions of the form $e^{ikx-i\omega t}$ which provide the dispersion relation

$$(\omega/c)^2 = k^2 + (2\pi/\lambda_c)^2 \tag{76}$$

between ω and k. The phase and group velocities

$$v_p = \omega/k \qquad v_g = d\omega/dk \tag{77}$$

both depend on k so a wave with a range of k values will spread out or disperse during propagation.

A non-linear form of the Klein–Gordon equation called the sine-Gordon equation

$$\frac{\partial^2}{\partial x^2}\psi - \frac{1}{c^2}\frac{\partial^2}{\partial t^2}\psi = (2\pi/\lambda_c)^2 \sin\psi \tag{78}$$

has soliton solutions, i.e., solitary waves capable of propagating over long distances before beginning to disperse. There are what are called kink solutions that represent a twist of the variable $\psi(x, t)$, and antikink solutions with the opposite twist.

CHAPTER 14

OPTICS

1 INTRODUCTION

We begin with a survey of geometical optics, including lenses and mirrors, then go into more detail about physical optics, covering polarization, interference, and diffraction. Some standard optics material has already been covered in other chapters, such as Chapter 13 on wave propagation.

2 INDEX OF REFRACTION

For most materials the index of refraction $n = (\varepsilon'/\varepsilon_0)^{1/2}$ is highest in value at very low frequencies and decreases to $n \approx 1$ in the x-ray region, where in this chapter we are concerned with the real part ε' of the complex dielectric constant $\varepsilon = \varepsilon' + i\varepsilon''$. We saw in Chapter 13 that over much of the frequency range the index of refraction actually increases as the frequency increases, and that the main decreases in n come about at resonances, as indicated in Fig. 13-3. For optical glasses the visible region of the spectrum is generally in the range below a resonance where n increases with the frequency. This behavior, whereby $dn/d\omega$ is positive, is referred to as normal dispersion; when $dn/d\omega$ is negative, the dispersion is called anomalous. Therefore n is larger at the blue end of the visible spectrum than it is at the red end, and a prism bends blue rays more than red rays. Colored glass or translucent materials have absorption bands in the visible owing to peaks in the $\varepsilon''(\omega)$ curve.

Some optical devices can compensate for dispersion. Two important prism combinations are a crown glass/flint glass arrangement which disperses light into colors without bending the beam, and an achromatic prism which bends light without dispersion, i.e., without color separation.

3 LENSES

A thin converging (i.e., positive) lense focuses parallel light at the focal point f, as indicated in Fig. 14-1. The lens formula

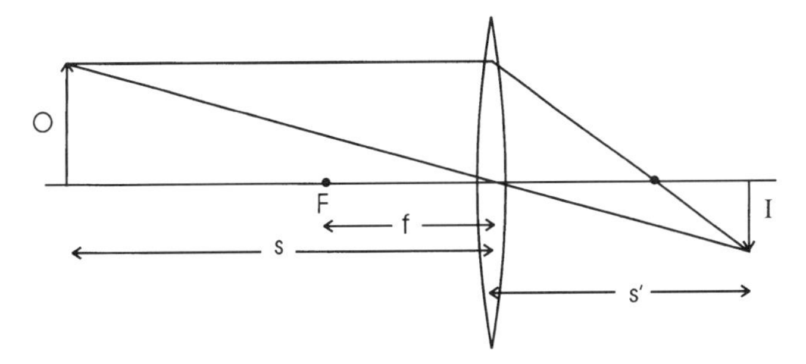

Fig. 14-1. Thin convergent lens showing an object O at a distance s to the left forming a real, inverted image I at a distance s' beyond the lens on the right. Note that for this case both s and s' are larger than the focal length f.

$$\frac{1}{s} + \frac{1}{s'} = \frac{1}{f} \tag{1}$$

gives the image distance s' for a particular object distance s. If both s and s' are greater than f, then the image is real (i.e., s' is positive) and inverted, as shown in Fig. 14-1, with the linear or lateral magnification m given by

$$m = -s'/s \tag{2}$$

where the negative sign indicates an inverted image. If the object distance is less than the focal length, $s < f$, then s' is negative and larger than f, thereby producing an image which is virtual, upright, and enlarged, as shown in Fig. 14-2.

A divergent lens has a negative focal length, and since the object distance is positive and the same lens formula applies, it follows that s' must be negative and smaller than s. The negative s' means that the image is virtual, and from Eq. (2) we have $0 < m < 1$ so the image is upright with a magnification less than 1, as shown in Fig. 14-3.

Since all quantities in the lens formula are reciprocals, optometrists find it convenient to use the units of diopters, or reciprocal meters. Lenses are rated by their power in diopters, the power being the reciprocal of the focal length in meters.

For thin lenses the lense makers' formula

$$\frac{1}{f} = (n-1)\left(\frac{1}{r_1} - \frac{1}{r_2}\right) \tag{3}$$

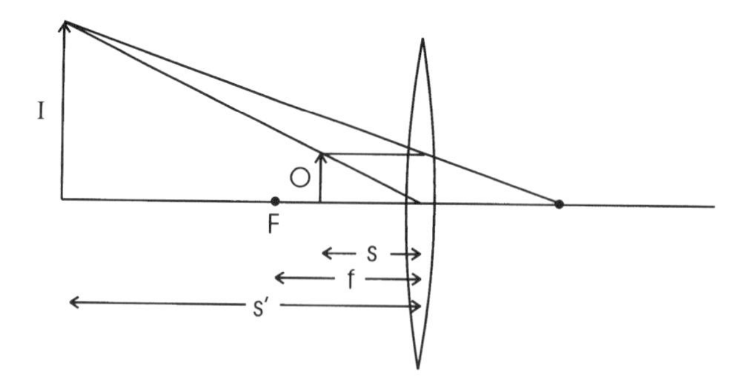

Fig. 14-2. Thin convex lens of Fig. 14-1 showing an object O inside the focal distance ($s < f$) forming a virtual, upright magnified image I at a distance s' to the left.

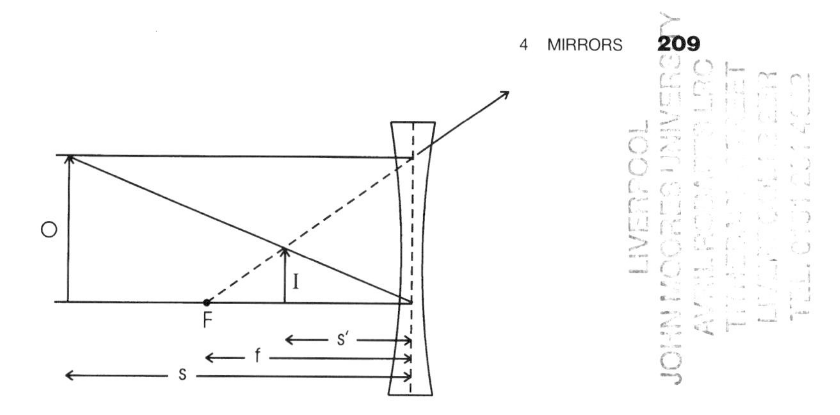

Fig. 14-3. Thin divergent lens showing an object O at a distance s to the left forming a virtual, upright image I at a distance s' with a magnification m less than 1.

provides the focal length, where r_1 is the first surface encountered by the incoming light ray; convex surfaces encountered have a positive r_j and negative ones a negative r_j. A typical converging lens has a positive r_1 and a negative r_2. For planar surface r_j is infinite and $1/r_j = 0$.

For a thin lens combination two lenses are next to each other. The lens formula (1) is applied sequentially to each with the image of the first lens acting as the object for the second.

Consider an object in a medium with an index of refraction n located a distance s from a convex surface of radius of curvature r beyond which the medium has an index of refraction n'. An image forms at a distance s' from the surface, where

$$\frac{n}{s} + \frac{n'}{s'} = \frac{n' - n}{r} \tag{4}$$

Thus, when the second medium is optically more dense than the first ($n' > n$) there is a real image in the second medium beyond the convex interface.

4 MIRRORS

A spherical mirror of radius r has a focal length f given by

$$f = -\tfrac{1}{2}r \tag{5}$$

and the mirror formula is the same as the lens formula (1)

$$\frac{1}{s} + \frac{1}{s'} = \frac{1}{f} \tag{6}$$

The object distance s and image distance s' are both positive when they are on the left-hand side of the mirror, corresponding to a real image. A mirror that is concave toward the object has, by convention, a positive focal length and a negative radius of curvature, and for this case the image is real and inverted for $s > f$.

Such a mirror focuses parallel incoming light ($s = \infty$) at the distance $s' = f$ in front of the mirror. A convex mirror, which has a negative focal length, always produces a virtual, upright image that is smaller than the object. The magnification m for both types of mirror is

$$m = -s'/s \tag{7}$$

For a planar mirror the focal length is infinite, so from Eq. (6) we see that $s = -s'$ and the image is virtual with no magnification.

5 POLARIZATION

Having finished this brief survey of geometrical optics we proceed to present some aspects of physical optics. Optics experiments are often carried out with narrow beams of plane waves with $\mathbf{E}$ vectors

$$\mathbf{E} = \mathbf{E}_0 \exp[i(\mathbf{k} \cdot \mathbf{r} - \omega t + \phi] \tag{8}$$

having polarizations, monochromaticities, amplitudes, and phases ϕ as uniform as possible. With the development of better and better lasers all of these factors have dramatically improved in recent years.

The polarization of an electromagnetic wave propagating in the z direction depends on the relative amplitudes and phases of the E_x and E_y components, A and B, respectively, of its electric field vector $\mathbf{E}$

$$\mathbf{E} = (A \exp(i\phi_x)\hat{\mathbf{i}} + B \exp(i\phi_y)\hat{\mathbf{j}}) \exp[i(kz - \omega t)] \tag{9}$$

Ordinarily it is more convenient to select the phase of the x component A as zero, and assign $e^{i\phi}$ to the y component B, where $\phi = \phi_y - \phi_x$ is the phase difference between the x and y components of $\mathbf{E}$. In Jones vector notation this gives

$$E_0 = \begin{vmatrix} A \\ Be^{i\phi} \end{vmatrix} \tag{10}$$

where the amplitudes are normalized to unity

$$A^2 + B^2 = 1 \tag{11}$$

For linear polarization $\phi = 0$ and the angle θ provides the direction of the vector $\mathbf{E}$

$$\tan\theta = B/A \tag{12}$$

For circular polarization $A = B = \sqrt{2}$ and the phase factor $\phi = \pm\pi/2$, which gives $e^{i\pi/2} = i$, and we have

$$\underset{\text{left circular (counterclockwise rotation)}}{E_0 = \frac{1}{\sqrt{2}}\begin{vmatrix} 1 \\ i \end{vmatrix}} \qquad \underset{\text{right circular (clockwise rotation)}}{E_0 = \frac{1}{\sqrt{2}}\begin{vmatrix} i \\ 1 \end{vmatrix}} \tag{13}$$

The corresponding elliptic polarizations are

$$\underset{\text{counterclockwise rotation}}{E_0 = \begin{vmatrix} A \\ iB \end{vmatrix}} \qquad \underset{\text{clockwise rotation}}{E_0 = \begin{vmatrix} iA \\ B \end{vmatrix}} \tag{14}$$

Examples of linear and circular polarization are given in Fig. 14-4.

We saw in Chapter 13 that unpolarized light incident on a dielectric medium at the angle of incidence i equal to Brewster's angle i_B defined by

$$\tan i_B = n \tag{15}$$

produces a reflected beam that is 100% plane polarized perpendicular to the plane of incidence. This is a convenient way to produce plane polarized light.

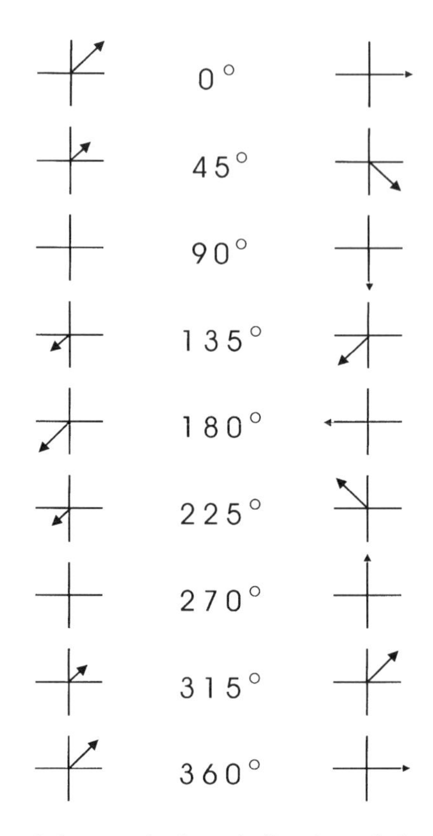

Fig. 14-4. Sketches of the magnitude and direction of the **E** vector as it varies throughout one cycle for the cases of linear (left) and circular (right) polarization.

6 INTERFERENCE AND DIFFRACTION

Two phenomena that reflect the wave nature of light are interference and diffraction. We discuss each in turn. Two beams of light that pass through each other emerge from the region of overlap without either of them being changed in any manner. However, in the region where they cross, the amplitudes of the electric and magnetic field vectors are superpositions from the two beams, a phenomenon called interference. When a light wave passes through an aperture or near edges of obstacles it becomes modified in the process, and this modification influences how is spreads into neighboring regions of space. The outgoing waveform is said to have undergone a process called diffraction.

There are two types of diffraction experiments that are often carried out. In Fraunhofer diffraction the source of light and the screen on which the diffraction pattern is observed are at great distances from the region of diffraction. Lenses can be employed to produce a parallel beam and to focus it on the screen. In Fresnel diffraction either the light source, the screen, or both are at lesser distances from the aperture, so lenses are not needed. Ordinarily Fresnel diffraction is easier to observe and Fraunhofer diffraction easier to explain. An example of the latter will be presented below when we examine diffraction at a wide slit.

Interference and diffraction can be analyzed by Huygens' principle whereby each point on a wavefront is regarded as a source of new wavelets. The next extension of a wavefront is generated by drawing small hemispheres, or in two dimensions small semicircles, along the front with radii proportional to the local velocity of light $v = c/n$ at that point. When a plane wave front passes obliquely into a second medium the Huygens' semicircles have different radii in the two media, and this accounts for the bending of the wave front at the interface. Radiating Huygens wavelets permit waves to enter regions shaded by obstacles, and to bend at the edges, thereby accounting for diffraction. Two cases of Huygens wavelet constructions for diffraction at multiple slits are given at Fig. 14-5.

7 INTERFERENCE

An example of interference is the reflection of light off the top and bottom faces of a glass plate, and the recombination of the two beams that emerge. We assume that the light is monochromatic and normally incident. The interference arises from the phase change that one of the beams aquires as it travels through the glass, and in addition one must take into account the extra phase change of π that occurs when a wave traveling in a medium of index of refraction n_1 reflects off a medium of index of refraction $n_2 > n_1$ and experiences a change in sign from the following expression (Eq. (22) in Chapter 13)

$$E_{\text{refl}} = \frac{n_1 - n_2}{n_1 + n_2} E_{\text{inc}} \tag{16}$$

The condition for constructive interference is

$$2nd = (m + \tfrac{1}{2})\lambda \qquad \text{(max)} \tag{17}$$

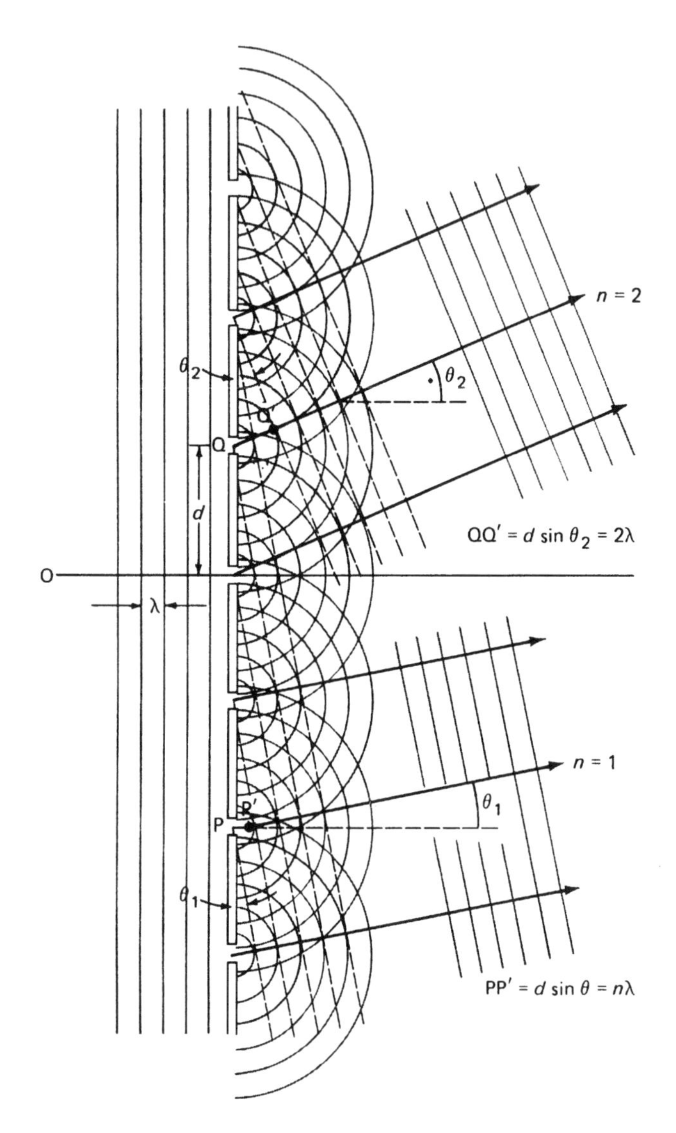

Fig. 14-5. Examples of the Huygens wavelet construction for diffraction at multiple slits. Cases of $d \sin \theta = n\lambda$ are shown for $n = 1$ and $n = 2$. (From J. P. McKelvey and H. Grotch, *Physics for Science and Engineering*, Harper and Row, New York, 1978, Fig. 26.34.)

where the total path length back and forth through the glass plate is $2d$, the wave length in the glass is λ/n, and the factor of $\frac{1}{2}$ corrects for the phase change arising from Eq. (16). Newton's rings are examples of this.

A Michelson interferometer splits a light beam into two beams, reflects each beam from a mirror, and then recombines them after each has traveled a comparable distance. Small adjustments in the mirror positions change the relative path lengths of the two beams to make the interference fringes appear light or dark.

8 SINGLE, DOUBLE, AND MULTIPLE SLITS

Consider a plane wave of light normally incident on an opaque screen containing two slits, S_1 and S_2, of width a separated by the distance d. Each slit acts like a source of light and the two emerging beams can interfere with each other. In addition the incident beam will be diffracted as it passes through each individual slit, and the result of these two factors will be a pattern of bright and dark regions seen on the screen at a distance D far beyond the slits. To explain the pattern on the screen we first consider what happens if the slits are very narrow, i.e., $a \ll d$, so that diffraction at the individual slits can be neglected. Then we examine the diffraction that occurs at a single slit, and finally the composite pattern arising from both factors acting together will be described.

The arrangement for what is called Young's experiment, sketched in Fig. 14-6, shows the beams from the two slits coming together at point P, which is a distance x along the screen from the midpoint P_0. We see from the figure that the extra path length $m\lambda$ between the distances from the two slits to P satisfies the expression

$$\sin\theta = m\lambda/d \tag{18}$$

and the distance x along the plate is given by

$$\tan\theta = x/D \tag{19}$$

Since for this experiment $x \ll D$, the angle θ is small and hence $\sin\theta$ can be equated to $\tan\theta$, to give the expression

$$x = m\lambda D/d \qquad \text{(max)} \tag{20}$$

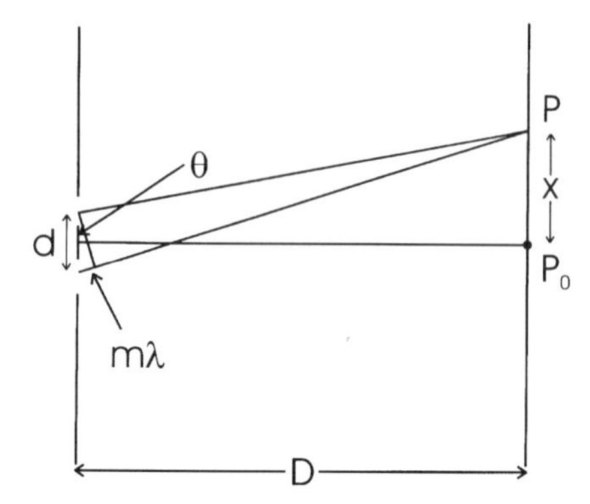

Fig. 14-6. Arrangement for Young's experiment showing the two slits separated by the distance d in the opaque screen on the left and light beams from each slit meeting at point P on the observing screen at a distance D to the right.

where m is an integer, as the condition for constructive interference and bright fringes whereby the **E** vectors from the two slits are in phase with each other and add. A similar argument gives

$$x = (m + \tfrac{1}{2})\lambda D/d \qquad \text{(min)} \tag{21}$$

as the condition for destructive interference and dark fringes.

Now that we have discussed the interference effects arising from the beams emerging from two narrow slits it is appropriate to examine the diffraction that occurs at a single slit. Consider the slit sufficiently wide so that light from different parts of it can interfere with each other and produce a pattern of maxima and minima on the screen. Parallel incoming light is used, as in the two-slit case, and the screen is far away so the conditions for Fraunhofer diffraction are present. If the diffraction angle θ is such that the light from one end of the slit takes a path to the observation point P which is one wavelength longer than the path from the other end, then the light on the screen will originate from an even distribution of all phases, half of which provide positively oriented **E** vectors and the other half negatively oriented **E** vectors. The result is destructive interference and a minimum in intensity. Thus, the condition for a minimum in the Fraunhofer diffraction of a wide slit is the same as the condition for a maximum in the two-slit interference case, and we have

$$x = m\lambda D/a \qquad \text{(min)} \tag{22}$$

for the mth minimum. Maxima will occur at

$$x = (m + \tfrac{1}{2})\lambda D/a \qquad \text{(max)} \tag{23}$$

where again m is an integer. The intensity I of a maximum depends on m and it is shown in optics books that

$$I = I_0\left(\frac{\sin\beta}{\beta}\right)^2 \tag{24}$$

with β given by

$$\beta = \frac{\pi a \sin\theta}{\lambda} \tag{25}$$

$$= (m + \tfrac{1}{2})\pi \qquad \text{(max)} \tag{26}$$

where we used Eq. (23) and the approximation $\sin\theta \approx \tan\theta \approx x/D$ as in the double slit case. There is a central bright spot of intensity I_0 for $\theta = \beta = 0$, and the remaining maxima have the intensities

$$I_{\mathrm{m}} = \frac{I_0}{(m + \frac{1}{2})^2\pi^2} \tag{27}$$

corresponding to the envelope curve of Fig. 14-7.

By way of summary we found from Eqs. (20) and (23) that the two-slit and single wide slit cases have the following respective separations between fringes on the screen

$$\Delta x = \lambda D/d \qquad \text{double slit} \tag{28}$$

$$\Delta x = \lambda D/a \qquad \text{single wide slit} \tag{29}$$

The pattern obtained with two wide slits is a combination of the two individual patterns with the diffraction pattern plus its associated intensities (27) acting as an envelope for the more closely spaced double slit interference fringes, since of course we have, by definition, $a < d$. Figure 14-7 gives an example of such a composite diffraction pattern for the case $d = 3.5a$.

When many equally spaced slits are used, the result is a diffraction grating. As the number of slits increases, the lines in the pattern become narrower. For normal incidence on the grating ($i = \pi/2$) the principal maximum occurs at the diffraction angle θ given by

$$d\sin\theta = m\lambda \tag{30}$$

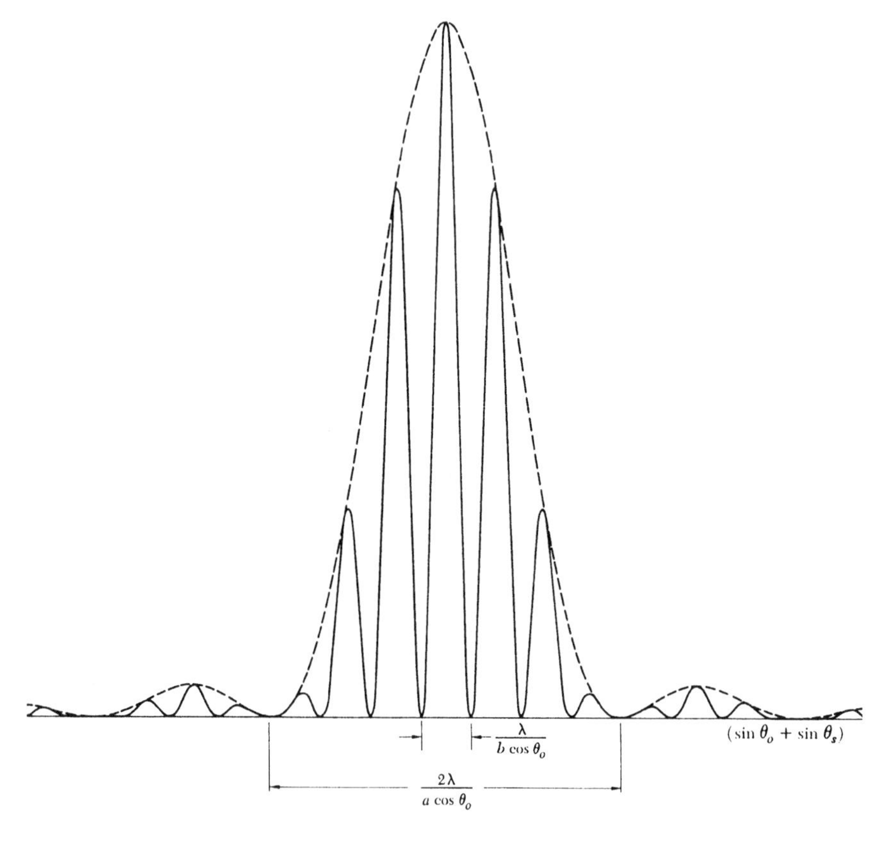

Fig. 14-7. Fraunhofer diffraction pattern from two slits of width a and separation $d = 3.5a$. The double slit interference fringes of relative spacing λ/d have their intensities determined by the envelope of relative spacing λ/a arising from the single slit diffraction. (From W. C. Elmore and M. A. Heald, *Physics of Waves*, McGraw-Hill, New York, 1969, Fig. 10.6.2, p. 367.)

For incidence at another angle i we have the grating equation

$$d(\sin\theta + \sin i) = m\lambda \tag{31}$$

A transmission grating has lines ruled on a plane glass surface and a reflection grating has lines ruled on a polished mirror.

CHAPTER 15

RADIATION

1 INTRODUCTION

The chapter begins with a general non-relativistic approach to radiation from charge and current distributions, and then presents examples of charges in motion. It concludes with a discussion of radiation from relativistically accelerated charges, Cherenkov radiation, and transition radiation.

2 RADIATING CHARGE AND CURRENT DISTRIBUTIONS

We start with a non-relativistic description of radiation from localized harmonically oscillating charge and current densities

$$\rho(\mathbf{r}', t) = \rho(\mathbf{r}')e^{-i\omega t} \tag{1}$$

$$\mathbf{J}(\mathbf{r}', t) = \mathbf{J}(\mathbf{r}')e^{-i\omega t} \tag{2}$$

These two sources are generally known quantities, and they are related to each other by the continuity equation

$$\nabla \cdot \mathbf{J} + \frac{\partial \rho}{\partial t} = 0 \tag{3}$$

which gives

$$\nabla \cdot \mathbf{J} = i\omega\rho \tag{4}$$

The oscillating current density produces a vector potential

$$\mathbf{A}(\mathbf{r}) = \frac{\mu_0}{4\pi} \int \frac{\mathbf{J}(\mathbf{r}')e^{ik(\mathbf{r}-\mathbf{r}')} d^3r'}{|\mathbf{r} - \mathbf{r}'|} \tag{5}$$

which can be integrated to provide us with the radiated magnetic field

$$\mathbf{B} = \nabla \times \mathbf{A} \tag{6}$$

and with the aid of Maxwell's relation

$$\nabla \times \mathbf{H} = \frac{\partial \mathbf{D}}{\partial t} = -i\omega\varepsilon_0 \mathbf{E} \tag{7}$$

we can take the curl of Eq. (6), recalling that $\mu_0\varepsilon_0 = 1/c^2$, to obtain the electric field of the radiation

$$\mathbf{E} = \frac{ic}{k} \nabla \times \mathbf{B} \tag{8}$$

where **B** and **E**, calculated from Eqs. (5), (6), and (8), are outside the region of the source where $\mathbf{J} = 0$. Thus, the integral (5) over the current density provides us with the electric and magnetic fields in the radiation pattern.

3 RADIATION ZONES

An important case to consider is that in which the radiating charge and current distributions are localized within a region of dimension d which is much less than the wavelength, $d \ll \lambda$. This restriction enables us to make some approximations which simplify the evaluation of the integral (5) of the vector potential over the sources. There are three zones at different distances r from the source where field patterns can be calculated:

$$d \ll r \ll \lambda \quad \text{near or static zone} \tag{9a}$$

$$d \ll r \approx \lambda \quad \text{intermediate or induction zone} \tag{9b}$$

$$d \ll \lambda \ll r \quad \text{far or radiation zone} \tag{9c}$$

Since $k = 2\pi/\lambda$ it follows from Eq. (9a) that in the near zone $\mathbf{k} \cdot \mathbf{r} \ll 1$, the exponential can be approximated by unity, and the $|\mathbf{r} - \mathbf{r}'|^{-1}$ term can be expanded in tesseral (or spherical) harmonics $Z_{lM}^{S,C}$, as is explained in Chapter 28.

$$\mathbf{A}(\mathbf{r}) = \frac{\mu_0}{4\pi} \sum \frac{2\pi}{2\ell + 1} \frac{Z_{\ell M}^{S,C}(\theta, \phi)}{r^{\ell+1}} \int \mathbf{J}(\mathbf{r}') Z_{\ell M}^{S,C}(\theta', \phi') r'^{\ell} d^3 r' \tag{10}$$

The near field has a static form with an harmonic oscillation, a state that can be called quasi-stationary. The factor $r^{\ell+1}$ in the denominator causes the non-vanishing lower order moments to dominate.

In the intermediate or induction zone where $\lambda \sim r$ and $kr \sim 2\pi$ all powers of $\mathbf{k} \cdot \mathbf{r}$ must be retained, and the situation is much more complicated. A multipole expansion can still be made, but we will not be concerned with this zone because it is of little practical importance.

The far or radiation zone is ordinarily the one of greatest interest. Here $\lambda \ll r$ so $\mathbf{k} \cdot \mathbf{r} \gg 1$, and the exponential oscillates rapidly. Since $r' \ll r$ we write $\mathbf{r} = \hat{\mathbf{n}} r$ and make the approximation

$$|\mathbf{r} - \mathbf{r}'| \approx r - \hat{\mathbf{n}} \cdot \mathbf{r}' \tag{11}$$

in the exponential to remove the term e^{ikr} from the integral. The distance $|\mathbf{r} - \mathbf{r}'|$ in the denominator is removed as $1/r$, and since $\mathbf{k} \cdot \mathbf{r}' \ll 1$ the exponential $e^{1k\hat{\mathbf{n}}\cdot\mathbf{r}'}$ is expanded in a rapidly converging power series, to give

$$\mathbf{A}(\mathbf{r}) = \frac{\mu_0}{4\pi} \frac{e^{ikr}}{r} \sum \frac{(-ik)^n}{n!} \int \mathbf{J}(\mathbf{r}') \, (\hat{\mathbf{n}} \cdot \mathbf{r}')^n d^3r' \tag{12}$$

so the emitted radiation arises mainly from the first non-vanishing term in the series.

4 ELECTRIC DIPOLE RADIATION

For electric dipole radiation $n = 0$ we can make use of Eq. (4), integrate Eq. (12) by parts, and obtain the vector potential $\mathbf{A}(\mathbf{r})$ as an integral over the electric dipole moment density $\rho(x', y', z')\mathbf{r}'$, which gives

$$\mathbf{A}(\mathbf{r}) = -\frac{\mu_0}{4\pi} ikc\mathbf{p} \frac{e^{ikr}}{r} \tag{13}$$

and the use of Eqs. (6) and (8) provides the radiation fields

$$\mathbf{B} = \frac{\mu_0}{4\pi} k^2 c(\hat{\mathbf{n}} \times \mathbf{p}) \frac{e^{ikr}}{r} \tag{14}$$

$$\mathbf{E} = -\frac{1}{4\pi\varepsilon_0} k^2 (\hat{\mathbf{n}} \times \mathbf{p}) \times \hat{\mathbf{n}} \frac{e^{ikr}}{r} \tag{15}$$

If the dipole moment $\mathbf{p}$ is taken in the z direction, then these fields expressed in spherical coordinates have the forms

$$\mathbf{B} = -\frac{\mu_0}{4\pi} k^2 cp \sin\theta \frac{e^{ikr}}{r} \hat{\boldsymbol{\phi}}_0 \tag{16}$$

$$\mathbf{E} = -\frac{1}{4\pi\varepsilon_0} k^2 p \sin\theta \frac{e^{ikr}}{r} \hat{\boldsymbol{\theta}}_0 \tag{17}$$

indicated in Fig. 15-1, which also shows the directions of the unit vectors $\hat{\theta}_0$ and $\hat{\phi}_0$. It is easy to show that Poynting's vector

$$\mathbf{E} \times \mathbf{H} = (\text{const}) \frac{p^2 \sin^2\theta}{r^2} \hat{\mathbf{n}} \tag{18}$$

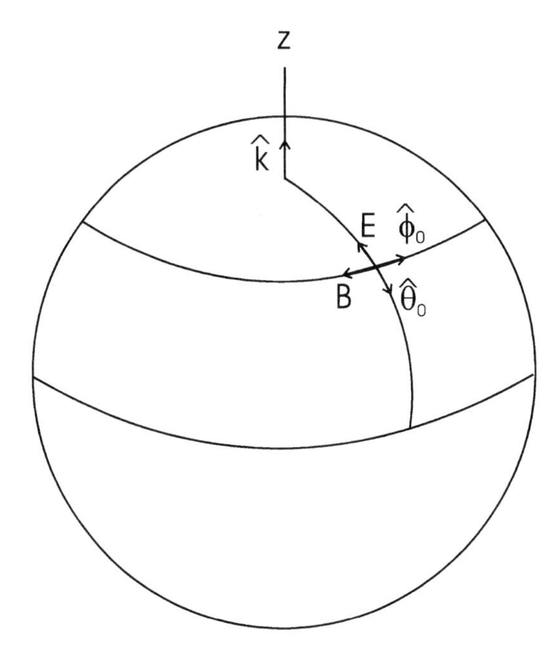

Fig. 15-1. Radiation fields **E** and **B** of an oscillating electric dipole located at the center and aligned along the z direction expressed in spherical coordinates θ, ϕ. The associated unit vectors are indicated.

in the radiation region falls off with distance as $1/r^2$, is axially symmetric (no ϕ dependence), has a $\sin^2\theta$ angular dependence, and is a maximum in the x, y plane, perpendicular to the direction of the dipole.

For far zone radiation the magnetic dipole moment $\boldsymbol{\mu}$ of Eq. (31) in Chapter 11

$$\boldsymbol{\mu} = \tfrac{1}{2}\int \mathbf{r}' \times \mathbf{J}(\mathbf{r}')d^3r' \tag{19}$$

produces the vector potential

$$\mathbf{A}(\mathbf{r}) = -\frac{\mu_0}{4\pi}(ik\hat{\mathbf{n}} \times \boldsymbol{\mu})\frac{e^{ikr}}{r} \tag{20}$$

which has the same form as the magnetic field **B** for the electric dipole case (14) with $\hat{\mathbf{n}} \times \mathbf{p}$ replaced by $\hat{\mathbf{n}} \times \boldsymbol{\mu}$. Hence the **B** field radiation pattern of a magnetic dipole is the same as the **E** field pattern of an electric dipole.

5 CHARGE IN UNIFORM MOTION

Uniformly moving charges carry along their electric and magnetic fields, while in addition accelerated charges radiate. In this section we examine the fields of a uniformly moving charge, and in the next section we treat accelerated charges.

Consider a charge q moving along a straight trajectory in the z, z' direction past an observation point P which is a distance b from the line in the y direction, as indicated in Fig. 15-2. The line from the charge to the observation point subtends an angle θ with respect to the direction of motion, so the motion starts with the charge at $z = -\infty$, $\theta = 0$, and ends with $z \Rightarrow +\infty$ and $\theta = \pi$, as is clear from Fig. 15-2, and we take $t = 0$ at the point of closest approach where $\theta = \pi/2$.

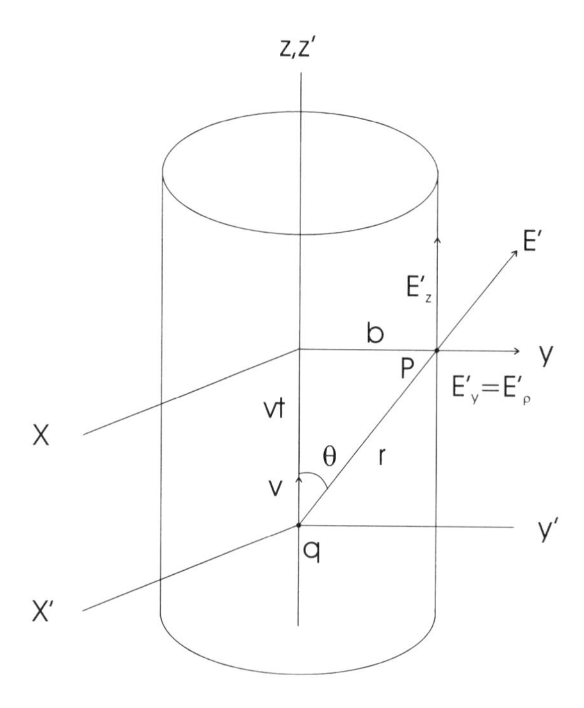

Fig. 15-2. Coordinate systems for calculating the electric and magnetic fields at a point P arising from an electric charge q in uniform motion at the velocity v along the z direction. The cartesian coordinates in the laboratory system (x, y, z) and in the rest system of the charge (x', y', z') are indicated.

At any instant the fields at the point P in the charge's rest system are all electric and radial, with the following components given by J. P. Jackson (*Classical Electrodynamics*, Wiley, New York, 1975, p. 553):

$$E'_z = E' \cos\theta = \frac{-qvt'}{4\pi\varepsilon_0 r'^3} = -\frac{1}{4\pi\varepsilon_0} \frac{q\gamma vt}{(b^2 + \gamma^2 v^2 t^2)^{3/2}} \tag{21}$$

$$E'_\rho = E'_y = E' \sin\theta = \frac{qb}{4\pi\varepsilon_0 r'^3} = \frac{1}{4\pi\varepsilon_0} \frac{qb}{(b^2 + \gamma^2 v^2 t^2)^{3/2}} \tag{22}$$

with $E'_\phi = 0$, and no B fields are present in the charge's rest system. When these fields are transformed to the observer's frame of reference we find the following three non-vanishing components:

$$E_z = -\frac{1}{4\pi\varepsilon_0} \frac{q\gamma vt}{(b^2 + \gamma^2 v^2 t^2)^{3/2}} \tag{23}$$

$$E_\rho = E_y = \frac{1}{4\pi\varepsilon_0} \frac{q\gamma b}{(b^2 + \gamma^2 v^2 t^2)^{3/2}} \tag{24}$$

$$B_\phi = B_x = \beta E_y / c \tag{25}$$

The electric fields cluster transverse to the direction of motion, and the B fields are circles around this direction, as indicated in Fig. 15-3.

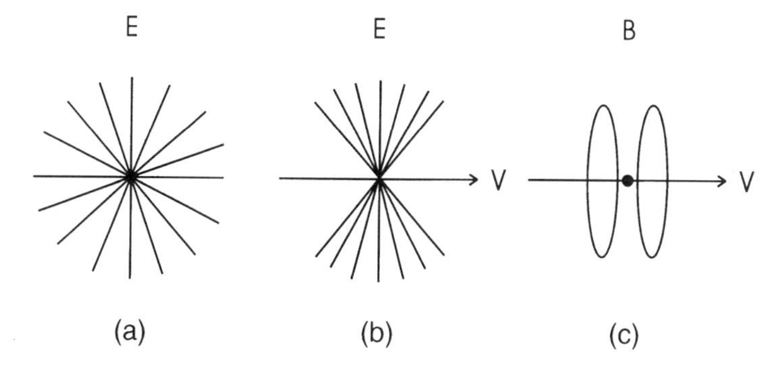

Fig. 15-3. Electric field configuration around (a) a stationary electric charge, and (b) a charge moving at a relativistic speed. The magnetic field is shown (c) around a moving charge. A stationary charge has no magnetic field.

6 ACCELERATED CHARGE

A relativistic approach to calculating the radiation of fast moving charges modifies the non-relativistic potentials by replacing the stationary charges and currents by charges in motion $e\mathbf{v}$. In addition we must take into account the fact that the potentials at a point $\mathbf{r}$ in space and time t are due to the position of the charge at an earlier time t' called the retarded time

$$t' = t - |\mathbf{r} - \mathbf{r}'|/c \tag{26}$$

and not at the time t. This is done by writing (see Jackson, *Classical Electrodynamics*, 1975, Chapter 14)

$$\mathbf{A}(\mathbf{r},t) = \frac{\mu_0}{4\pi}\left(\frac{e\boldsymbol{\beta}}{(1 - \boldsymbol{\beta}\cdot\hat{\mathbf{n}})R}\right)_{\text{ret}} \tag{27}$$

$$\Phi(\mathbf{r},t) = \frac{1}{4\pi\varepsilon_0}\left(\frac{e}{(1 - \boldsymbol{\beta}\cdot\hat{\mathbf{n}})R}\right)_{\text{ret}} \tag{28}$$

where R is the observer-to-charge distance, and $\hat{\mathbf{n}}$ is a unit vector in the $\mathbf{R}$ direction. After considerable work one obtains for the electric field

$$\mathbf{E}(\mathbf{r},t) = \frac{e}{4\pi\varepsilon_0}\left(\frac{\hat{\mathbf{n}} - \boldsymbol{\beta}}{\gamma^2(1 - \boldsymbol{\beta}\cdot\hat{\mathbf{n}})^3R^2}\right)_{\text{ret}} + \frac{e}{4\pi\varepsilon_0 c}\left(\frac{\hat{\mathbf{n}} \times \{(\hat{\mathbf{n}} - \boldsymbol{\beta}) \times \dot{\boldsymbol{\beta}}\}}{(1 - \boldsymbol{\beta}\cdot\hat{\mathbf{n}})^3R}\right)_{\text{ret}} \tag{29}$$

and for the magnetic field

$$\mathbf{B} = (1/c)[\hat{\mathbf{n}} \times \mathbf{E}]_{\text{ret}} \tag{30}$$

The velocity fields of the first term fall off as $1/R^2$ and do not contribute to the radiation, but the acceleration fields arising from the second term fall off as $1/R$, corresponding to radiation fields.

7 SLOWLY MOVING ACCELERATED CHARGE

For a slowly moving accelerated charge, $\beta \ll 1$, the radiation field term of Eq. (29) reduces to

$$\mathbf{E}(\mathbf{r},t) \approx \frac{e}{4\pi\varepsilon_0 c}\,\frac{\hat{\mathbf{n}} \times (\hat{\mathbf{n}} \times \dot{\boldsymbol{\beta}})}{R} \tag{31}$$

with the magnitude

$$E(\mathbf{r},t) \approx \frac{e}{4\pi\varepsilon_0 c} \frac{\dot{\beta}\sin\theta}{R} \tag{32}$$

where θ is the angle between $\dot{\boldsymbol{\beta}}$ and the direction of the radiation. The power radiated per unit solid angle is

$$\frac{dP}{d\Omega} = \frac{e^2\dot{\beta}^2\sin^2\theta}{16\pi^2\varepsilon_0 c} \tag{33}$$

and integration gives, for the total radiated power,

$$P = \frac{e^2\dot{\beta}^2}{6\pi\varepsilon_0 c} \tag{34}$$

a result due to Larmor.

8 RELATIVISTICALLY ACCELERATED CHARGE

At relativistic speeds the Larmor approximation (33) is not valid, and if the acceleration occurs over a short enough time so that $\boldsymbol{\beta}$ and $\dot{\boldsymbol{\beta}}$ remain constant in direction and magnitude, and the observation is made so far away that $\hat{\mathbf{n}}$ and R do not change appreciably during the acceleration, then we obtain for the radiation pattern

$$\frac{dP}{d\Omega} = \frac{e^2}{16\pi^2\varepsilon_0 c} \frac{|\hat{\mathbf{n}} \times \{(\hat{\mathbf{n}} - \boldsymbol{\beta}) \times \dot{\boldsymbol{\beta}}\}|^2}{(1 - \boldsymbol{\beta}\cdot\hat{\mathbf{n}})^5} \tag{35}$$

There are two cases to consider: linear acceleration and transverse acceleration.

For linear acceleration, in which $\boldsymbol{\beta}$ and $\dot{\boldsymbol{\beta}}$ are parallel, we obtain

$$\frac{dP}{d\Omega} = \frac{e^2\dot{\boldsymbol{\beta}}^2}{16\pi^2\varepsilon_0 c} \frac{\sin^2\theta}{(1 - \boldsymbol{\beta}\cdot\hat{\mathbf{n}})^5} \tag{36}$$

In the limit as $\beta \Rightarrow 1$ the denominator becomes very small, and there is a maximum at the angle

$$\theta_{\max} \approx \frac{1}{2\gamma} \tag{37}$$

as shown in Fig. 15-4 so the radiation predominates in a narrow cone in the forward direction. The total radiated power is

$$P = \frac{e^2\dot{\beta}^2\gamma^6}{6\pi\varepsilon_0 c} = \frac{e^2}{6\pi\varepsilon_0 c^3}\frac{\dot{p}^2}{m^2} \tag{38}$$

For uniform circular motion in which the acceleration is transverse to the direction of motion we obtain what is called synchrotron radiation, which has the radiation pattern

$$\frac{dP}{d\Omega} = \frac{e^2}{16\pi^2\varepsilon_0 c}\frac{\dot{\beta}^2}{(1-\beta\cos\theta)^3}\left[1 - \frac{\sin^2\theta\cos^2\phi}{\gamma^2(1-\beta\cos\theta)^2}\right] \tag{39}$$

This has a maximum in the forward direction ($\theta = 0$), goes through a zero, and then has a secondary maximum. The beam is emitted in the plane of the orbit. It has a cross-section shaped somewhat like an oval or ellipse with its long axis in the direction of the orbit axis ($\phi = \frac{1}{2}\pi$, angular width $\approx 2/\gamma$) and its short axis in the orbit plane ($\phi = 0$, angular width $\approx 1/\gamma$) as shown in Fig. 15-4. The total radiated power is

$$P = \frac{e^2\dot{\beta}^2\gamma^4}{6\pi\varepsilon_0 c} = \frac{e^2}{6\pi\varepsilon_0 c^3}\frac{\dot{p}^2\gamma^2}{m^2} \tag{40}$$

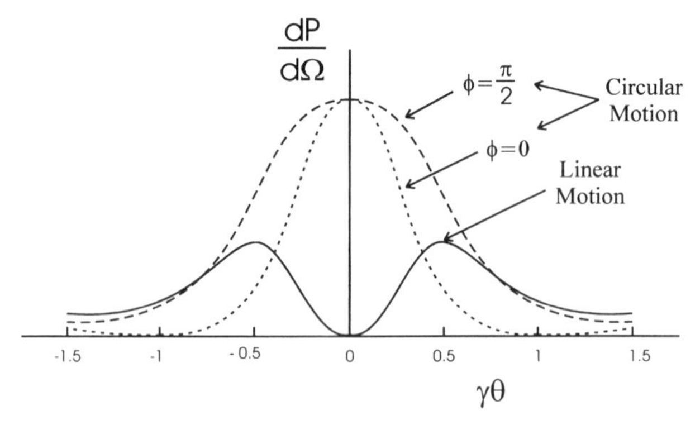

Fig. 15-4. Angular dependences of radiation beams emitted by relativistically accelerated charges undergoing linear and circular motions.

Comparing the expressions for the linear and transverse acceleration we see that for the same $\dot{p}$ (same applied force) the radiation is γ^2 times stronger for the transverse case, while for constant $\dot{\beta}$ it is γ^2 larger for the parallel case. This difference arises because the time derivatives of the momentum $p = \gamma m v$ in these two cases are given by

$$\begin{aligned} \dot{p} &= \gamma^3 m \dot{v} \quad \text{linear motion} \\ \dot{p} &= \gamma m \dot{v} \quad \text{circular motion} \end{aligned} \tag{41}$$

as may be checked by direct differentiation, recalling that $\gamma = [1 - (v^2/c^2)]^{-1/2}$ remains constant in time for the circular case, and hence is not differentiated.

The light radiated by the circular motion of the electron is detected in the plane of the orbit as successive pulses, once every cycle. We assume $\beta \approx 1$, and the beam's angular width $\theta \approx 1/\gamma$ is observed while the particle goes a distance $\rho\theta$ in the orbit, where ρ is the radius of the orbit. Pulses of width Δt in time

$$\Delta t \approx \rho / 2\gamma^3 c \tag{42}$$

are observed at intervals T_0

$$T_0 = 2\pi\rho / c \tag{43}$$

where $2\pi\rho$ is the circumference of the orbit. Figure 15-5 gives details of this. A Fourier decomposition of the pulse shows that the beam is broad and weak in intensity for low frequencies, and narrow and weak for very high frequencies, relative to the more intense beam near the critical frequency ω_c given by

$$\omega_c = 3\gamma^3 c / \rho \tag{44}$$

as shown in Fig. 15-6, where ω_c is close to the reciprocal of the pulse width Δt mentioned above (42).

An extremely relativistic charge experiencing a very short duration arbitrary acceleration emits synchrotron radiation comparable to that from a charge moving instantaneously in a circular orbit at constant speed.

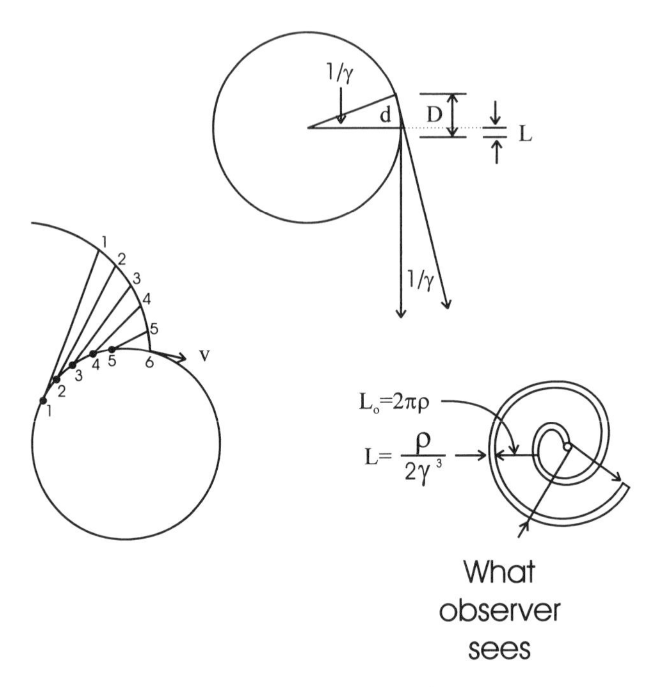

Fig. 15-5. Pulses of radiation of length $\Delta t = \rho/2c\gamma^3$ spaced at intervals $T_0 = 2\pi\rho/c$ emitted in the orbital plane by a charge moving at a uniform relativistic speed $v \sim c$ in a circular orbit of radius ρ.

9 CHERENKOV RADIATION

Another type of radiation by fast moving charged particles occurs when they move through a medium at a speed that exceeds the speed of light $c/n = c/[\varepsilon(\omega)/\varepsilon_0]^{1/2}$ in that medium, where n is the index of refraction. This can be analyzed qualitatively in terms of the wavefronts sketched in Fig. 15-7. For particle speeds v less than c/n the wavefronts of the E and B fields always stay ahead of the particle, as shown in part (a) of the figure, so there is no radiation since particles moving at constant velocity do not radiate. However, when the particle speed exceeds that of light in the medium, then it is always ahead of its fields, as shown in part (b) of the figure. There is an angle θ_c, called the Cherenkov angle, also shown, associated with wavefronts moving outward within this angle. The cosine of this angle is defined by the ratio of the distances

$$\cos\theta_c = \frac{v_{\text{light}}t}{v_{\text{particle}}t} \tag{45}$$

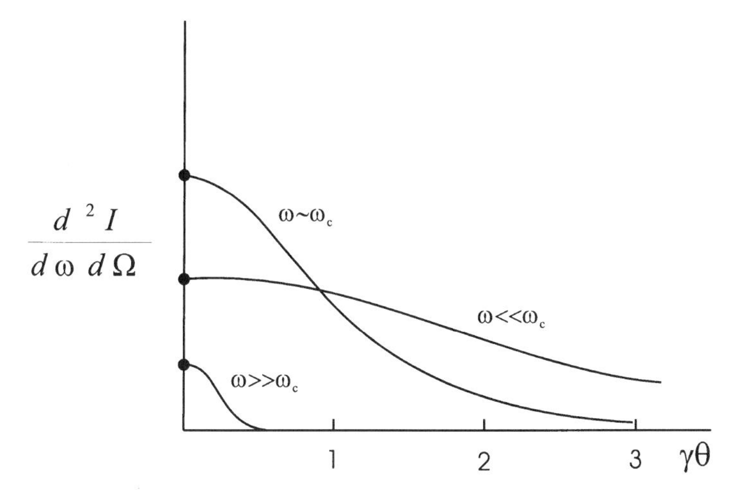

Fig. 15-6. Angular dependence of synchrotron radiation spectrum for frequencies ω much less than, comparable with, and much greater than the critical frequency $\omega_c = 3\gamma^3 c/\rho$.

and this gives

$$\cos\theta_c = \frac{1}{\beta[\varepsilon(\omega)/\varepsilon_0]^{1/2}} \tag{46}$$

The radiation is emitted at the Cherenkov angle θ_c, and a measurement of this angle plus a knowledge of the dielectric constant ε permits the determination of the electron velocity. Since $\varepsilon(\omega)$ is frequency dependent, light of different colors is emitted at somewhat different angles. This phenomenon is easily seen as a faint glow in the water around nuclear reactors. The emitted light is linearly polarized in the plane containing the particle path and the observation direction.

10 TRANSITION RADIATION

We have seen that a charged particle moving at a fast speed in a medium carries along electric and magnetic fields that are characteristic of its speed and of the medium. When such a particle passes abruptly from one medium to another, the fields must reorganize themselves during the approach to and movement past the interface. Within a certain depth D along the path at the

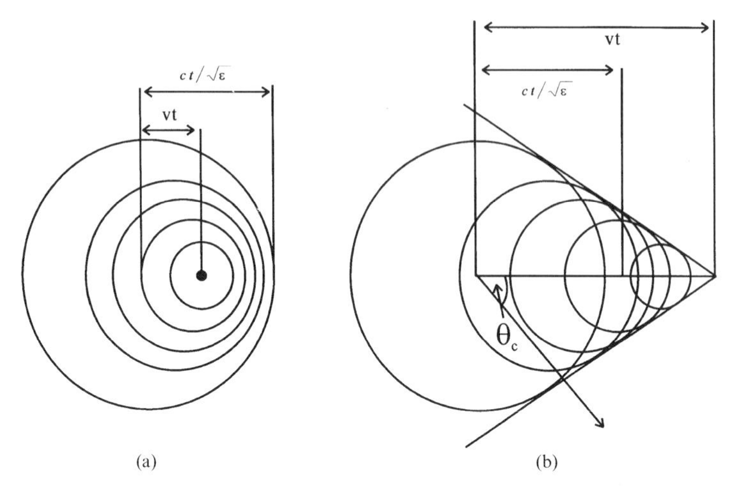

Fig. 15-7. Cherenkov radiation. Wave fronts for a charged particle moving slower than the speed of light in a medium ($v < c/n$) are shown in (a), and wave fronts for the charge exceeding the speed of light in the medium ($v > c/n$) are shown in (b). For the latter case light is emitted at the Cherenkov angle θ_c defined by Eq. (46).

interface, called the formation length, fields from different points in space combine coherently and radiate in the forward direction, mainly within an angular range $\theta \leq 1/\gamma$. The formation depth is

$$D \approx \gamma c/\omega_p \approx \gamma\lambda_p/2\pi \tag{47}$$

and the spectrum contains frequencies up to $\omega \approx \omega_p$, where ω_p is the plasma frequency, and λ_p is the plasma wavelength.

CHAPTER 16

COLLISIONS

1 INTRODUCTION

When heavy charged particles move through matter they interact with the electronic charge clouds around nuclei without much loss in energy per electron encounter, and with negligible deflection from their path. Nevertheless most of the energy lost by the particles is the accumulation from many electron encounters, although very little deviation in path results from these successive energy decrements. The major mechanism for the wide angle scattering of these particles is collisions with the more massive nuclei. A great deal of energy is lost by radiation produced by accelerations that charged particles undergo during their passage through matter. This can be the principal mechanism of energy loss for highly relativistic charged particles.

2 COULOMB COLLISION

Consider a fast moving incoming particle of charge ze and momentum $p = \gamma M v$ which approaches a stationary electron of mass m with the impact parameter b, as shown in Fig. 16-1. The electric field E_ρ of the charge, obtained from Eq. (24) in Chapter 15, gives a momentum impulse Δp to the electron in the transverse direction

$$\Delta p = \int_{-\infty}^{\infty} eE_\perp(t)dt = \frac{ze^2}{2\pi\epsilon_0 bv} \tag{1}$$

and we can estimate that this impulse also brings about the transfer of energy

$$\Delta E(b) = \frac{(\Delta p)^2}{2m} = \frac{2z^2e^4}{(4\pi\epsilon_0)^2 mv^2}\frac{1}{b^2} \tag{2}$$

to the electron. A more exact treatment gives

$$\Delta E(b) = \frac{(\Delta p)^2}{2m} = \frac{2z^2e^4}{(4\pi\epsilon_0)^2 mv^2}\frac{1}{b^2 + b_{\min}^2} \tag{3}$$

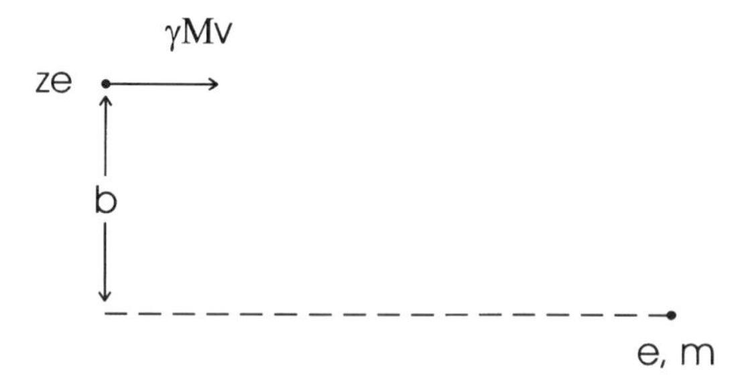

Fig. 16-1. Impact parameter b for a fast moving incoming particle of charge ze and momentum $\gamma M v$ scattering off an electron of charge e and mass m.

where the minimum impact parameter b_{min} is obtained by equating the electron energy with the Coulomb energy at the closest approach

$$\gamma m v^2 = \frac{ze^2}{4\pi\epsilon_0 b_{\text{min}}} \tag{4}$$

This is a classical result. The quantum mechanical minimum impact parameter is obtained by equating the angular momentum of the collision with Planck's constant

$$\gamma m v b_{\text{min}} = \hbar \tag{5}$$

The larger of these two minimum impact parameters (4) and (5) should be used in Eq. (3). They are the same for the speed $\beta = 2\alpha = z/137$.

When the collision time is longer than the orbital period $\tau = 1/\omega$ of the electron, the effect of the electric field of the incoming particle on the motion of the electron tends to average out, and very little energy transfer occurs. This averaging process becomes important for impact parameters greater than b_{max} defined by the expression

$$\gamma v = \omega b_{\text{max}} \tag{6}$$

Figure 16-2 shows how the energy transfer to the electron depends on the impact parameter. We see from the figure that in the middle of the range $b_{\text{min}} < b < b_{\text{max}}$ the plot of $\log \Delta E$ versus $\log b$ is a straight line of slope -2 corresponding to Eq. (2). The Coulomb collisions that occur for $b < b_{\text{max}}$ have appreciable energy transfer, whereas for larger impact parameters, $b > b_{\text{max}}$, the collisions are adiabatic with negligible energy transfer.

Experimental measurements usually determine the energy loss per unit distance dE/dx in a material, and for N atoms of atomic number Z per unit volume we can integrate over impact parameters to obtain

$$\frac{dE}{dx} = NZ \frac{z^2 e^4}{4\pi\epsilon_0^2 m v^2} \log(b_{\text{max}}/b_{\text{min}}) \tag{7}$$

Figure 16-3 shows how dE/dx varies with the ratio of the kinetic energy E_{K} to the rest energy Mc^2, where the kinetic energy E_{K} is defined as the total energy γMc^2 minus the rest energy Mc^2

$$E_{\text{K}} = (\gamma - 1)Mc^2 \tag{8}$$

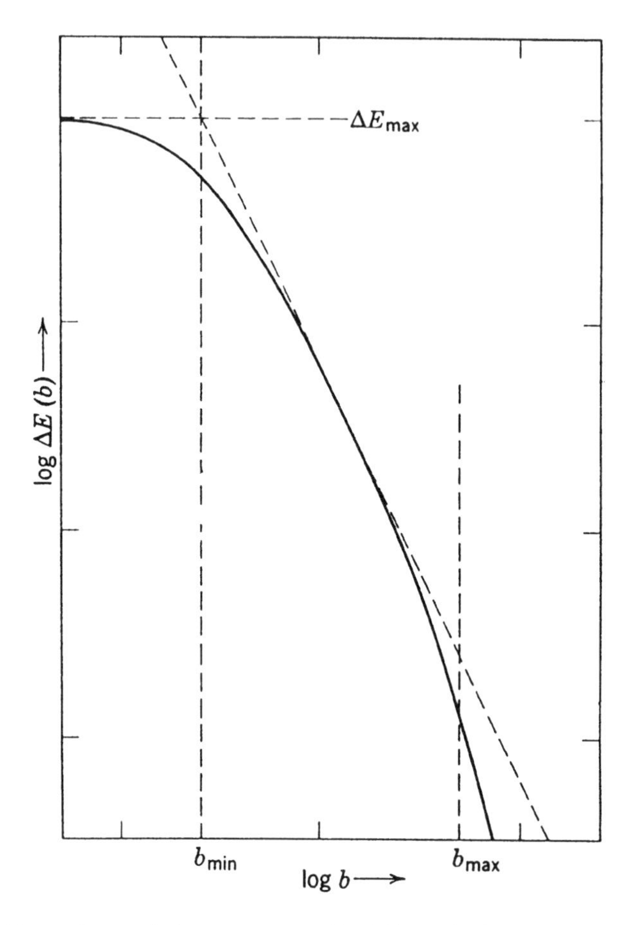

Fig. 16-2. Dependence of the energy transfer ΔE on the impact parameter b for a Coulomb collision. (From J. D. Jackson, *Classical Electrodynamics*, Wiley, New York, 1975, Fig. 13-2.)

so

$$(\gamma - 1) = E_K/Mc^2 \tag{9}$$

We see that $(\gamma - 1)$ has two limiting behaviors

$$(\gamma - 1) \approx \begin{cases} \frac{1}{2}\beta^2 & \beta \ll 1, \gamma \sim 1 \\ \gamma & \beta \sim 1, \gamma \gg 1 \end{cases} \tag{10}$$

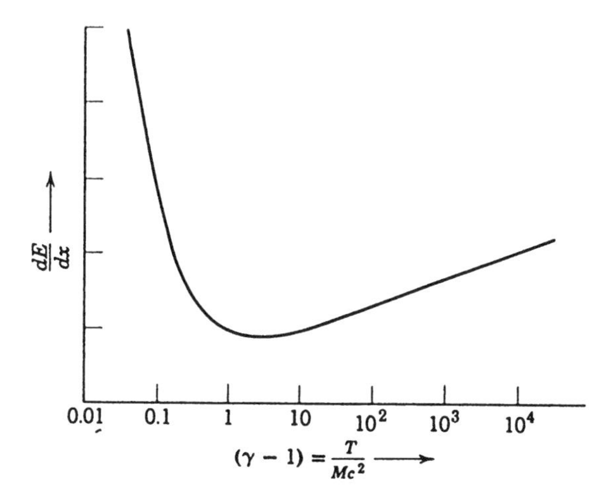

Fig. 16-3. Dependence of the energy loss in Coulomb collisions on the relativistic kinetic energy, where $E_K = Mc^2(\gamma - 1)$. (From J. D. Jackson, *Classical Electrodynamics*, Wiley, New York, 1975, Fig. 13-4.)

The figure does not take into account the so-called density effect whereby the polarization of the medium alters the particle's fields and decreases the energy loss of ultravrelativistic particles ($\gamma \gg 1$).

3 COLLISIONS WITH NUCLEI

Collisions with atoms, or more precisely with the nuclei of atoms, are the principal events responsible for the large angle scattering of incoming charged particles. Collisions with electrons determine the energy loss, but have very little effect on the scattering. Neglecting the screening effect of the atomic electrons, the angular deflection θ of the fast incoming particle of momentum $p = \gamma Mv$ and charge ze moving past a nucleus of charge Ze at an impact parameter b may be estimated as

$$\theta \approx \Delta p/p \tag{11}$$

and with the aid of Eq. (1) this gives the scattering angle θ

$$\theta = \frac{zZe^2}{2\pi\epsilon_0 pvb} \tag{12}$$

The classical differential scattering cross-section $d\sigma/d\Omega$ of Eq. (45) in Chapter 2

$$\frac{d\sigma}{d\Omega} = \frac{b}{\sin\theta}\left|\frac{db}{d\theta}\right| \tag{13}$$

has the following form for the small angle approximation $\sin\theta \cong \theta$:

$$\frac{d\sigma}{d\Omega} \approx \left(\frac{zZe^2}{2\pi\epsilon_0 pv}\right)^2 \frac{1}{\theta^4} \tag{14}$$

For non-relativistic particles and wide angles the classical Rutherford formula, Eq. (46) in Chapter 2, applies

$$\frac{d\sigma}{d\Omega} \approx \left(\frac{zZe^2}{8\pi\epsilon_0 Mv^2}\right)^2 \operatorname{cosec}^4 \tfrac{1}{2}\theta \tag{15}$$

even for quantum mechanical cases.

For large impact parameters where the scattering angle is small, the atomic electrons shield the nucleus, and a correction factor prevents the cross-section from becoming infinite as θ approaches zero

$$\frac{d\sigma}{d\Omega} \approx \left(\frac{zZe^2}{2\pi\epsilon_0 pv}\right)^2 \frac{1}{(\theta^2 + \theta_{\rm min}^2)^2} \tag{16}$$

where

$$\theta_{\rm min} = \begin{cases} \dfrac{zZe^2}{2\pi\epsilon_0 pva} & \text{classically} \\ \hbar/pa & \text{quantum mechanically} \end{cases} \tag{17}$$

where $a \approx 1.4a_0/Z^{1/3}$ is the radius of the atom and $a_0 = 4\pi\epsilon_0\hbar^2/m_e e^2$ is the Bohr radius.

For small impact parameters where the scattering is into large angles it is necessary to take into account the finite radius R of the nucleus of atomic 'weight' A

$$R = \tfrac{1}{2} r_0 A^{1/3} \tag{18}$$

where r_0 is the classical electron radius obtained by equating the electron's rest energy $m_e c^2$ and its Colulomb self-energy $e^2/4\pi\epsilon_0 r_0$

$$r_0 = \frac{e^2}{4\pi\epsilon_0 m_e c^2} = \alpha^2 a_0 = 2.28 \times 10^{-15}\,\text{m} \tag{19}$$

where $\alpha = e^2/4\pi\epsilon_0\hbar c \sim 1/137$ is the dimensionless fine structure constant. The finite nuclear radius (18) provides a maximum limit to the scattering from the quantum mechanical expression

$$\theta_{\max} = \frac{\hbar}{pR} = \frac{274}{\beta\gamma A^{1/2}} \tag{20}$$

Large angle Rutherford scattering events are rare. Most particles traversing matter emerge with small deflections arising from many successive minor Rutherford scatterings by nuclei. A particle emerging with a large deflection will most likely have only had one scattering encounter to give it that large deflection.

4 BREMSSTRAHLUNG

We have already mentioned that the principal mechanism for energy loss by particles moving through matter is interactions with atomic electrons. We have seen how energy is passed on to the electrons via the transfer of momentum through an impulse. For relatively slow moving particles, $v \ll c$, very little power P is radiated via Larmor's non-relativistic expression, Eq. (34) in Chapter 15

$$P = \frac{z^2 e^2 a^2}{6\pi\epsilon_0 c^3} \tag{21}$$

during the accelerations $\mathbf{a} = \dot{\beta} c$ that take place during the impulse. However, for relativistic particles the radiation that accompanies this impulse is the dominant mode of energy loss. This radiation is called Bremsstrahlung, the German word for breaking radiation.

The differential radiation intensity for a particle of charge ze moving at the relativistic velocity $\boldsymbol{\beta}$ and undergoing a small change in velocity $\Delta\boldsymbol{\beta}$ during the collision is, in the limit as $\omega \Rightarrow 0$, given by

$$\frac{d^2 I}{d\omega d\Omega} = \frac{z^2 e^2}{16\pi^3 \epsilon_0 c}\left[\epsilon^* \cdot \left(\frac{\Delta\boldsymbol{\beta} + \hat{\mathbf{n}} \times (\boldsymbol{\beta} \times \Delta\boldsymbol{\beta})}{(1 - \hat{\mathbf{n}} \cdot \boldsymbol{\beta})^2}\right)\right]^2 \tag{22}$$

where the radiation of polarization ϵ^* is emitted in the direction $\hat{\mathbf{n}}$. Explicit formulae (Jackson 5.10) can be obtained for polarization vectors parallel to and perpendicular to the plane containing $\boldsymbol{\beta}$ and $\hat{\mathbf{n}}$. A power series expansion shows that for radiation at small angles θ relative to the velocity direction $\boldsymbol{\beta}$ we can write for the denominator of Eq. (22)

$$1 - \hat{\mathbf{n}} \cdot \boldsymbol{\beta} = 1 - \beta \cos\theta \approx \frac{1 + \gamma^2 \theta^2}{2\gamma^2} \tag{23}$$

and the radiation is strongest for angles in the narrow range around the forward direction

$$\theta < 1/\gamma = Mc^2/E_{\mathrm{K}} \tag{24}$$

where use was made of Eq. (9) for the limit $\gamma \gg 1$.

In practice X-rays are generally produced by electron beams impinging on metallic targets such as tungsten, copper, or molybdenum. The electrons induce transitions between the low lying atomic levels of the tungsten, such as the $n = 1$ or $n = 2$ levels, and this produces sharp X-ray lines at the Rydberg energies characteristic of the atom. The incoming electrons also undergo rapid decelerations in the metal, producing a continuous background X-ray emission arising from the Bremsstrahlung radiation.

5 THOMSON SCATTERING

When a plane electromagnetic wave is incident on a free charge q of mass m the charge interacts with the electric field of the wave, is accelerated, and hence radiates. The power $dP/d\Omega$ radiated by the accelerated charge was given by Eq. (33) in Chapter 15 for a charge $q = e$

$$\frac{dP}{d\Omega} = \frac{e^2}{16\pi^2 \epsilon_0 c} \dot{\beta}^2 \sin^2\theta \tag{25}$$

Integration over θ gives Eq. (21). This can also be looked upon as a scattering of the incident radiation, and for unpolarized incident radiation we have the Thomson formula for the scattering cross-section

$$\frac{d\sigma}{d\Omega} = \tfrac{1}{2}(1 + \cos^2\theta) r_0^2 \tag{26}$$

where r_0 is the classical electron radius defined by Eq. (19). An angular integration of Eq. (26) provides the total Thomson cross-section

$$\sigma = (8\pi/3) r_0^2 \approx 0.66 \times 10^{-28}\ \mathrm{m}^2 \tag{27}$$

for the scattering of radiation by a free charge e.

CHAPTER 17

ANGULAR MOMENTUM

1 INTRODUCTION

Angular momentum is basically a classical mechanics concept, but its main importance is in the field of quantum mechanics. The operator approach is widely used in quantum mechanics where commutators and algebraic formalisms become important. In this chapter relationships involving the gradient and Laplacian operators, spherical harmonics, and symmetry will be examined. Orbital and spin angular momenta will be treated, as well as coupling schemes involving them. We will conclude with a discussion of Clebsch Gordan coefficients and irreducible tensor operators.

2 LINEAR AND ANGULAR MOMENTUM

Linear momentum p for a single particle is defined in relativity theory as

$$p = \gamma m v \tag{1}$$

where $\gamma = (1 - \beta^2)^{-1/2}$ and $\beta = v/c$, but in this chapter we treat the non-relativistic case $\beta \ll 1$ and $\gamma \approx 1$, so we have

$$p = mv \tag{2}$$

with the three cartesian components $p_i = mv_i$, $i = x, y, z$. For a free particle p_x is the momentum conjugate to the variable x.

For this same free particle in cylindrical and spherical coordinates we have the sets of conjugate variables shown in Table 17-1. Three of these conjugate momenta are angular momenta because they are conjugate to angular variables. In classical mechanics an angular momentum vector $\mathbf{L}$ may be written

$$\mathbf{L} = \mathbf{I} \cdot \omega = I\omega \tag{3}$$

where the moment of inertia I was defined in Chapter 3. Here we have, for example, $I = m\rho^2$ and $\omega = \dot{\phi}$ for p_ϕ in cylindrical coordinates, and $I = mr^2$ and $\omega = \dot{\theta}$ for p_0 in spherical coordinates.

A more basic definition of angular momentum is the cross product of the position vector $\mathbf{r}$ with the linear momentum $\mathbf{p}$

$$\mathbf{L} = \mathbf{r} \times \mathbf{p} \tag{4}$$

TABLE 17-1 Conjugate coordinates and momenta

Cylindrical coordinates		Spherical coordinates	
$\rho = (x^2 + y^2)^{1/2}$	$p_\rho = m\dot{\rho}$	$r = (x^2 + y^2 + z^2)^{1/2}$	$p_r = m\dot{r}$
$\phi = \tan^{-1}(y/x)$	$p_\phi = m\rho^2\dot{\phi}$	$\theta = \cos^{-1}(z/r)$	$p_\theta = mr^2\dot{\theta}$
z	$p_z = m\dot{z}$	$\phi = \tan^{-1}(y/x)$	$p_\phi = mr^2 \sin^2 \theta\dot{\phi}$

A torque $\mathbf{N}$ is defined by the expression

$$\mathbf{N} = \mathbf{r} \times \mathbf{F} \tag{5}$$

and we have the equations of motion for linear and angular cases

$$\frac{d\mathbf{p}}{dt} = \mathbf{F} \tag{6}$$

$$\frac{d\mathbf{L}}{dt} = \mathbf{N} \tag{7}$$

For constant mass and constant moment of inertia these equations simplify to

$$m\ddot{r} = \mathbf{F} \tag{8}$$

$$I\ddot{\theta} = \mathbf{N} \tag{9}$$

For systems of particles these expressions can be summed over the particles and over the applied and internal forces and torques. When there are no applied forces on a system the linear momentum $\mathbf{p}$ is conserved, and when there are no applied torques the angular momentum $\mathbf{L}$ is conserved.

3 MOMENTUM OPERATORS

In quantum mechanics both the linear momentum $\mathbf{p}$ and the angular momentum $\mathbf{L}$ are operators

$$\mathbf{p} = -i\hbar\nabla = -i\hbar\left\{\hat{\mathbf{i}}\frac{\partial}{\partial x} + \hat{\mathbf{j}}\frac{\partial}{\partial y} + \hat{\mathbf{k}}\frac{\partial}{\partial z}\right\} \tag{10}$$

$$\mathbf{L} = -i\hbar\mathbf{r} \times \nabla = i\hbar\left\{-\hat{\phi}_0\frac{\partial}{\partial\theta} + \hat{\theta}_0\frac{1}{\sin\theta}\frac{\partial}{\partial\phi}\right\} \tag{11}$$

and the latter is obtained from the angular part of the gradient operator in spherical coordinates

$$\nabla = \hat{\mathbf{r}}_0\frac{\partial}{\partial r} + \hat{\boldsymbol{\theta}}_0\frac{1}{r}\frac{\partial}{\partial\theta} + \hat{\boldsymbol{\phi}}_0\frac{1}{r\sin\theta}\frac{\partial}{\partial\phi} \tag{12}$$

with the aid of the unit vector relations $\hat{\mathbf{r}}_0 \times \hat{\theta}_0 = \hat{\boldsymbol{\phi}}_0$ and $\hat{\mathbf{r}}_0 \times \hat{\boldsymbol{\phi}}_0 = -\hat{\boldsymbol{\theta}}_0$, recalling that $\mathbf{r} = r\hat{\mathbf{r}}_0$. the z component of L_z is

$$L_z = -i\hbar\left\{x\frac{\partial}{\partial y} - y\frac{\partial}{\partial x}\right\} = -i\hbar\frac{\partial}{\partial\phi} \tag{13}$$

The components L_x and L_y have much more complicated forms in spherical coordinates. The square of the total angular momentum operator

$$L^2 = L_x^2 + L_y^2 + L_z^2 \tag{14}$$

given by

$$L^2 = -\hbar^2\left\{\frac{1}{\sin\theta}\frac{\partial}{\partial\theta}\left(\sin\theta\frac{\partial}{\partial\theta}\right) + \frac{1}{\sin^2\theta}\frac{\partial^2}{\partial\phi 2}\right\} \tag{15}$$

is related to the Laplacian ∇^2 in spherical coordinates

$$L^2 = -\hbar^2 r^2\nabla^2 + \hbar^2\frac{\partial}{\partial r}\left(r^2\frac{\partial}{\partial r}\right) \tag{16}$$

Another important relationship is

$$\mathbf{L} \times \mathbf{L} = i\hbar\mathbf{L} \tag{17}$$

Linear and angular momenta obey some commutator relations, where the commutator of A and B is defined by

$$[A, B] = AB - BA \tag{18}$$

For linear momentum we have the expressions

$$[q_i, q_j] = 0 \qquad [p_i, p_j] = 0 \tag{19}$$

$$[q_i, p_j] = i\hbar\delta_{ij} \tag{20}$$

Angular momentum components satisfy the following commutators:

$$[L_i, L_j] = i\hbar\varepsilon_{ijk}L_k \tag{21}$$

where the Levi–Civita symbol ε_{ijk} is +1 for an even permutation of indices, −1 for an odd permutation, and zero if any two indices are the same. Each component L_i of **L** commutes with the square of the total angular momentum L^2

$$[L^2, L_i] = 0 \tag{22}$$

The spherical harmonics $Y_{LM}(\theta, \phi)$ are eigenfunctions of the angular part of the Laplacian so they are also eigenfunctions of the angular momentum. Each spherical harmonic has the ϕ dependence $e^{iM\phi}$ and it is clear from Eq. (13) that

$$L_z Y_{LM} = M Y_{LM} \tag{23}$$

which means that the integer M is the eigenvalue of the operator L_z. The total angular momentum operator L^2 also satisfies a simple eigenvalue equation

$$L^2 Y_{LM} = L(L+1) Y_{LM} \tag{24}$$

The linear combinations L_+ and L_-, respectively, of the x and y components of **L**

$$L_+ = L_x + iL_y \qquad L_- = L_x - iL_y \tag{25a}$$

$$L_x = \tfrac{1}{2}(L_+ + L_-) \qquad L_y = -\tfrac{1}{2}i(L_+ - L_-) \tag{25b}$$

are called raising and lowering operators, or ladder operators, because they satisfy the recursion relation

$$\begin{aligned} L_\pm Y_{LM} &= [L(L+1) - M(M \pm 1)]^{1/2} Y_{LM\pm1} \\ &= [(L \mp M)(L \pm M + 1)]^{1/2} Y_{LM\pm1} \end{aligned} \tag{26}$$

with the spherical harmonics.

4 ORBITAL AND SPIN ANGULAR MOMENTUM

We have been discussing angular momentum $\hbar L$ which can be visualized as one mass moving in an orbit around another. It is confined to integral values of L which are the eigenvalues of the Laplacian operator. There is also an intrinsic or spin angular momentum $\hbar S$ which is visualized as a mass rotating on its axis, and it can assume either integral or half integral values: $S = 0, 1/2, 1, 3/2, \ldots$. When a particle has both orbital and spin motion, we can define an overall angular momentum vector **J** which is the sum of the orbital and spin components

$$\mathbf{J} = \mathbf{L} + \mathbf{S} \tag{27}$$

where $\hbar$ is omitted for convenience. The total angular momentum assumes values in the range

$$|L - S| \leq J \leq L + S \tag{28}$$

The configuration of an electron can be specified by the Russell Saunders symbol $^{2S+1}L_J$, where we use the spectroscopic notation S, P, D, F, G, H,... for orbital states $L = 0, 1, 2, 3, 4, 5, \ldots$. Thus, for the case $S = 1/2, L = 2$ we can have, via Eq. (28), the values $J = 3/2$ and $J = 5/2$, and the $J = 3/2$ configuration, for example, is designated by $^2D_{3/2}$.

We have been discussing a single electron. In a many-electron, or more generally a many-particle system, each particle has an orbital L_i and a spin S_i contribution to its individual angular momentum J_i, and there are two commonly used methods for calculating the total angular momentum **J** of the system. One method, called Russell Saunders coupling, involves calculating the total orbital and spin angular momenta for the system

$$\mathbf{L} = \sum \mathbf{L}_i \tag{29}$$

$$\mathbf{S} = \sum \mathbf{S}_i \tag{30}$$

and then adding them vectorially via Eq. (27), subject to the limitations of Eq. (28), to obtain the allowed overall angular momenta **J**.

The other method for calculating total **J**, called the *jj* coupling scheme, involves starting with the individual $\mathbf{J}_i$ values of each particle

$$\mathbf{J}_i = \mathbf{L}_i + \mathbf{S}_i \tag{31}$$

and adding them vectorially to obtain total **J**

$$\mathbf{J} = \sum \mathbf{J}_i \tag{32}$$

In practice there are several final **J** values, and they may be determined through either coupling scheme. Russell Saunders coupling is generally found in atoms, especially light atoms, while *jj* coupling is characteristic of heavy atoms and nuclei.

5 SPIN–ORBIT INTERACTION

We have been considering angular momentum coupling schemes without suggesting how they might come about. The simplest one involves the spin–orbit interaction $\lambda\mathbf{L}\cdot\mathbf{S}$, where λ is the spin–orbit coupling constant. This interaction can split a degenerate orbital state ${}^{2S+1}L$ into J states of different energy. To determine this splitting we write the vector identity

$$\mathbf{J}^2 = (\mathbf{L}+\mathbf{S})^2 = \mathbf{L}^2 + 2\mathbf{L}\cdot\mathbf{S} + \mathbf{S}^2 \tag{33}$$

We know from Eq. (24) that the operator $\mathbf{L}^2$ has the eigenvalue $L(L+1)$, and in like manner $\mathbf{S}^2$ and $\mathbf{J}^2$ have the respective eigenvalues $S(S+1)$ and $J(J+1)$, and this permits us to write from Eq. (33)

$$\lambda\mathbf{L}\cdot\mathbf{S} = \tfrac{1}{2}\lambda[J(J+1) - S(S+1) - L(L+1)] \tag{34}$$

For the particular case $L = 1$, $S = 1/2$ we have two J states with the symbols ${}^2\mathrm{P}_{3/2}$ and ${}^2\mathrm{P}_{1/2}$, and the energies

$$\lambda\mathbf{L}\cdot\mathbf{S} = \begin{cases} \tfrac{1}{2}\lambda & \text{for } J = 3/2 \quad ({}^2\mathbf{P}_{3/2},\ \text{four } \mathbf{M}_\mathrm{J} \text{ states}) \\ -\lambda & \text{for } J = 1/2 \quad ({}^2\mathbf{P}_{1/2},\ \text{two } \mathbf{M}_\mathrm{J} \text{ states}) \end{cases} \tag{35}$$

We see that the center of gravity of the energy is preserved, a result typical for splittings of angular momentum levels.

An interaction analogous to $\lambda\mathbf{L}\cdot\mathbf{S}$ is the hyperfine interaction T between an electron spin S and a nuclear spin I. This can be isotropic with the form $T\mathbf{S}\cdot\mathbf{I}$, or it can be anisotropic with the form of a tensor, and in the principal axis system we have the operator forms

$$\mathbf{S}\cdot\mathbf{T}\cdot\mathbf{I} = T_{xx}S_xI_x + T_{yy}S_yI_y + T_{zz}S_zI_z \tag{36}$$

$$= \tfrac{1}{4}T_{xx}(S_+ + S_-)(I_+ + I_-) - \tfrac{1}{4}T_{yy}(S_+ - S_-)(I_+ - I_-) + T_{zz}S_zI_z \tag{37}$$

where use was made of Eqs. (25). If orbital motion is also present, then we can define a total angular momentum **F**

$$\mathbf{F} = \mathbf{L} + \mathbf{S} + \mathbf{I} \tag{38}$$

to take this into account. There are several coupling schemes available for the determination of **F**.

6 WAVE FUNCTIONS AND CLEBSCH GORDAN COEFFICIENTS

There are two principal ways to express the wave functions of the two $^2\mathrm{P_J}$ electron configurations of Eq. (35) for which $L = 1$ and $S = 1/2$. One way is to assign eigenkets $|M_L, M_S\rangle$ to the six states in the $\mathrm{M_L, M_S}$ quantization scheme

$$|M_L, M_s\rangle : \ |-1, \tfrac{1}{2}\rangle,\ |-1, -\tfrac{1}{2}\rangle,\ |0, \tfrac{1}{2}\rangle,\ |0, -\tfrac{1}{2}\rangle,\ |1, \tfrac{1}{2}\rangle,\ |1, -\tfrac{1}{2}\rangle \tag{39}$$

and the other way is to assign eigenkets $|J, M\rangle$ in the J, M scheme, where

$$\begin{aligned} -J \le M \le +J \\ M = M_L + M_S \end{aligned} \tag{40}$$

and M is written instead of M_J. For this case we can have $J = 1/2$ or $J = 3/2$, which gives a doublet and a quartet with the six eigenkets $|J, M\rangle$

$$\begin{aligned} \text{doublet } ^2\mathrm{P}_{1/2} \quad & |\tfrac{1}{2}, -\tfrac{1}{2}\rangle,\ |\tfrac{1}{2}, \tfrac{1}{2}\rangle \\ \text{quartet } ^2\mathrm{P}_{3/2} \quad & |\tfrac{3}{2}, -\tfrac{3}{2}\rangle,\ |\tfrac{3}{2}, -\tfrac{1}{2}\rangle,\ |\tfrac{3}{2}, \tfrac{1}{2}\rangle,\ |\tfrac{3}{2}, \tfrac{3}{2}\rangle \end{aligned} \tag{41}$$

The total degeneracy is the same in both quantization schemes, so we have a total of N states given by

$$\sum (2J + 1) = (2S + 1)(2L + 1) = N \tag{42}$$

where the summation is over J states, and for the $^2\mathrm{P_J}$ case this expression gives

$$2 + 4 = 2 \times 3 = 6 \tag{43}$$

for the six states.

The wave functions $|J, M\rangle$ for the spin–orbit split states $|\frac{1}{2}, M\rangle$ and $|\frac{3}{2}, M\rangle$ can be written as linear combinations of the six $|M_L, M_S\rangle$ states

$$|J, M\rangle = \sum a_i |M_{Li}, M_{Si}\rangle \tag{44}$$

where $\sum a_i^2 = 1$, and we have for the $^2\mathrm{P}_{1/2}$ doublet case $|\frac{1}{2}, \mathrm{M}\rangle$

$$\begin{aligned} |\tfrac{1}{2}, \tfrac{1}{2}\rangle &= -\left(\tfrac{2}{3}\right)^{1/2}|1, -\tfrac{1}{2}\rangle + \left(\tfrac{1}{3}\right)^{1/2}|0, \tfrac{1}{2}\rangle \\ |\tfrac{1}{2}, -\tfrac{1}{2}\rangle &= \left(\tfrac{2}{3}\right)^{1/2}|-1, \tfrac{1}{2}\rangle - \left(\tfrac{1}{3}\right)^{1/2}|0, -\tfrac{1}{2} \end{aligned} \tag{45}$$

and for the $^2\mathrm{P}_{3/2}$ quartet case $|\frac{3}{2}, M\rangle$

$$\begin{aligned} |\tfrac{3}{2}, \tfrac{3}{2}\rangle &= |1, \tfrac{1}{2}\rangle \\ |\tfrac{3}{2}, \tfrac{1}{2}\rangle &= \left(\tfrac{1}{3}\right)^{1/2}|1, -\tfrac{1}{2}\rangle + \left(\tfrac{2}{3}\right)^{1/2}|0, \tfrac{1}{2}\rangle \\ |\tfrac{3}{2}, -\tfrac{1}{2} &= \left(\tfrac{1}{3}\right)^{1/2}|-1, \tfrac{1}{2}\rangle + \left(\tfrac{2}{3}\right)^{1/2}|0, -\tfrac{1}{2} \\ |\tfrac{3}{2}, -\tfrac{3}{2}\rangle &= |-1, -\tfrac{1}{2}\rangle \end{aligned} \tag{46}$$

The coefficients in these expressions, which we will denote by $\langle \mathrm{LSM_L M_S|JM}\rangle$, or by $\langle \mathrm{M_L M_S|JM}\rangle$ in shorthand notation, are called Clebsch Gordan coefficients. For example, we have from above

$$\begin{aligned} \langle 1\tfrac{1}{2}|\tfrac{3}{2}\tfrac{3}{2}\rangle &= 1 \\ \langle -1\tfrac{1}{2}|\tfrac{3}{2} - \tfrac{1}{2}\rangle &= 1/\sqrt{3} \end{aligned} \tag{47}$$

A coefficient is zero unless conditions (40) are fulfilled. The coefficients are all real, and satisfy the orthogonality relations

$$\begin{aligned} \sum \langle M_L M_S|JM\rangle\langle M_L M_S|J'M'\rangle &= \delta_{JJ'}\delta_{MM'} \\ \sum \langle M_L M_S|JM\rangle\langle M_L' M_S'|JM\rangle &= \delta_{M_L M_L'}\delta_{M_S M_S'} \end{aligned} \tag{48}$$

where the first summation is over M_L, M_S, and the second is over J, M. The non-zero coefficients for the present case can be written in the following convenient array, where the only states which exist are those for which $M_S = M - M_L = \pm\frac{1}{2}$:

$$
\begin{array}{lccc}
 & M_L = 1 & M_L = 0 & M_L = -1 \\
J = \frac{3}{2}, M = -\frac{3}{2} & - & - & 1 \\
J = \frac{3}{2}, M = -\frac{1}{2} & - & \left(\frac{2}{3}\right)^{1/2} & \left(\frac{1}{3}\right)^{1/2} \\
J = \frac{3}{2}, M = +\frac{1}{2} & \left(\frac{1}{3}\right)^{1/2} & \left(\frac{2}{3}\right)^{1/2} & - \\
J = \frac{3}{2}, M = +\frac{3}{2} & 1 & - & - \\
J = \frac{1}{2}, M = -\frac{1}{2} & - & -\left(\frac{1}{3}\right)^{1/2} & \left(\frac{2}{3}\right)^{1/2} \\
J = \frac{1}{2}, M = +\frac{1}{2} & -\left(\frac{2}{3}\right)^{1/2} & \left(\frac{1}{3}\right)^{1/2} & -
\end{array} \tag{49}
$$

which agrees with the coefficients in Eqs. (45) and (46). This array (49) can be summarized as follows, where M takes on values $M_L \pm \frac{1}{2}$:

$$
\begin{array}{lccc}
 & M_L = 1 & M_L = 0 & M_L = -1 \\
J = \frac{3}{2} & \left\{\frac{(\frac{1}{2}+M)(\frac{3}{2}+M)}{6}\right\}^{1/2} & \left\{\frac{(\frac{3}{2}-M)(\frac{3}{2}+M)}{3}\right\}^{1/2} & \left\{\frac{(\frac{1}{2}-M)(\frac{3}{2}-M)}{6}\right\}^{1/2} \\
J = \frac{1}{2} & -\left\{\frac{(\frac{1}{2}+M)(\frac{3}{2}-M)}{\frac{3}{2}}\right\}^{1/2} & \frac{2M}{[3]^{1/2}} & \left\{\frac{(\frac{1}{2}-M)(\frac{3}{2}+M)}{\frac{3}{2}}\right\}^{1/2}
\end{array} \tag{50}
$$

Table 17-2 presents general expressions for the Clebsch Gordan coefficients $\langle j_1 j_2 M_1 M_2 | JM \rangle$ corresponding to the cases $j_2 = 1$, and other authors have given expressions for higher order cases. Table 17-3 provides specific coefficients for particular cases. Related coefficients have been proposed by Wigner (3-j symbols) and by Racah (V coefficients).

7 IRREDUCIBLE TENSORS

Many operators in quantum mechanics can be expressed as irreducible tensor operators, sometimes called irreducible spherical tensors, with the symbol T_k^q, where

$$-k \leq q \leq +k \tag{51}$$

For example, a vector operator V_1^q with $k = 1$ and $q = 0, \pm 1$ has the irreducible form

TABLE 17-2. General forms of the Clebsch Gordan coefficients for the particular cases $j_2 = 1/2$ and $j_2 = 1$ using the Condon-Shortley sign convention

$J_2 = 1/2$		
$m_2 \backslash J =$	$j_1 + \frac{1}{2}$	$j_1 = \frac{1}{2}$
$+\frac{1}{2}$	$\sqrt{\frac{1}{2} + \frac{M}{2j_1+1}}$	$-\sqrt{\frac{1}{2} - \frac{M}{2j_1+1}}$
$-\frac{1}{2}$	$\sqrt{\frac{1}{2} - \frac{M}{2j_1+1}}$	$\sqrt{\frac{1}{2} + \frac{M}{2j_1+1}}$

$J_2 = 1$			
$m_2 \backslash J =$	$j_1 + 1$	j_1	$j_1 - 1$
$+1$	$\sqrt{\frac{\{(j_1 + M)(j_1 + M + 1)}{2(j_1 + 1)(2j_1 + 1)}}$	$-\sqrt{\frac{(j_1 + M)(j_1 - M + 1)}{2j_1(j_1 + 1)}}$	$-\sqrt{\frac{(j_1 - M)(j_1 - M + 1)}{2j_1(2j_1 + 1)}}$
0	$\sqrt{\frac{(j_1 - M + 1)(j_1 + M + 1)}{(j_1 + 1)(2j_1 + 1)}}$	$\frac{M}{\sqrt{j_1(j_1 + 1)}}$	$-\sqrt{\frac{(j_1 - M)(j_1 + M)}{j_1(2j_1 + 1)}}$
-1	$\sqrt{\frac{(j_1 - M)(j_1 - M + 1)}{2(j_1 + 1)(2j_1 + 1)}}$	$\sqrt{\frac{(j_1 - M)(j_1 + M + 1)}{2j_1(j_1 + 1)}}$	$\sqrt{\frac{(j_1 + M)(j_1 + M + 1)}{2j_1(2j_1 + 1)}}$

Source: E. U. Condon and G. H. Shortley, *The Theory of Atomic Spectra*, Cambridge University Press, 1953, p. 76. This classic treatise gives analogous expressions for $j_2 = \frac{3}{2}$ and $j_2 = 2$.

$$\begin{aligned} V_1^1 &= -\frac{V_x + iV_y}{\sqrt{2}} \\ V_1^0 &= V_z \\ V_1^{-1} &= \frac{V_x - iV_y}{\sqrt{2}} \end{aligned} \tag{52}$$

and the electric quadrupole tensor operator Q_2^q, which has $k = 2$ and $q = 0, \pm 1, \pm 2$, is given by

$$\begin{aligned} Q_2^0 &= 2A[3I_z^2 - I(I+1)] \\ Q_2^{\pm 1} &= \sqrt{6}A[I_z(I_x \pm iI_y) + (I_x \pm iI_y)I_z] \\ Q_2^{\pm 2} &= \sqrt{6}A(I_x \pm iI_y)^2 \end{aligned} \tag{53}$$

TABLE 17-3. Clebsch Gordan coefficients for various specific cases using the Condon–Shortley sign convention

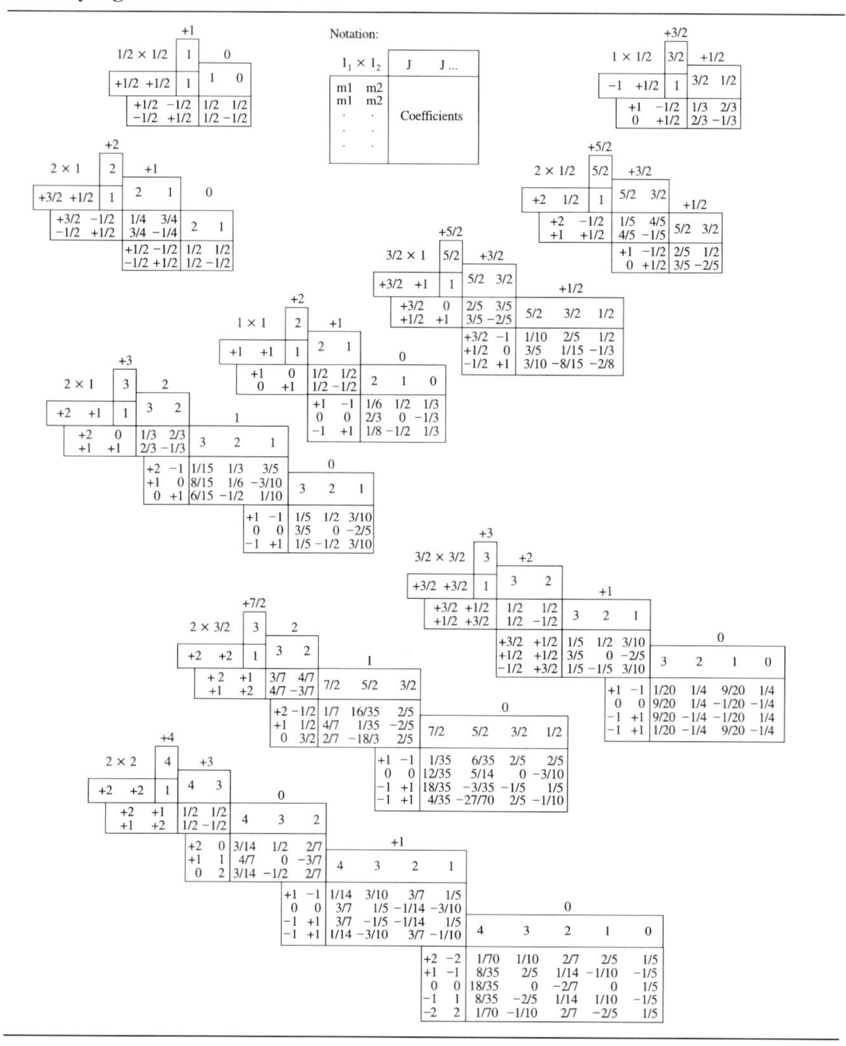

Note: To conserve space without sacrificing reliability, only the coefficients for $M \geq 1/2$ are shown. Those for $M < 0$ may be obtained from the symmetry relation

$$(J, -M|j_1, -m_1; j_2, -m_2) = (-1)^{j_1+j_2-J}(J, M|j_1, m_1; j_2, m_2)$$

The square root sign is understood for every entry, as in Table 17-2, thus $-4/5$ should be read as $-2/\sqrt{5}$.

Source: E. R. Cohen (ed.), *The Physics Quick Reference Guide*, AIP Press, New York, 1996, p. 180.

where I_i is the nuclear spin angular momentum operator and the coefficient $A = eQ/4I(2I-1)$.

An irreducible tensor operator T_k^q satisfies angular momentum like commutation rules

$$\begin{aligned}&[J_z, T_k^q] = qT_k^q \\ &[J_\pm, T_k^q] = [k(k+1) - q(q\pm 1)]^{1/2} T_k^{q\pm 1} = [(k\mp q)(k\pm q+1)]^{1/2} T_k^{q\pm 1}\end{aligned} \tag{54}$$

The Wigner–Eckart theorem tells us that the matrix element $\langle \tau JM | T_k^q | \tau' J' M' \rangle$ of an irreducible operator can be written in the form

$$\langle \tau JM | T_k^q | \tau' J' M' \rangle = \langle \tau J \| T_k \| \tau' J' \rangle \langle J' k M' q | J' k JM \rangle \tag{55}$$

where the right-hand side depends only on a Clebsch Gordan coefficient $\langle J' k M' q | j' k JM \rangle$ and a reduced matrix element $\langle \tau J \| T_k \| \tau' J' \rangle$ which is characteristic of the operator, and independent of M, M', and q. The additional quantum number τ only affects the reduced matrix element.

CHAPTER 18

MATRIX FORMULATION OF QUANTUM MECHANICS

1 INTRODUCTION

There are several ways to approach the subject of quantum mechanics. One way is in terms of differential equations, and another emphasizes operators. In this chapter we adopt a matrix approach to the subject and develop it in terms of angular momentum matrices.

2 ANGULAR MOMENTUM MATRICES

In Chapter 3 we discussed some of the properties of the Pauli spin matrices. These provide the matrices associated with the three cartesian components of a spin $\frac{1}{2}$ particle

$$J_x = \tfrac{1}{2}\begin{pmatrix} 0 & 1 \\ 1 & 0 \end{pmatrix} \qquad J_y = \tfrac{1}{2}\begin{pmatrix} 0 & -i \\ i & 0 \end{pmatrix} \qquad J_z = \tfrac{1}{2}\begin{pmatrix} 1 & 0 \\ 0 & -1 \end{pmatrix} \tag{1}$$

The matrices for higher order spins are generated with the aid of the coefficient $[(J \mp M)(J \pm M + 1)]^{1/2}$ in Eq. (26) in Chapter 17

$$J_{\pm}|M\rangle = [(J \pm M)(J \mp M + 1)]^{1/2}|M \pm 1\rangle \tag{2}$$

This coefficient permits us to generate the following triangular array which is symmetrical about its vertical axis:

$$\begin{array}{lc}
J = \frac{1}{2} & \sqrt{1} \\
J = 1 & \sqrt{2}\ \sqrt{2} \\
J = \frac{3}{2} & \sqrt{3}\ \sqrt{2 \cdot 2}\ \sqrt{3} \\
J = 2 & \sqrt{4}\ \sqrt{3 \cdot 2}\ \sqrt{2 \cdot 3}\ \sqrt{4} \\
J = \frac{5}{2} & \sqrt{5}\ \sqrt{4 \cdot 2}\ \sqrt{3 \cdot 3}\ \sqrt{2 \cdot 4}\ \sqrt{5} \\
J = 3 & \sqrt{6}\ \sqrt{5 \cdot 2}\ \sqrt{4 \cdot 3}\ \sqrt{3 \cdot 4}\ \sqrt{2 \cdot 5}\ \sqrt{6} \\
J = \frac{7}{2} & \sqrt{7}\ \sqrt{6 \cdot 2}\ \sqrt{5 \cdot 3}\ \sqrt{4 \cdot 4}\ \sqrt{3 \cdot 5}\ \sqrt{2 \cdot 6}\ \sqrt{7}
\end{array} \tag{3}$$

The row for a general spin J is

$$\sqrt{(2J) \cdot (1)}\ \sqrt{(2J - 1 \cdot (2)]}\ \sqrt{(2J - 2) \cdot (3)}$$
$$\sqrt{(2J - 3) \cdot (4)} \ldots \sqrt{(1) \cdot (2J)}$$

The numbers in this triangle constitute the coefficients once removed from the diagonal of the $J = \frac{1}{2}$ and higher order J_x and Jy matrices, with $\pm i$ inserted in the case of J_y. The J_z matrices have the M values along the diagonal. The matrices for $J = \frac{1}{2}$, 1 and $\frac{3}{2}$, taken from C. P. Poole, Jr and H. A. Farach, *Theory of Magnetic Resonance* (Wiley, New York, 1987), are as follows:

$$\text{unit matrix} \rightarrow \begin{pmatrix} 1 & 0 \\ 0 & 1 \end{pmatrix} \rightarrow \begin{pmatrix} 1 & 0 & 0 \\ 0 & 1 & 0 \\ 0 & 0 & 1 \end{pmatrix} \rightarrow \begin{pmatrix} 1 & 0 & 0 & 0 \\ 0 & 1 & 0 & 0 \\ 0 & 0 & 1 & 0 \\ 0 & 0 & 0 & 1 \end{pmatrix} \tag{4a}$$

$$\vec{J}_x \rightarrow \frac{1}{2}\begin{pmatrix} 0 & 1 \\ 1 & 0 \end{pmatrix} \rightarrow \frac{1}{2}\begin{pmatrix} 0 & \sqrt{2} & 0 \\ \sqrt{2} & 0 & \sqrt{2} \\ 0 & \sqrt{2} & 0 \end{pmatrix} \rightarrow \frac{1}{2}\begin{pmatrix} 0 & \sqrt{3} & 0 & 0 \\ \sqrt{3} & 0 & 2 & 0 \\ 0 & 2 & 0 & \sqrt{3} \\ 0 & 0 & \sqrt{3} & 0 \end{pmatrix} \tag{4b}$$

$$\vec{J}_y \rightarrow \frac{1}{2}\begin{pmatrix} 0 & -i \\ i & 0 \end{pmatrix} \rightarrow \frac{1}{2}\begin{pmatrix} 0 & -i\sqrt{2} & 0 \\ i\sqrt{2} & 0 & -i\sqrt{2} \\ 0 & i\sqrt{2} & 0 \end{pmatrix} \rightarrow \frac{1}{2}\begin{pmatrix} 0 & -\sqrt{3}i & 0 & 0 \\ \sqrt{3}i & 0 & -2i & 0 \\ 0 & 2i & 0 & -\sqrt{3}i \\ 0 & 0 & \sqrt{3}i & 0 \end{pmatrix} \tag{4c}$$

$$\vec{J}_z \rightarrow \frac{1}{2}\begin{pmatrix} 1 & 0 \\ 0 & -1 \end{pmatrix} \rightarrow \begin{pmatrix} 1 & 0 & 0 \\ 0 & 0 & 0 \\ 0 & 0 & -1 \end{pmatrix} \rightarrow \frac{1}{2}\begin{pmatrix} 3 & 0 & 0 & 0 \\ 0 & 1 & 0 & 0 \\ 0 & 0 & -1 & 0 \\ 0 & 0 & 0 & -3 \end{pmatrix} \tag{4d}$$

$$\vec{J}^2 \rightarrow \frac{3}{4}\begin{pmatrix} 1 & 0 \\ 0 & 1 \end{pmatrix} \rightarrow 2\begin{pmatrix} 1 & 0 & 0 \\ 0 & 1 & 0 \\ 0 & 0 & 1 \end{pmatrix} \rightarrow \frac{15}{4}\begin{pmatrix} 1 & 0 & 0 & 0 \\ 0 & 1 & 0 & 0 \\ 0 & 0 & 1 & 0 \\ 0 & 0 & 0 & 1 \end{pmatrix} \tag{4e}$$

$$\vec{J}^+ \rightarrow \begin{pmatrix} 0 & 1 \\ 0 & 0 \end{pmatrix} \rightarrow \begin{pmatrix} 0 & \sqrt{2} & 0 \\ 0 & 0 & \sqrt{2} \\ 0 & 0 & 0 \end{pmatrix} \rightarrow \begin{pmatrix} 0 & \sqrt{3} & 0 & 0 \\ 0 & 0 & 2 & 0 \\ 0 & 0 & 0 & \sqrt{3} \\ 0 & 0 & 0 & 0 \end{pmatrix} \tag{4f}$$

$$\vec{J}^- \rightarrow \begin{pmatrix} 0 & 0 \\ 1 & 0 \end{pmatrix} \rightarrow \begin{pmatrix} 0 & 0 & 0 \\ \sqrt{2} & 0 & 0 \\ 0 & \sqrt{2} & 0) \end{pmatrix} \rightarrow \begin{pmatrix} 0 & 0 & 0 & 0 \\ \sqrt{3} & 0 & 0 & 0 \\ 0 & 2 & 0 & 0 \\ 0 & 0 & \sqrt{3} & 0 \end{pmatrix} \tag{4g}$$

These angular momentum matrices are hermitian, $M_{ij} = M_{ji}^*$, and have zero trace. Also shown here are the corresponding unit matrices, and the matrices for the raising and lowering operators, J_+ and J_-, respectively, defined by the expressions

$$\begin{aligned} J_+ &= J_x + iJ_y \\ J_- &= J_x - iJ_y \end{aligned} \tag{5}$$

They are real matrices obtained from J_x and J_y by matrix addition, as explained in Chapter 27, Section 5, and may also be constructed from the triangle, (3). We see that the matrix for J^2 obtained from the expression

$$J^2 = J_x^2 + J_y^2 + J_z^2 \tag{6}$$

by matrix multiplication [i.e., $J_x^2 = (J_x)(J_x)$] followed by matrix addition (6) equals the unit matrix times $J(J+1)$, which is the eigenvalue for J^2. The angular momentum matrices obey the standard commutation rules

$$[J_i, J_j] = iJ_k \quad (i, j, k \text{ cyclic}) \tag{7}$$

$$[J^2, J_k] = 0 \quad (k = x, y, z) \tag{8}$$

as can be demonstrated by carrying out some matrix multiplications.

3 ADDITION OF ANGULAR MOMENTUM

In Chapter 17 we discussed the addition of two angular momenta

$$\mathbf{J} = \mathbf{J}_1 + \mathbf{J}_2 \tag{9}$$

and in the matrix representation this is carried out with the aid of the direct product expansion described in Section 5 of Chapter 27. For each cartesian component, for example

$$J_x = J_{1x} + J_{2x} \tag{10}$$

we form the direct products with the aid of unit matrices U_1 and U_2 by writing

$$J_x = (J_{1x}) \times (U_2) + (U_1) \times (J_{2x}) \tag{11}$$

where U_i denotes a unit matrix the size of J_{xi}. Expanding for the case $J_1 = \frac{1}{2}$ and $J_2 = \frac{1}{2}$ we have

$$J_x = \frac{1}{4}\begin{pmatrix} 0 & 1 \\ 1 & 0 \end{pmatrix} \times \begin{pmatrix} 1 & 0 \\ 0 & 1 \end{pmatrix} + \frac{1}{4}\begin{pmatrix} 1 & 0 \\ 0 & 1 \end{pmatrix} \times \begin{pmatrix} 0 & 1 \\ 1 & 0 \end{pmatrix} \tag{12}$$

and carrying out the direct product expansion first and then the matrix addition we obtain

$$J_x = \frac{1}{4}\begin{pmatrix} 0 & 0 & 0 & 1 \\ 0 & 0 & 1 & 0 \\ 0 & 1 & 0 & 0 \\ 1 & 0 & 0 & 0 \end{pmatrix} + \frac{1}{4}\begin{pmatrix} 0 & 1 & 0 & 0 \\ 1 & 0 & 0 & 0 \\ 0 & 0 & 0 & 1 \\ 0 & 0 & 1 & 0 \end{pmatrix} = \frac{1}{4}\begin{pmatrix} 0 & 1 & 0 & 1 \\ 1 & 0 & 1 & 0 \\ 0 & 1 & 0 & 1 \\ 1 & 0 & 1 & 0 \end{pmatrix} \tag{13}$$

The 4×4 matrix on the right is the direct product matrix representation for J_x.

4 ZEEMAN EFFECT OF HYDROGEN ATOM

As an example of the direct product formalism that we have presented, consider a hydrogen atom in its ground electronic state with $L = 0$ and $S = \frac{1}{2}$. There is also a nuclear spin $I = \frac{1}{2}$ on the proton, so in our notation we are dealing with the system $J_1 = S = \frac{1}{2}$ and $J_2 = I = \frac{1}{2}$. We write the Hamiltonian for the high field Zeeman effect in which we assume quantization of the spins along the magnetic field direction B_z.

The Hamiltonian $\mathcal{H}$ for the electronic spin magnetic moment $g\mu_B\mathbf{S}$ and the nuclear spin magnetic moment $-g_N\mu_S\mathbf{I}$ in the magnetic field $\mathbf{B} = \mathbf{k}B$ is

$$\mathcal{H} = g\mu_B S_z B - g_N\mu_N I_z B + T\mathbf{S} \cdot \mathbf{I} \tag{14}$$

where the first two terms on the right are the electronic and nuclear Zeeman terms, respectively, with $g\mu_B/g\mu_N = 660$, and $T\mathbf{S} \cdot \mathbf{I}$ is the hyperfine interaction between the electronic and nuclear spins that was introduced in Section 5 of Chapter 17. The Hamiltonian in direct product matrix form is

$$\begin{aligned}\mathcal{H} = g\mu_B(S_z) \times (U) - g_N\mu_N B(U) \times (I_z) + T[(S_x) \times (I_x) \\ + (S_y) \times (I_y) + (S_z) \times (I_z)]\end{aligned} \tag{15}$$

where the matrices are indicated by parentheses, and U denotes the 2×2 unit matrix. We see that each S_i matrix is on the left of a direct product pair, and each I_i matrix is on the right. When one of these is missing, as in the Zeeman terms, it is replaced by a unit matrix, as shown. When we insert the actual matrices, we have

$$\begin{aligned}\mathcal{H} = \tfrac{1}{2}g\mu_B B\begin{pmatrix}1 & 0\\ 0 & 1\end{pmatrix} \times \begin{pmatrix}1 & 0\\ 0 & 1\end{pmatrix} - \tfrac{1}{2}g_N\mu_N B\begin{pmatrix}1 & 0\\ 0 & 1\end{pmatrix} \times \begin{pmatrix}1 & 0\\ 0 & -1\end{pmatrix}\\ + \tfrac{1}{4}T\left[\begin{pmatrix}0 & 1\\ 1 & 0\end{pmatrix} \times \begin{pmatrix}0 & 1\\ 1 & 0\end{pmatrix} + \begin{pmatrix}0 & -i\\ i & 0\end{pmatrix} \times \begin{pmatrix}0 & -i\\ i & 0\end{pmatrix}\right.\\ \left. + \begin{pmatrix}1 & 0\\ 0 & -1\end{pmatrix} \times \begin{pmatrix}1 & 0\\ 0 & -1\end{pmatrix}\right]\end{aligned} \tag{16}$$

Working out the direct products and matrix additions we obtain the following 4×4 Hamiltonian matrix:

$$\begin{pmatrix}\tfrac{1}{2}g\mu_B B - \tfrac{1}{2}g_N\mu_N B + \tfrac{1}{4}T & 0 & 0 & 0\\ 0 & \tfrac{1}{2}g\mu_B B + \tfrac{1}{2}g_N\mu_N B - \tfrac{1}{4}T & \tfrac{1}{2}T & 0\\ 0 & \tfrac{1}{2}T & -\tfrac{1}{2}g\mu_B B - \tfrac{1}{2}g_N\mu_N B - \tfrac{1}{4}T & 0\\ 0 & 0 & 0 & -\tfrac{1}{2}g\mu_B B + \tfrac{1}{2}g_N\mu_N B + \tfrac{1}{4}T\end{pmatrix} \tag{17}$$

which gives two energies directly, and a 2×2 submatrix equivalent to a quadratic equation to solve for the other two energies. The result is

$$\begin{aligned}E_1 &= \tfrac{1}{2}g\mu_B B - \tfrac{1}{2}g_N\mu_N B + \tfrac{1}{4}T\\ E_2 &= -\tfrac{1}{4}T + \tfrac{1}{2}[(g\mu_B + g_N\mu_N B)^2 B^2 + T^2]^{1/2}\\ E_3 &= -\tfrac{1}{4}T - \tfrac{1}{2}[(g\mu_B + g_N\mu_N B)^2 B^2 + T^2]^{1/2}\\ E_4 &= -\tfrac{1}{2}g\mu_B B + \tfrac{1}{2}g_N\mu_N B + \tfrac{1}{4}T\end{aligned} \tag{18}$$

The wave functions for the first and fourth energies are known exactly, whereas the other two states have mixed wave functions

$$\begin{aligned}|E_1\rangle &= |\tfrac{1}{2}\tfrac{1}{2}\rangle\\ |E_2\rangle &= \alpha|\tfrac{1}{2} - \tfrac{1}{2}\rangle + \gamma| - \tfrac{1}{2}\tfrac{1}{2}\rangle\\ |E_3\rangle &= \gamma^*|\tfrac{1}{2} - \tfrac{1}{2}\rangle - \alpha^*| - \tfrac{1}{2}\tfrac{1}{2}\rangle\\ |E_1\rangle &= | - \tfrac{1}{2} - \tfrac{1}{2}\rangle\end{aligned} \tag{19}$$

with the normalization condition

$$\alpha\alpha^* + \gamma\gamma^* = 1 \tag{20}$$

The quantities α and γ are Clebsch Gordan coefficients, and they are also the coefficients in the unitary matrix U that diagonalizes the Hamiltonian matrix

$$U = \begin{pmatrix} 1 & 0 & 0 & 0 \\ 0 & \alpha & \gamma & 0 \\ 0 & \gamma^* & -\alpha^* & 0 \\ 0 & 0 & 0 & 1 \end{pmatrix} \tag{21}$$

Figures 18-1 and 18-2 show the dependence of the energies E_i and the wavefunction coefficients α and γ on the ratio $g\mu_B B/T$. We see that near zero field α and γ approach $\sqrt{1/2}$, and at high field $\alpha \Rightarrow 1$ and $\gamma \ll 1$. The

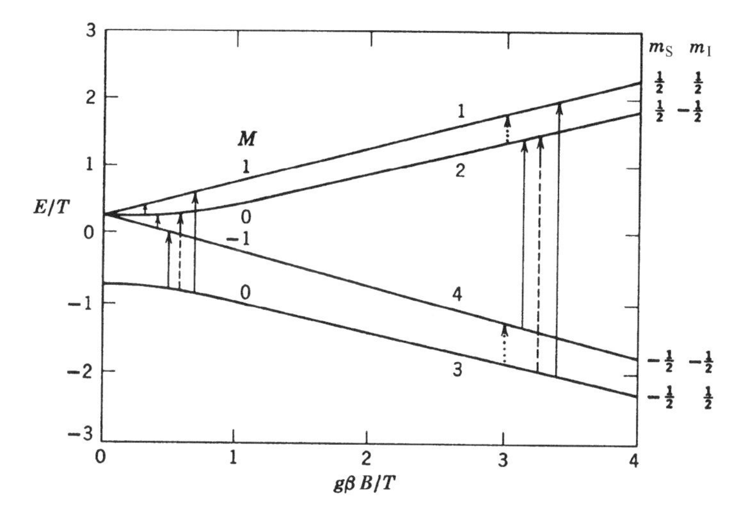

Fig. 18-1. Zeeman energies (Eqs. (18)] of a hydrogen atom as a function of an applied magnetic field B. The strong allowed (—) and weak forbidden (- - -) electron spin resonance (ESR) transitions as well as the high field nuclear magnetic resonance (NMR, $\cdots$) transitions are indicated. The low field lines are labeled by the $|J, M\rangle$ wave function scheme, and the high field lines by the $|m_S, m_I\rangle$ scheme, where $M = m_S + m_I$. On the figure the symbol β is used for the Bohr magneton μ_B. (From C. P. Poole, Jr and H. A. Farach, *Theory of Magnetic Resonance*, 2nd Ed, Wiley, New York, 1987, p. 48.)

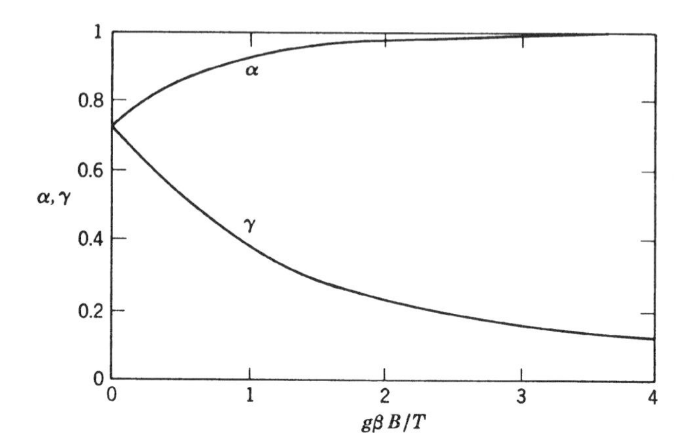

Fig. 18-2. Magnitudes of the wave function coefficients α and γ of Eqs. (19) as a function of the dimensionless ratio $g\mu_B B/T$. On the figure the symbol β is used for the Bohr magneton μ_B. (From C. P. Poole, Jr and H. A. Farach, *Theory of Magnetic Resonance*, 2nd Ed, Wiley, New York, 1987, p. 51.)

selection rules for allowed electron spin resonance (ESR) transitions at low field (levels $3 \Rightarrow 4$ and $3 \Rightarrow 1$) are $\Delta M = \pm 1$, and at high field ($3 \Rightarrow 1$ and $4 \Rightarrow 2$) they are $\Delta m_S = \pm 1$, $\Delta m_I = 0$, where $M = m_S + m_I$. For nuclear magnetic resonance (NMR) transitions ($2 \Rightarrow 1$ and $3 \Rightarrow 4$) the selection rules are $\Delta m_S = 0$, $\Delta m_I = \pm 1$.

5 GENERAL HAMILTONIAN MATRIX

We have been discussing the formation of Hamiltonian matrices by the direct product method. When we wish to form such a matrix in a more general case we will usually begin with a Hamiltonian $\mathcal{H}$ consisting of a dominant part $\mathcal{H}_0$ for which we already know the eigenfunctions $\phi_{oj} = |j\rangle$ and the eigenvalues E_{oj}, and a smaller part $\mathcal{H}'$ to be evaluated

$$\mathcal{H} = \mathcal{H}_0 + \mathcal{H}' \tag{22}$$

On this basis $|j\rangle$ the dominant or zero order Hamiltonian $\mathcal{H}_0$ is diagonal

$$\langle \mathrm{i}|\mathcal{H}_0|\,\mathrm{j}\rangle = E_{0j}\delta_{ij} \tag{23}$$

and it is necessary to calculate the diagonal and the off-diagonal matrix elements $\langle i|\mathcal{H}'|j\rangle$ of the secondary Hamiltonian $\mathcal{H}'$. The matrix for the total Hamiltonian $\mathcal{H}$ is as follows, assuming that there are only four terms in the basis set:

$$\begin{pmatrix} E_{01} + \langle 1|\mathcal{H}'|1\rangle & \langle 1|\mathcal{H}'|2\rangle & \langle 1|\mathcal{H}'|3\rangle & \langle 1|\mathcal{H}'|4\rangle \\ \langle 2|\mathcal{H}'|1\rangle & E_{02} + \langle 2|\mathcal{H}'|2\rangle & \langle 2|\mathcal{H}'|3\rangle & \langle 2|\mathcal{H}'|4\rangle \\ \langle 3|\mathcal{H}'|1\rangle & \langle 3|\mathcal{H}'|2\rangle & E_{03} + \langle 3|\mathcal{H}'|3\rangle & \langle 3|\mathcal{H}'|4\rangle \\ \langle 4|\mathcal{H}'|1\rangle & \langle 4|\mathcal{H}'|2\rangle & \langle 4|\mathcal{H}'|3\rangle & E_{04} + \langle 4|\mathcal{H}'|4\rangle \end{pmatrix} \tag{24}$$

where use was made of Eq. (23). This hermitian matrix can be diagonalized by a unitary transformation, or it can be solved as a determinant by putting $-\lambda$ on the diagonals, and this is equivalent to solving a quartic equation. Perturbation theory can also be applied to obtain an approximate solution, as will be explained in Chapter 22. Quite commonly the trace will be zero for the zero order energies

$$\sum E_{0i} = E_{01} + E_{02} + E_{03} + E_{04} = 0 \tag{25}$$

and also for the first-order (diagonal) perturbation terms

$$\sum \langle \mathrm{i}|\mathcal{H}'|\mathrm{i}\rangle = \langle 1|\mathcal{H}'|1\rangle + \langle 2|\mathcal{H}'|2\rangle + \langle 3|\mathcal{H}'|3\rangle + \langle 4|\mathcal{H}'|4\rangle = 0 \tag{26}$$

and this can be of help in finding the solution. The off-diagonal terms $\langle \mathrm{i}|\mathcal{H}'|\mathrm{j}\rangle$ provide second-order corrections to the energy.

CHAPTER 19

SCHRÖDINGER EQUATION

1 INTRODUCTION

One of the principal ways to do quantum mechanics is to write the Hamiltonian $\mathcal{H}$ for a system, express it in operator form, and then solve the associated differential equation, called the Schrödinger equation. We start with a brief discussion of time dependence, and then confine our attention to the time independent Schrödinger equation. The capabilities of this equation are illustrated by examining the square well and the harmonic oscillator potentials in three dimensions. In several places we draw upon results presented in other chapters. We conclude with a discussion of orbital degeneracies.

2 THE HAMILTONIAN

We begin with a time dependent wavefunction $\Psi(\mathbf{r}, t)$ which is an eigenfunction of the Hamiltonian $\mathcal{H}(\mathbf{r}, t)$ and satisfies the time dependent Schrödinger equation

$$\mathcal{H}\psi = i\hbar\frac{\partial}{\partial t}\psi \tag{1}$$

The absolute value squared of the wavefunction $|\psi|^2$ is the probability density. It is associated with a probability current density $\mathbf{J}_\Psi$, and together they satisfy a continuity equation for the conservation of probability

$$\frac{\partial}{\partial t}|\Psi|^2 + \nabla \cdot \mathbf{J}_\psi = 0 \tag{2}$$

This current density $\mathbf{J}_\psi$ is related to the wavefunction Ψ through the velocity operator $-i(\hbar/m)\nabla$ as follows:

$$\mathbf{J}_\psi = \frac{\hbar}{2im}[\psi^*\nabla\psi - (\nabla\psi^*)\psi] \tag{3}$$

The time dependence of an operator $\mathbf{A}$ is related to the Hamiltonian $\mathcal{H}$ through the following expression:

$$\frac{d\mathbf{A}}{dt} = \frac{\partial \mathbf{A}}{\partial t} - \frac{1}{i\hbar}[\mathcal{H}, \mathbf{A}] \tag{4}$$

where the notation $[\mathcal{H}, \mathbf{A}]$ denotes the commutator $\mathcal{H}\mathbf{A} - \mathbf{A}\mathcal{H}$. The expectation value $\langle \mathbf{A} \rangle$ of an operator may be obtained by integration over space

$$\langle \mathbf{A} \rangle = \int \psi^* \mathbf{A} \psi d\tau \tag{5}$$

To obtain the time independent Schrödinger equation we assume an harmonic time dependence $\psi(\mathbf{r}, t) = \psi(\mathbf{r})e^{-iEt/\hbar}$ in Eq. (1) which gives

$$\mathcal{H}\psi = E\psi \tag{6}$$

For a single particle the Hamiltonian becomes

$$\mathcal{H} = \frac{p^2}{2m} + V(\mathbf{r}) \tag{7}$$

and writing the momentum p in operator form $\mathbf{p} = -i\hbar\nabla$ provides the Schrödinger equation in its usual form

$$-\frac{\hbar^2}{2m}\nabla^2\psi + V(\mathbf{r})\psi = E\psi \tag{8}$$

which is solved to obtain the energies and the eigenfunctions. In the present chapter we are particularly interested in potentials $V(r)$ which are functions of the radial variable r in spherical polar coordinates

$$r = (x^2 + y^2 + z^2)^{1/2} \tag{9}$$

Examples are the Coulomb $-Ze^2/4\pi\varepsilon_0 r$, the isotropic three-dimensional harmonic oscillator $\frac{1}{2}kr^2$, and the three-dimensional square well potentials. The next chapter deals with various types of one-dimensional potentials.

3 LAPLACIAN OPERATOR AND RADIAL EQUATION

Margeneau and Murphy, in their monograph *The Mathematics of Physics and Chemistry*, (D. van Nostrand, NY, 1943), mention 11 coordinate systems in which the Schrödinger equation is separable so the solution can be written as the product of individual solutions, one for each variable. These include, of course, cartesian, polar, and cylindrical coordinates. We examine the spherical polar case. For the present, spin will not be taken into account.

Chapter 28 gives the Laplacian operator ∇^2 in spherical coordinates, and substituting this into the Schrödinger equation, (Eq. 8), gives

$$-\frac{\hbar^2}{2m}\left[\frac{1}{r^2}\frac{\partial}{\partial r}\left(r^2\frac{\partial}{\partial r}\right) + \frac{1}{r^2\sin\theta}\frac{\partial}{\partial\theta}\left(\sin\theta\frac{\partial}{\partial\theta}\right) + \frac{1}{r^2\sin^2\theta}\frac{\partial^2}{\partial\phi^2}\right]\psi + V(r)\psi = E\psi \tag{10}$$

The wave function is separable, i.e., $\psi(r, \theta, \phi) = R(r)Y_{L,M}(\theta, \phi)$, where the angular part $Y_{L,M}(\theta, \phi)$ is a spherical harmonic which is a product of an associated Legendre polynomial $P_L^M(\cos\theta)$ and the exponential function $e^{iM\phi}$. Substituting these into Eq. (10) gives the radial equation

$$-\frac{\hbar^2}{2m}\left[\frac{1}{r^2}\frac{\partial}{\partial r}\left(r^2\frac{\partial}{\partial r}\right)-\frac{L(L+1)}{r^2}\right]R+V(r)R=ER \tag{11}$$

where $R(r)$ is the radial wave function. The solutions of this equation depend on the form of the potential $V(r)$.

As an example consider the hydrogen atom with the Coulomb potential

$$V(r)=-\frac{Ze^2}{4\pi\varepsilon_0 r} \tag{12}$$

We explain in Chapter 21 that the wave function solutions are products of the exponential term $e^{-1/2\rho}$ and a Laguerre polynomial, where $\rho = 2Zr/na_0$, n is the principal quantum number, and $a_0 = 4\pi\varepsilon_0\hbar^2/me^2$ is the Bohr radius. The energies

$$E_n=-\frac{Z^2me^4}{8\varepsilon_0^2h^2}\frac{1}{n^2} \tag{13}$$

are negative and depend inversely on the square of the quantum number n. We show in Chapter 21 that there is also a second-order dependence of the energy on the orbital quantum number L. For each n level there are L values in the range

$$L=0,1,\ldots,n-1 \tag{14}$$

so the n levels do not have orbital parity.

4 THREE-DIMENSIONAL SQUARE WELL

The simplest radial potential is the three-dimensional square well for which $V(r) = -V_0$ inside and $V(r) = 0$ outside

$$V(r)=\begin{cases}-V_0 & r<a\\ 0 & r>a\end{cases} \tag{15}$$

The radial part of the Schrödinger equation in these two regions becomes

$$\frac{\hbar^2}{2m}\left[\frac{1}{r^2}\frac{\partial}{\partial r}\left(r^2\frac{\partial}{\partial r}\right)-\frac{L(L+1)}{r^2}\right]R+(V_0+E)R=0 \quad r<a \tag{16}$$

$$\frac{\hbar^2}{2m}\left[\frac{1}{r^2}\frac{\partial}{\partial r}\left(r^2\frac{\partial R}{\partial r}\right)-\frac{L(L+1)}{r^2}\right]R+ER=0 \quad r>a \tag{17}$$

To simplify this equation it is convenient to define the quantities

$$\alpha=\left(\frac{2m(V_0-|E|)}{\hbar^2}\right)^{1/2} \tag{18}$$

$$\beta=\left(\frac{2m|E|}{\hbar^2}\right)^{1/2} \tag{19}$$

where the absolute value sign is used because the energy E is negative.

The solutions inside the well are spherical Bessel functions $j_L(\alpha r)$ which are related to ordinary Bessel functions $J_{L+1/2}(\alpha r)$ through the expression

$$j_L(r)=[\pi/2\alpha r]^{1/2}J_{L+1/2}(\alpha r) \tag{20}$$

and the two lowest order ones are

$$j_0(\alpha r)=\frac{\sin\alpha r}{\alpha r} \qquad j_1(\alpha r)=\frac{\sin\alpha r}{\alpha^2 r^2}-\frac{\cos\alpha r}{\alpha r} \tag{21}$$

The solutions outside the well are spherical Hankel functions h_L which decay with distance faster than $e^{-\beta r}$. These inside and outside solutions are matched at the boundary of the well by requiring that $(1/R)dR/dr$ be continuous at $r=a$, and the result is a transcendental equation which must be solved numerically. For example, for $L=0$ we have

$$\alpha\cot\alpha a=-\beta \tag{22}$$

For a shallow well

$$V_0<\pi^2\hbar^2/2ma^2 \tag{23}$$

there is only one solution for $L = 0$ and none for higher L values. The deeper the well, i.e. the greater the magnitude of V_0, the larger the number of solutions. This behavior is analogous to that of the one-dimensional square well that will be examined in the next chapter.

5 INFINITE SQUARE WELL

For a very deep well, $V_0 \gg \pi^2\hbar^2/2ma^2$, it is more convenient to take the zero of energy at the bottom, corresponding to

$$V(r) = \begin{cases} 0 & r < a \\ +V_0 & r > a \end{cases} \tag{24}$$

and for this case α and β become

$$\alpha = \left(\frac{2mE}{\hbar^2}\right)^{1/2} \tag{25}$$

$$\beta = \left(\frac{2m(V_0 - E)}{\hbar^2}\right)^{1/2} \tag{26}$$

where the energy E is now positive.

If we let the potential V_0 become arbitrarily large, then β does also ($E \ll V_0$), the wave function $\approx e^{-\beta r}$ drops precipitously to zero immediately outside the well, and the boundary condition requires the interior solution to vanish at the edge of the well

$$j_L(\alpha a) = 0 \tag{27}$$

Therefore αa corresponds to a root of j_L

$$\alpha a = \gamma_{Ln} \tag{28}$$

where γ_{Ln} is the nth root of the Lth order spherical Bessel function. The energies from Eq. (25) are

$$E_{Ln} = \frac{\hbar^2 \gamma_{Ln}^2}{2ma^2} \tag{29}$$

For the $L = 0$ state we see from Eqs. (21) and (28) that the root γ_{on} is

$$\gamma_{on} = n\pi \tag{30}$$

to give for the energy

$$E_{on} = \frac{\pi^2 \hbar^2 n^2}{2ma^2} \tag{31}$$

The situation for higher L can be examined with the aid of the asymptotic behavior of $j_L(\alpha r)$ in the limit $n \gg L$

$$j_L(\alpha r) \sim \frac{1}{\alpha r} \sin\left(\alpha r - \frac{L\pi}{2}\right) \tag{32a}$$

$$\sim \frac{1}{\alpha r}\left[\sin(\alpha r)\cos\left(\frac{L\pi}{2}\right) - \cos(\alpha r)\sin\left(\frac{L\pi}{2}\right)\right] \tag{32b}$$

which has the roots

$$\gamma_{Ln} = \frac{1}{2} L\pi + \pi(\text{integer}) \tag{33}$$

Thus there is an odd-L case and an even-L case.

6 THREE-DIMENSIONAL HARMONIC OSCILLATOR

The three-dimensional harmonic oscillator has the potential $\frac{1}{2}m\omega^2 r^2$ which is independent of angle, and the Schrödinger equation is separable in spherical coordinates. The wave function is the product of a radial part $R(r)$ and a spherical harmonic $Y_{LM}(\theta, \phi)$, as in the square well case. The radial equation is the same as (16) with the harmonic oscillator potential inserted in place of $-V_0$

$$\frac{\hbar^2}{2m}\left[\frac{1}{r^2}\frac{\partial}{\partial r}\left(r^2 \frac{\partial R}{\partial r}\right) - \frac{L(L+1)}{r^2}\right]R + \left(-\frac{1}{2}m\omega^2 r^2 + E\right)R = 0 \tag{34}$$

Instead of solving this equation we discuss its solutions and then compare them with the solutions obtained in cartesian coordinates.

Solutions of the radial equation (34) involve powers of r times the factor $e^{-\Gamma r^2/2}$, where

$$\Gamma = m\omega/\hbar \tag{35}$$

ω being the classical oscillator frequency, and the energy has the form

$$E = (n + \tfrac{3}{2})\hbar\,\omega \tag{36}$$

where n is zero or a positive integer, $n = 0, 1, 2, \ldots$. These harmonic oscillator energies are independent of the orbital quantum number L. The harmonic oscillator has the same parity as the square well, with the L values given by

$$\begin{aligned} L &= 0, 2, \ldots, n \quad n \text{ even} \\ L &= 1, 3, \ldots, n \quad n \text{ odd} \end{aligned} \tag{37}$$

The energy level numbering scheme is displaced by 1 relative to the square well case. Again the lowest level is an s-state ($L = 0$), the next higher level is a p-state ($L = 1$), the third ($n = 2$) has s- and d-states, the fourth ($n = 3$) has p and f states, etc.

The harmonic oscillator potential written in cartesian coordinates

$$V(r) = \tfrac{1}{2}m\omega^2 r^2 = V(x, y, z) = \tfrac{1}{2}m\omega^2(x^2 + y^2 + z^2) \tag{38}$$

has a Schrödinger equation separable into three uncoupled equations, one for each cartesian coordinate x, y, z, all of the form

$$-\frac{\hbar^2}{2m}\frac{\partial^2\Psi}{\partial x^2} + \tfrac{1}{2}m\omega^2 x^2\Psi = E\Psi \tag{39}$$

If we define the dimensionless energy parameter α and the dimensionless variable u

$$\alpha = 2E/\hbar\omega \qquad u = (m\omega/\hbar)^{1/2}x \tag{40}$$

then the Schrödinger equation assumes the form

$$\frac{d^2\Psi}{du^2} + (\alpha - u^2)\Psi = 0 \tag{41}$$

If we try the solution

$$\Psi(u) = \mathrm{e}^{-u^2/2} H_n(u) \tag{42}$$

we obtain Hermite's equation

$$\frac{d^2 H_n}{du^2} - 2u \frac{dH_n}{du} + (\alpha - 1)H_n = 0 \tag{43}$$

where $H_n(u)$ is a Hermite polynomial. The four lowest order polynomials are

$$\begin{aligned} H_0(u) &= 1 \\ H_1(u) &= 2u \\ H_3(u) &= 4u^2 - 2 \\ H_4(u) &= 8u^3 - 12u \end{aligned} \tag{44}$$

and the energies are

$$E_n = \left(n_x + \tfrac{1}{2}\right) \hbar\omega \tag{45}$$

where ω is the classical oscillator frequency as in Eq. (36) above.

The total wavefunction is the product of the three wavefunctions of the type (42) for the three directions

$$\Phi(x, y, z) = \mathrm{e}^{-\Gamma r^2/2} H_n(u_x) H_n(u_y) H_n(u_z) \tag{46}$$

The energies for the three directions are additive

$$E_{n_x n_y n_z} = \left(n_x + n_y + n_z + \tfrac{3}{2}\right) \hbar\omega \tag{47}$$

where by comparison with Eq. (36) the principal quantum number n is

$$n = n_x + n_y + n_z \tag{48}$$

The lowest energy E_{000} with $n = 0$ is a singlet (s-state) since the only choice is $n_x = n_y = n_z = 0$. The next level is three-fold degenerate since there are three choices of n_x, n_y, and n_z to make $n = 1$, with the energies

$$E_{100} = E_{010} = E_{001} \tag{49}$$

and the Hermite polynomial H_1 wave functions x, y, and z are linear combinations of the p-state, $L = 1$, spherical harmonics $x \pm iy$ and z, as explained in Chapter 28.

For the $n = 2$ state there are the following six possibilities for $(n_x n_y n_z)$:

$$\begin{array}{llll} (110) & xy/r^2 & (200) & 2[x^2 - r^2]/r^2 \\ (101) & xz/r^2 & (020) & 2[y^2 - r^2]/r^2 \\ (011) & yz/r^2 & (002) & 2[z^2 - r^2]/r^2 \end{array} \tag{50}$$

We see from the T^j_{2M} Tesseral harmonic expressions of Eq. (58) in Chapter 10 that the three functions of the left are second rank ($L = 2$) Tesseral harmonics, and the three on the right are linear combinations of the remaining two T^j_{2M} plus the s-state $[x^2 + y^2 + z^2]/r^2$ function. Thus, the $n = 2$ state is a combination of an s-state and a d-state.

We analyze the $n = 3$ case from a slightly different point of view. There are 10 possibilities for $(n_x n_y n_z)$, three of the type (300), six of the type (120), and the tenth equal to (111). There are also 10 third-rank products of x, y, z given by

$$x^3,\ y^3,\ z^3,\ x^2y,\ x^2z,\ y^2x,\ y^2z,\ z^2x,\ z^2y,\ xyz \tag{51}$$

Seven of these can be written as linear combinations of the $L = 3$ Tesseral harmonics which form the f-state, and the remaining three linearly independent functions can be written in the form r^2x, r^2y, and r^2z which has a p-state angular dependence. Therefore the 10-fold degenerate $n = 3$ state is composed of a p-state and an f-state.

We can see from Eqs. (37) that the energy states and the wave functions of the three dimensional harmonic oscillator have parity, i.e. they are either even or odd under the parity operator P in accordance with the expression

$$P\Psi(x, y, z) = \Psi(-x, -y, -z) = \pm\Psi(x, y, z) \tag{52}$$

where the plus sign is for even parity and the minus sign for odd parity. The degeneracy of each orbital level is $2L + 1$, so the overall degeneracy for each case is the summation

$$\begin{array}{llll} \text{degeneracy} = \sum(2L+1) & L = 0, 2, \ldots, n & n \text{ even} \\ \text{degeneracy} = \sum(2L+1) & L = 1, 3, \ldots, n & n \text{ odd} \end{array} \tag{53}$$

The total degeneracy for both parities can be written as a sum over successive integers m

$$\text{degeneracy} = \sum m = 1, 3, 6, 10, \ldots \quad \text{for } n = 0, 1, 2, 3, \ldots \tag{54}$$

from $m = 1$ to $m = n + 1$. The result is shown in Table 19-1, where we use the spectroscopic notation s, p, d, f, ... for successive L states, as defined in Chapter 17 after Eq. (28). Other interactions such as spin–orbit coupling or the Zeeman effect can raise these degeneracies

TABLE 19-1 Parities and degeneracies of harmonic oscillator energy states

n	Parity	States	L-values	Degeneracy
0	even	s	0	1
1	odd	p	1	3
2	even	s, d	0, 2	6
3	odd	p, f	1, 3	10
4	even	s, d, g	0, 2, 4	15
5	odd	p, f, h	1, 3, 5	21

CHAPTER 20

ONE-DIMENSIONAL QUANTUM SYSTEMS

1 INTRODUCTION

To acquire an understanding of quantum mechanics it can be helpful to examine the properties of one-dimensional systems because this simplifies the mathematics, and makes it easy to learn general principles without becoming bogged down with details. These systems illustrate the propagation and reflection of waves, boundary conditions at interfaces, the confinement of particles in potential wells, bonding and antibonding states, etc. All of this serves to highlight many of the important ideas about quantum mechanics.

2 SCHRÖDINGER EQUATION

The time dependent Schrödinger equation (see Chapter 19, Eq. (1))

$$\mathcal{H}\Psi = i\hbar \frac{\partial \Psi}{\partial t} \tag{1}$$

with a harmonic time dependence $\exp[-iEt/\hbar]$ becomes (see Chapter 19, Eq. (6))

$$\mathcal{H}\Psi = E\Psi \tag{2}$$

And if the Hamiltonian $\mathcal{H}$ contains just the kinetic energy and a one-dimensional potential $V(x)$, then we obtain

$$-\frac{\hbar^2}{2m}\frac{\partial^2 \Psi}{\partial x^2} + V(x)\Psi = E\Psi \tag{3}$$

This can be written in the simplified form

$$\Psi'' + [\varepsilon - U(x)]\Psi = 0 \tag{4}$$

by making use of the following definitions for the reduced potential $U(x)$ and the reduced energy ε, respectively

$$V = \frac{\hbar^2}{2m} U \tag{5a}$$

$$E = \frac{\hbar^2}{2m} \varepsilon \tag{5b}$$

In some cases we use potentials that are constant over certain regions of space, as inside a square well, while other potentials, like that of the harmonic oscillator $\frac{1}{2}kx^2$, depend directly on the position.

Solutions of the Schrödinger equation (4) for various potentials $V(x)$ when the total energy E exceeds the potential energy $U(x)$ constitutes the main topic of the present chapter. We are concerned with both traveling wave and standing wave solutions

$$\Psi(x) = Ae^{ikx-i\omega t} + Be^{-ikx-i\omega t} \quad \text{traveling wave} \tag{6}$$

$$\left.\begin{aligned}\Psi(x) &= C\sin(kx) + D\cos(kx) \\ \Psi(x) &= F\sin(kx+\phi)\end{aligned}\right\} \text{standing wave} \tag{7, 8}$$

for cases where E exceeds $V(x)$. The traveling wave variety (6) is found when the particle is not restricted to a particular region of space, but rather can move to the left or to the right, and standing wave solutions (7) or (8) occur where the particle is confined. The two standing wave solutions are equivalent to each other with $C = F\cos\phi$ and $D = F\sin\phi$. The coefficients A, B, C, D, F, and ϕ are evaluated from the boundary conditions at locations x, where the potential changes while the wave function and its derivative (or its logarithmic derivative Ψ'/Ψ) remain continuous.

When the total energy E is less than the potential energy, exponential growth and decay solutions are obtained

$$\Psi(x) = Ae^{\kappa x} + Be^{-\kappa x} \tag{9}$$

or equivalently

$$\Psi(x) = C\sinh(\kappa x) + D\cosh(\kappa x) \tag{10}$$

These reflect the penetration of the wave function into what is a classically forbidden region.

If there is no potential, $U = 0$, then we have a free particle, and the solution (6) is a plane wave traveling to the left or to the right

$$\Psi(x) = e^{ikx-i\omega t} \quad \text{wave traveling to the right} \tag{11}$$

$$\Psi(x) = e^{-ikx-i\omega t} \quad \text{wave traveling to the left} \tag{12}$$

The energy is entirely kinetic with the value

$$E = \frac{\hbar^2 k^2}{2m} \tag{13}$$

where the momentum $\mathbf{p}$ is

$$\mathbf{p} = m\mathbf{v} = \hbar\mathbf{k} \tag{14}$$

and $\mathbf{k}$ is the wave vector with the magnitude $k = 2\pi/\lambda$.

3 REFLECTION AT THE POTENTIAL STEP

Let us consider the case in which the reduced potential has the positive value U_1 in the region where $x < 0$, and the positive value U_2 for $x > 0$, where $U_1 < U_2$, as shown in Fig. 20-1. For the case $\varepsilon < U_1$ there is no physically meaningful solution since the kinetic energy cannot be negative. Thus, there are two cases to consider: reduced energies ε between U_1 and U_2, and reduced energies greater than U_2. The former case is examined in this section, and the latter in the next section.

The wave function in the $x < 0$ region for $U_1 < \varepsilon < U_2$ is an incident plane wave $I\,e^{ikx}$ plus a reflected plane wave $R\,e^{-1kx}$, while that in the forbidden region on the right is an exponential decay $D\,e^{-\kappa x}$

$$\Psi(x) = I\,e^{ikx} + R\,e^{-ikx} \qquad x < 0 \tag{15a}$$

$$\Psi(x) = De^{-\kappa x} \qquad x > 0 \tag{15b}$$

The boundary conditions for the continuity of Ψ and Ψ' at $x = 0$ give

$$\begin{aligned} I + R &= D \\ ik(I - R) &= -\kappa D \end{aligned} \tag{16}$$

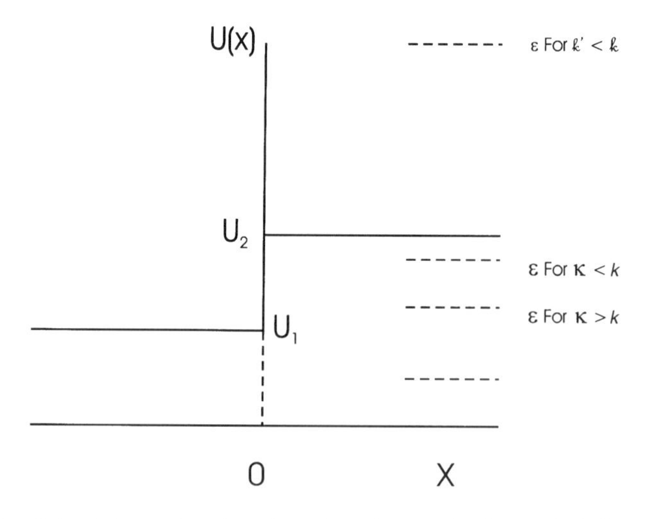

Fig. 20-1. Potential step showing the reduced potential U_1 to the left ($x < 0$) of the step and the reduced potential $U_2 > U_1$ to the right ($x > 0$). Several energies are indicated by dashed horizontal lines.

Solving for the coefficients R and D in terms of I we find

$$R = \frac{ik + \kappa}{ik - \kappa} I \tag{17}$$

$$D = \frac{2ik}{ik - \kappa} I \tag{18}$$

Equation (17) indicates that the amplitude R of the reflected wave is the same as I, the amplitude of the incident wave, but with a phase shift.

To determine the wavelength in the left-hand region we substitute the wavefunction (15a) into the Schrödinger equation (4) to obtain

$$k^2 = \varepsilon - U_1 = (2\pi/\lambda)^2 \tag{19}$$

and this gives for the wavelength to the left of the potential step

$$\lambda = \frac{2\pi}{(\varepsilon - U_1)^{1/2}} \tag{20}$$

Thus, the wavelength is increased by the presence of the potential U_1.

To determine the decay factor κ for $x > 0$ we substitute wave function (15b) into Eq. (4) to obtain

$$\kappa = (U_2 - \varepsilon)^{1/2} \tag{21}$$

and $1/\kappa$ is the distance into the forbidden region where the amplitude decays to $1/e$ of its value at the interface $x = 0$. We see from Eq. (18) that the amplitude D in the forbidden medium becomes very small for $\kappa \gg k$, and the wave function decays rapidly. For $\kappa \ll k$, on the other hand, the amplitude in the forbidden region is large, and decays slowly. These results are shown in Fig. 20-2.

4 TRANSMISSION PAST THE POTENTIAL STEP

We will now consider the case of a potential step with the total energy greater than both potentials, i.e., $U_1 < U_2 < \varepsilon$. There will be an incident and a reflected wave from the left, as before, and a transmitted wave to the right in the second medium, with reflection coefficient R/I and transmission coefficient T/I, respectively. The waves are as follows:

$$\Psi(x) = I\,\mathrm{e}^{ik_1 x} + R\,\mathrm{e}^{-ik_1 x} \quad x < 0 \tag{22a}$$

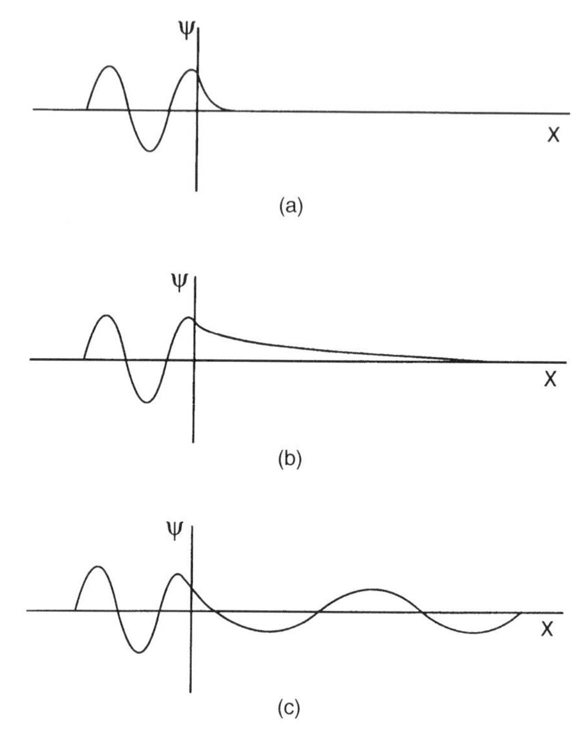

Fig. 20-2. Sketch of wave functions for the reduced energy ε in the range $\varepsilon > U_1$ of the potential step of Fig. 20-1. The solutions of the Schrödinger equation are traveling waves to the left of the step ($x < 0$) and a decaying exponential $e^{-\kappa x}$ [(a) and (b)] or traveling wave $\sin k'x$ (c) to the right ($x > 0$). These correspond, respectively, to the cases of ε close to U_1 (a), of ε midway between U_1 and U_2 (b), and of ε above U_2 (c).

$$\Psi(x) = T\, e^{ik_2 x} \qquad x > 0 \tag{22b}$$

where from Eq. (19) the propagation constants are

$$k_1 = (\varepsilon - U_1)^{1/2} \qquad k_2 = (\varepsilon - U_2)^{1/2} \tag{23}$$

which means that, since $U_1 < U_2$, for this case $k_1 > k_2$ and $\lambda_1 < \lambda_2$. The boundary conditions for continuity of Ψ and Ψ' at $x = 0$ give

$$\begin{aligned} I + R &= T \\ ik_1(I - R) &= ik_2 T \end{aligned} \tag{24}$$

Solving for the coefficients R and D in terms of I we find

$$R = \frac{k_1 - k_2}{k_1 + k_2} I \tag{25a}$$

$$T = \frac{2k_1}{k_1 + k_2} I \tag{25b}$$

Figure 20-2(c) plots the passage of the wave past the potential step for the case $U_2 < \varepsilon$. From Eq. (23) we see that the wavelength is longer for positive x, after passing the potential step, and the transmitted amplitude is less than the incident amplitude because there is a reflected wave that keeps some of the incident energy on the left-hand side.

The squares of the coefficients I, R, and T give the energy density in each component of the wave, and they are multiplied by the respective wave velocities to provide the energy flow balance

$$v_1 I^2 - v_1 R^2 = v_2 T^2 \tag{26}$$

which gives

$$I^2 = R^2 + (k_2/k_1) T^2 \tag{27}$$

since the velocity in a medium is given by

$$v_i = p_i/m = \hbar k_i/m \tag{28}$$

from Eq. (14).

5 POTENTIAL BARRIER

Next we consider the case of a potential barrier of positive reduced energy U_0 with a wave coming from a potential free region ($U_1 = 0$) on the left, and passing through the barrier into the potential free region ($U_3 = 0$) on the right. The barrier, which is sketched in Fig. 20-3, has a potential which extends over the range $-\frac{1}{2}L < x < +\frac{1}{2}L$. We consider the two cases of the reduced energy less than and greater than U_0, the former giving growth–decay solutions in the barrier region, and the latter propagation solutions through the barrier region. In both cases there is a transmitted wave that passes through or over the barrier to enter the zero potential region on the

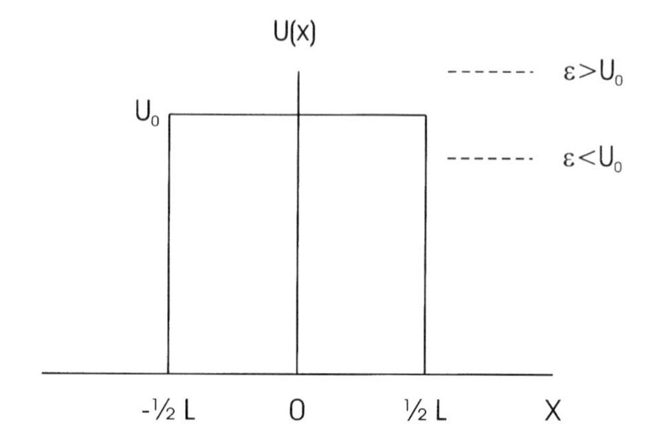

Fig. 20-3. Potential barrier of height U_0 extending over the range $-\frac{1}{2}L < x < +\frac{1}{2}L$. The wave function in the barrier region is the traveling wave type for $\varepsilon > U_0$, and the growth–decay type for $\varepsilon < U_0$.

right. The wave functions have the following forms in both free space regions:

$$\Psi(x) = I\,\mathrm{e}^{ik_1 x} + R\,\mathrm{e}^{-ik_1 x} \qquad x < -\tfrac{1}{2}L \tag{29}$$

$$\Psi(x) = T\,\mathrm{e}^{ik_1 x} \qquad x > \tfrac{1}{2}L \tag{30}$$

and in the region of the barrier $-\frac{1}{2}L < x < +\frac{1}{2}L$. There are two cases:

$$\Psi(x) = A\,\mathrm{e}^{\kappa x} + B\,\mathrm{e}^{-\kappa x} \qquad \varepsilon < U_0 \tag{31}$$

$$\Psi(x) = C\,\mathrm{e}^{ikx} + D\,\mathrm{e}^{-ikx} \qquad \varepsilon > U_0 \tag{32}$$

where

$$\kappa = (U_0 - \varepsilon)^{1/2} \tag{33}$$

$$k = (\varepsilon - U_0)^{1/2} \tag{34}$$

The waves are sketched in Figs. 20-4(a) and 20-4(b), respectively, for $\varepsilon < U_0$ and $\varepsilon > U_0$. If the boundary conditions are matched in the manner described above, then the transmission coefficient T can be determined, and it is given by (A. Messiah, *Méchanique Quantique*, Dunod, Paris, 1964, vol. 1, p. 82)

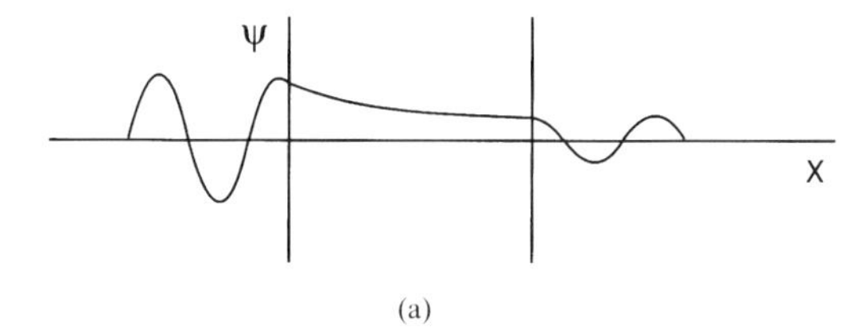

(a)

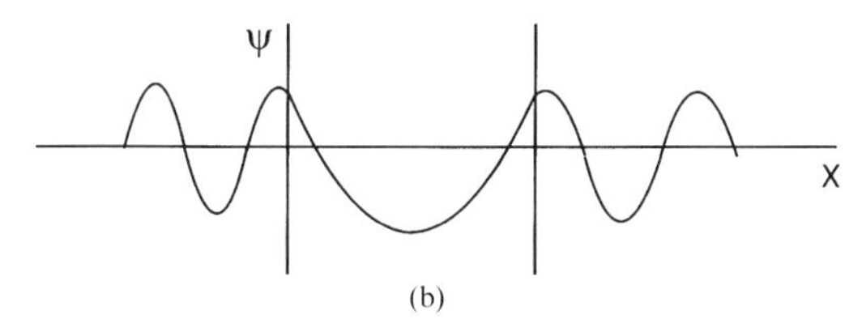

(b)

Fig. 20-4. Wave function in the neighborhood of the potential barrier of Fig. 20-3 showing the exponential decay for $\varepsilon < U_0$ (a) and traveling wave behavior for $\varepsilon > U_0$ (b).

$$T^2 = \frac{\varepsilon(U_0 - \varepsilon)}{\varepsilon(U_0 - \varepsilon) + \frac{1}{4} U_0^2 \sinh^2 \kappa L} \qquad \varepsilon < U_0 \tag{35}$$

$$T^2 = \frac{\varepsilon(\varepsilon - U_0)}{\varepsilon(\varepsilon - U_0) + \frac{1}{4} U_0^2 \sin^2 kL} \qquad \varepsilon > U_0 \tag{36}$$

Figure 20-5 shows a plot of T^2 versus ε/U_0 for the condition $U_0 L^2 = 40$. The oscillations arise for $\varepsilon > U_0$ because, as we see from Eq. (36), $T^2 = 1$ every time $kL = n\pi$, where n is an integer. The dashed line is a plot for the condition $\sin kL = 1$, where minima in T occur, but does not correspond to actual transmission coefficients except at these minima.

6 SQUARE WELL POTENTIAL

We now consider a square well of width a and depth V_0 below the surface. It is convenient to select the zero of energy at the bottom of the well so the potential at the top is V_0, and to let $x = 0$ in the center and $x = \pm\frac{1}{2}a$ at the edges, as shown in Fig. 20-6. We are only interested in energies below the

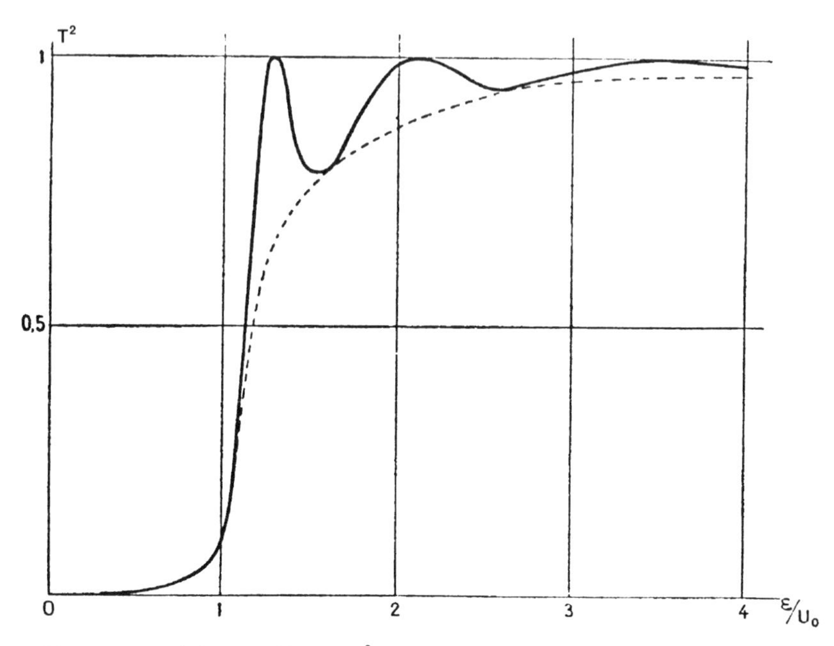

Fig. 20-5. Transmission coefficient T^2 through the potential barrier of Fig. 20-3 as a function of the normalized energy ε/U_0. It reaches the value $T^2 = 1$ when $kL = n\pi$. The dashed curve is drawn tangent to the $T(\varepsilon)$ curve near the successive minima at $kL = (n + \frac{1}{2})\pi$. (From Messiah, *Méchanique quantique*, Dunod, Paris, 1964, p. 82.)

top of the well, meaning $E < V_0$, so the solutions are standing waves inside, either sines or cosines, and decaying exponentials outside. The approach is similar to that of the three-dimensional square well treated in Section 4 of Chapter 19 except that now we select the zero of the potential at the bottom. We have, in particular

$$\Psi(x) = A\,e^{\kappa x} \qquad x < -\tfrac{1}{2}a \tag{37a}$$

$$\Psi(x) = B\sin(kx) + C\cos(kx) \qquad -\tfrac{1}{2}a < x < +\tfrac{1}{2}a \tag{37b}$$

$$\Psi(x) = D\,e^{-\kappa x} \qquad +\tfrac{1}{2}a < x \tag{37c}$$

$$k = (\varepsilon)^{1/2} \tag{38a}$$

$$\kappa = (U_0 - \varepsilon)^{1/2} \tag{38b}$$

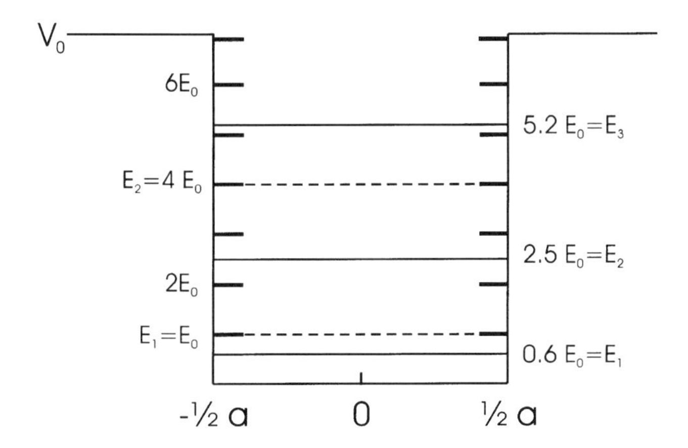

Fig. 20-6. One-dimensional potential well showing how the energies E_n for the finite well (right-hand side, solid horizontal lines) lie below their infinite well counterparts (left-hand side, dashed lines).

where U_0 and ε are defined in terms of V_0 and E by Eqs. (5). For the case of a sine wave solution inside the well, $\Psi = \sin(kx)$, the boundary conditions give the expression

$$k \cot(\tfrac{1}{2}ka) = -\kappa \quad \text{odd parity} \tag{39a}$$

and for a cosine solution $\Psi = \cos(kx)$ we have the condition

$$k \tan(\tfrac{1}{2}ka) = \kappa \quad \text{even parity} \tag{39b}$$

These transcendental equations cannot be solved in closed form.

To obtain a graphical solution we define the angle $\mathcal{E}$

$$\mathcal{E} = \tfrac{1}{2}ak = \tfrac{1}{2}a(\varepsilon)^{1/2} \tag{40}$$

which gives for Eq. (38b)

$$\tfrac{1}{2}a\kappa = [(\tfrac{1}{2}a^2)U_0 - \mathcal{E}^2]^{1/2} \tag{41}$$

and the two transcendental equations become

$$-\mathcal{E} \cot \mathcal{E} = [(\tfrac{1}{2}a)^2 U_0 - \mathcal{E}^2]^{1/2} \qquad \text{odd parity} \tag{42a}$$

$$\mathcal{E} \tan \mathcal{E} = [(\tfrac{1}{2}a)^2 U_0 - \mathcal{E}^2]^{1/2} \qquad \text{even parity} \tag{42b}$$

The solutions $\mathcal{E}$ are at the intersections of plots of the functions

$$r(\mathcal{E}) = -\mathcal{E} \cot \mathcal{E} \qquad \text{odd parity} \tag{43a}$$

$$p(\mathcal{E}) = \mathcal{E} \tan \mathcal{E} \qquad \text{even parity} \tag{43b}$$

and the function

$$q(\mathcal{E}) = [(\tfrac{1}{2}a)^2 U_0 - \mathcal{E}^2]^{1/2} \tag{44}$$

and Figs. 20-7(a) and 20-7(b) show examples of such intersections. Equation (44) rewritten in the form

$$q^2 + \mathcal{E}^2 = (\tfrac{1}{2}a)^2 U_0 \tag{45}$$

demonstrates that it is a circle characterized by

$$\text{radius} = \tfrac{1}{2}a\sqrt{U_0} \tag{46}$$

in the q, $\mathcal{E}$ plane, as shown in Fig. 20-7. To obtain an expression for the energy we combine Eqs. (5b) and (40) for values of $\mathcal{E}$ which are solutions of one or the other of the transcendental equations (42)

$$E = \frac{\hbar^2}{2ma^2}(2\mathcal{E})^2 \tag{47}$$

We see from Fig. 20-7 that (i) there is always at least one allowed energy, (ii) the lowest energy is an even parity cosine, (iii) the even and odd parity levels alternate in energy, and (iv) there is only a finite number of them for each particular choice of the well parameter $U_0 a^2$. The number of energy levels increases with $U_0 a^2$ because from Eq. (47) increasing a brings the allowed energies closer together, and from Eq. (46) and Fig. 20-7 increasing U_0 sets a higher maximum value on the range of allowed energies. Therefore the dimensionless parameter $U_0 a^2$ may be considered as a measure of the strength of a square well. Only one bound state exists for a circle (45) of

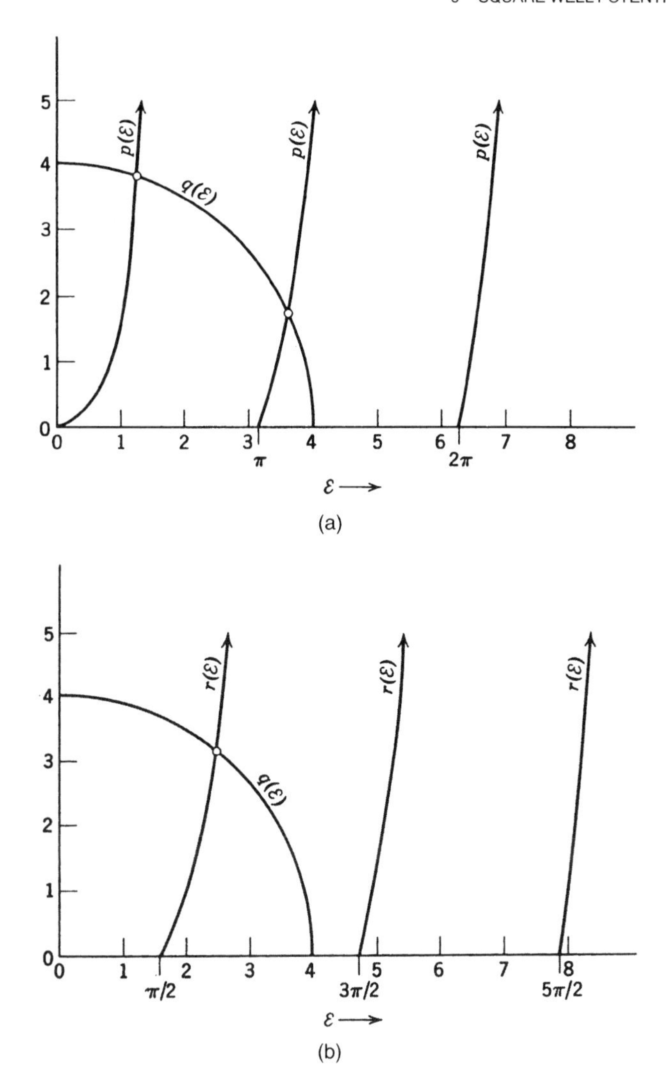

Fig. 20-7. Plot (b) of the function Eq. (43a) and (a) of the function Eq. (43b) for the one-dimensional square well. The roots $\mathcal{E}$ that give the energies E of Eq. (47) are at the intersections of Eqs. (43) with the circles $q(\mathcal{E})$ of Eq. (45). Two roots are shown for the upper even parity case $p(\mathcal{E})$ and one root for the lower odd parity case $r(\mathcal{E})$ for the circle $q(\mathcal{E})$ of radius $\frac{1}{2}a\sqrt{U_0} = 4$. (From R. M. Eisberg, *Fundamentals of Modern Physics*, Wiley, New York, 1961, Figs. 8-15 and 8-16.)

radius $\frac{1}{2}a\sqrt{U_0} < \frac{1}{2}\pi$, and the condition for the onset of a second bound state is a radius of $\frac{1}{2}a\sqrt{U_0} = \frac{1}{2}\pi$ to give for Eq. (44)

$$q(\mathcal{E}) = [(\tfrac{1}{2}\pi)^2 - \mathcal{E}^2]^{1/2} \tag{48}$$

This onset occurs for the potential

$$V_0 = \frac{\hbar^2\pi^2}{2ma^2} \tag{49}$$

as can be seen from Eq. (5a).

7 INFINITE SQUARE WELL

To solve the problem of an infinite square well we again take the zero of energy at the bottom of the well so the potential outside is infinite. We proceed as with the three-dimensional infinite well of Chapter 19, Section 5. The value of κ in Eqs. (37a) and (37c) becomes infinite so the wave function does not extend outside the well. The solutions are still given by Eq. (37b), but with the following boundary conditions:

$$\cos(\pm k_n a/2) = 0 \quad k_n = n\pi/a \quad n = 1, 3, 5, \ldots \quad \text{even parity} \tag{50a}$$

$$\sin(\pm k_n a/2) = 0 \quad k_n = n\pi/a \quad n = 2, 4, 6, \ldots \quad \text{odd parity} \tag{50b}$$

Thus there are two kinds of solutions, even solutions and odd solutions, as sketched in Fig. 20-8. They both give the same expression for the energy, but with different n values. Since the potential is zero inside the well the energy is entirely of the kinetic type, given by Eq. (13) subject to the conditions of Eqs. (50)

$$E_n = \frac{\hbar^2 k_n^2}{2m} = \frac{\pi^2\hbar^2 n^2}{2ma^2} \tag{51}$$

We can define what is called the zero point energy E_0, the lowest possible energy of a particle in an infinite square well, the energy corresponding to $n = 1$

$$E_0 = \frac{\pi^2\hbar^2}{2ma^2} \tag{52}$$

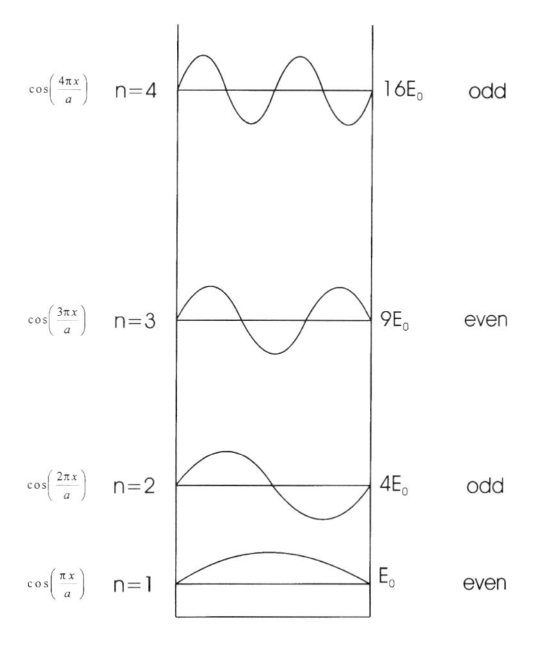

Fig. 20-8. Sketch of the wave functions for the four lowest energy levels ($n = 1$ to 4) of the one-dimensional infinite square well.

The first five energies written in terms of E_0

$$\begin{array}{ccccc} E_1 = E_0 & E_2 = 4E_0 & E_3 = 9E_0 & E_4 = 16E_0 & E_5 = 25E_0 \\ \text{even} & \text{odd} & \text{even} & \text{odd} & \text{even} \end{array} \tag{53}$$

alternate in parity as in the finite well. These energies can be obtained from the construction of Figs. 20-7(a) and 20-7(b) by using a circle of infinite radius. We can see from Figs. 20-6 and 20-7 that the energies of a finite square well are always lower than their infinite well counterparts.

CHAPTER 21

ATOMS

1 INTRODUCTION

In Chapter 19 we gave the solution to the Schrödinger equation for the hydrogen atom. In this chapter we review the hydrogen atom solution and apply it to helium as well as to much larger atoms. The splitting of the levels brought about by the spin-orbit interaction is explained, and the Hartree–Fock approach for treating large atoms is outlined. The construction of the periodic table and the variation of the energies and sizes of atoms with atomic number are discussed.

2 THE HYDROGEN ATOM

A hydrogenic atom, which means an atom with nuclear charge $+Ze$ and a single orbiting electron, has the Coulomb potential $V(\mathbf{r}) = Ze^2/4\pi\varepsilon_0 r$ and the Schrödinger equation

$$-\frac{\hbar^2}{2m}\nabla^2\Psi - \frac{Ze^2}{4\pi\varepsilon_0 r}\Psi = E\Psi \tag{1}$$

In Chapter 19 we wrote down this equation in spherical coordinates

$$\frac{-\hbar^2}{2m}\left\{\frac{1}{r^2}\frac{\partial}{\partial r}\left(r^2\frac{\partial}{\partial r}\right) + \frac{1}{r^2\sin\theta}\frac{\partial}{\partial\theta}\left(\sin\theta\frac{\partial}{\partial\theta}\right) + \frac{1}{r^2\sin^2\theta}\frac{\partial^2}{\partial\phi^2}\right\}\Psi - \frac{Ze^2}{4\pi\varepsilon_0 r}\Psi = E\Psi \tag{2}$$

and for a spherically symmetric potential like the Coulomb potential we noted that the Schrödinger equation is separable with the following product solution:

$$\Psi(r,\theta,\phi) = R(r)Y_{LM}(\theta,\phi) \tag{3}$$

where the spherical harmonic $Y_{LM}(\theta,\phi)$ is an eigenfunction of the angular part of the Laplacian, and the term $R(r)$ satisfies the following radial equation:

$$-\frac{\hbar^2}{2m}\left\{\frac{1}{r^2}\frac{\partial}{\partial r}\left(r^2\frac{\partial}{\partial r}\right) - \frac{L(L+1)}{r^2}\right\}\Psi - \frac{Ze^2}{4\pi\varepsilon_0 r}\Psi = E\Psi \tag{4}$$

To solve this second-order homogeneous differential equation we change to a dimensionless variable ρ and we define a dimensionless energy parameter n

$$\rho = \frac{2Zr}{\mathrm{n}a_0} \qquad = \frac{Ze^2}{4\pi\varepsilon_0\hbar}\left(\frac{m}{2|E|}\right)^{1/2} \tag{5}$$

where a_0 is the Bohr radius

$$a_0 = \frac{4\pi\varepsilon_0\hbar^2}{me^2} = 0.0529 \text{ nm} = 0.529 \text{ Å} \tag{6}$$

and we write

$$R(r) = e^{-\rho/2}\rho^L L(\rho) \tag{7}$$

to obtain the following differential equation for $L(\rho)$:

$$\rho L'' + [2(L+1) - \rho]L' + [n - L - 1]L = 0 \tag{8}$$

The related associated Laguerre differential equation

$$\rho L'' + [m + 1 - \rho]L' + pL = 0 \tag{9}$$

has solutions that are associated Laguerre polynomials $L_p^m(\rho)$ of order n and degree m, where p and m are integers. Comparing Eqs. (8) and (9) we conclude that the hydrogen atom radial function (7) contains Laguerre polynomials L_p^m of integer order $p = n - L - 1$ and integer degree $m = 2L + 1$. The total wave function of the Schrödinger equation is therefore

$$\Psi(r, \theta, \phi) = e^{-\rho/2}\rho^L L_{n-L-1}^{2L+1}(\rho) Y_{LM}(\theta, \phi) \tag{10}$$

For the hydrogen atom in its ground state, $n = 1$ and $L = 0$, there is no angular dependence, and the wave function is

$$\Psi(r, \theta, \phi) = (Z^3/\pi a^3)^{1/2} e^{-Zr/a_0} \tag{11}$$

where we include the nuclear charge Z for later use, and of course $Z = 1$ for hydrogen.

To obtain an expression for the energy we note that n in Eq. (5) is an integer, and we identify it with the principal quantum number so Eq. (5) gives the familiar Rydberg formula for the energy

$$E_n = -\frac{Z^2 me^4}{32\pi^2\varepsilon_0^2\hbar^2}\frac{1}{n^2} = \frac{Z^2 e^2}{8\pi\varepsilon_0 a_0}\frac{1}{n^2} \tag{12}$$

where a_0 is given by Eq. (6) and for the ground state $E_1 = -13.6\,\text{eV}$. The energy does not depend on the orbital quantum number L so there can be more than one L value for the same energy, with the possibilities

$$L = 0, 1, 2, \ldots, (n-1) \tag{13}$$

Therefore parity is not a good quantum number for the hydrogen atom since the parity of an L state is $(-1)^L$, and both odd and even L states can occur for the same n. More precisely, the L states corresponding to the four lowest values of n are

$$\begin{array}{ll} n = 1 & \text{s state only} \\ n = 2 & \text{s and p states} \\ n = 3 & \text{s, p and d states} \\ n = 4 & \text{s, p, d, and f states} \end{array} \tag{14}$$

where $L = 0, 1, 2, 3$, for s, p, d, and f states, respectively. These hydrogen energy levels are plotted in Fig. 21-1.

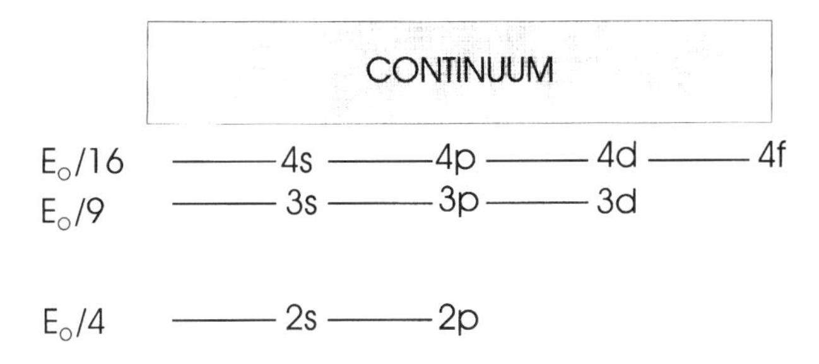

Fig. 21-1. Zero-order energy levels $E_n = E_0/n^2$ of the hydrogen atom.

If we make use of the dimensionless fine structure constant α

$$\alpha = \frac{e^2}{4\pi\varepsilon_0 \hbar c} \approx \frac{1}{137} \tag{15}$$

we obtain for the energies

$$E_n = \frac{\frac{1}{2} mc^2 \alpha^2}{n^2} \tag{16}$$

where mc^2 is the relativistic rest energy, and the reduced mass m given by

$$\frac{1}{m} = \frac{1}{m_e} + \frac{1}{m_p} \tag{17}$$

is very close to the electron mass m_e since it is so much smaller than the proton mass m_p, where $m_p/m_e \approx 1837$.

The square of the wave function $|\Psi(\rho, \theta, \phi)|^2$ is proportional to the electron density at the position r, θ, ϕ. Of greater interest is the probability of finding an electron at a particular distance r from the nucleus, and this is given by $4\pi r^2[R(r)]^2$, where $4\pi r^2 \Delta r$ is the volume of a spherical shell of thickness Δr at a distance r from the nucleus. Figure 21-2 shows plots of $4\pi r^2[R(r)]^2$ as a function of r, and we see from these plots that the electron density is concentrated at larger values of r for larger principal quantum numbers n. The average distance $\langle r \rangle$, the rms distance $[\langle r^2 \rangle]^{\frac{1}{2}}$ and $\langle 1/r \rangle$, depend on n and L, but not on M_L in the following way:

$$\langle r \rangle = \frac{n^2 a_0}{Z} \left\{ 1 + \tfrac{1}{2}\left(1 - \frac{L(L+1)}{n^2} \right) \right\} \tag{18}$$

$$\langle r^2 \rangle^{\frac{1}{2}} = \frac{n^2 a_0}{Z} \left\{ 1 + \tfrac{3}{2}\left(1 - \frac{L(L+1) - \frac{1}{3}}{n^2} \right) \right\}^{\frac{1}{2}} \tag{19}$$

$$\langle 1/r \rangle = Z/a_0 n^2 \tag{20}$$

the latter value being independent of L.

The Laguerre polynomials are not applicable to the treatment of heavy atoms. The spherical harmonics, on the other hand, provide the angular dependence for atoms of all sizes.

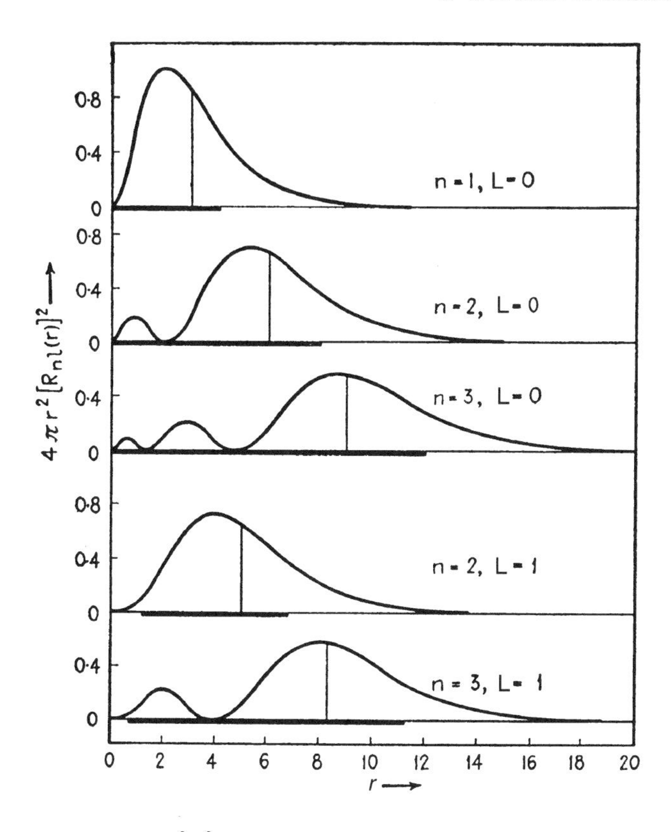

Fig. 21-2. Probability $4\pi r^2 R_n^2(r)$ of finding an electron a distance r from the hydrogen atom nucleus calculated for five hydrogen atom wave functions with the indicated n and L values. (From Pauling and Wilson, McGraw Hill, New York, 1935 p. 143.)

3 SPIN–ORBIT INTERACTION

If the spin–orbit interaction is taken into account, then there is an additional energy splitting ΔE_{SO} given by

$$\Delta E_{\mathrm{SO}} = \frac{1}{2m^2c^2}\left(\frac{1}{r}\frac{dV(r)}{dr}\right)\mathbf{S}\cdot\mathbf{L} \tag{21}$$

$$= \frac{\hbar^2}{4m^2c^2}\left(\frac{1}{r}\frac{dV(r)}{dr}\right)[J(J+1) - L(L+1) - S(S+1)] \tag{22}$$

Section 5 in Chapter 17 explains how to derive Eq. (22) from Eq (21). The relativistic theory of Dirac gives a better approximation to this splitting

$$E = -\frac{me^4}{2(4\pi\varepsilon_0\hbar n)^2}\left\{1 + \frac{a^2Z^2}{n}\left(\frac{1}{J+\frac{1}{2}} - \frac{3}{4n}\right)\right\} \tag{23}$$

which is in close agreement with experiment. Figure 21-3 shows the spin–orbit splittings for the lowest hydrogen ($Z = 1$) levels.

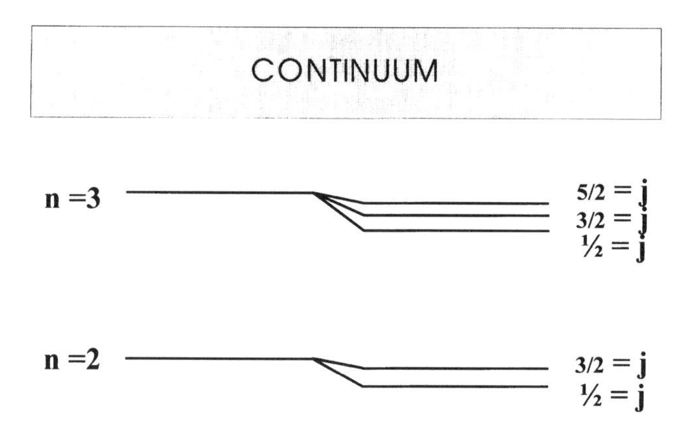

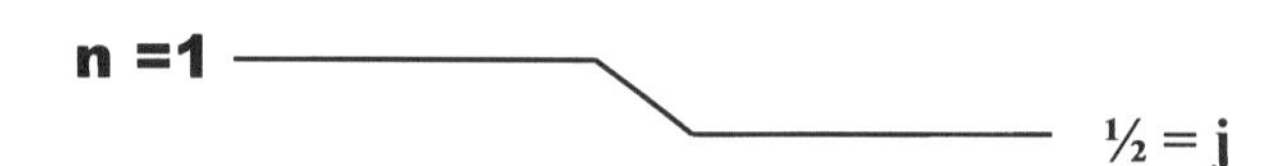

Fig. 21-3. Effect of spin–orbit coupling on the splitting of the three lowest ($n = 1, 2, 3$) hydrogen atom energy levels.

4 THE HELIUM ATOM

The helium atom has atomic number $Z = 2$ and two electrons at positions r' and r'', respectively, and its Hamiltonian $\mathcal{H}$ for the approximation of Eq. (1) gives the Schrödinger equation

$$-\frac{\hbar^2}{2m}(\nabla'^2 + \nabla''^2)\Psi - \frac{2e^2}{4\pi\varepsilon_0}\left(\frac{1}{r'} + \frac{1}{r''}\right)\Psi + \frac{e^2}{4\pi\varepsilon_0|\mathbf{r}' - \mathbf{r}''|}\Psi = E\,\Psi \qquad (24)$$

where the wave function $\Psi = \Psi(\mathbf{r}', s', \mathbf{r}'', s'')$ depends on the space ($\mathbf{r}$) and spin ($\mathbf{s}$) coordinates of the two electrons. The inclusion of the term for Coulombic repulsion of the two electrons makes the helium atom problem much more difficult to solve than that of the hydrogen atom. The Hamiltonian is especially complex for heavy elements because it includes the repulsive interaction for all pairs of electrons.

The spins of the two electrons of helium can be either antiparallel or parallel, so there are both singlet states and triplet states, and transitions between these two sets of states are forbidden. Helium atoms in singlet states are called parahelium and those in triplet states form orthohelium.

To a first approximation we can assume that the wave function of helium is the product of two hydrogen atom wave functions (10), one for each electron. This means that the wave function of the ground state of helium is the product of two ground state hydrogen wave functions (11) with $Z = 2$

$$\Psi_0(r', r'') = (Z^3/\pi a_0^3)\mathrm{e}^{-Z(r'+r'')/a_0} \qquad (25)$$

If the interelectronic term $e^2/4\pi\varepsilon_0|\mathbf{r}' - \mathbf{r}''|$ is neglected to zeroth order, then the Schrödinger equation breaks up into two uncoupled hydrogenic equations

$$-\left(\frac{\hbar^2}{2m}\nabla'^2 + \frac{2e^2}{4\pi\varepsilon_0 r'}\right)\Psi - \left(\frac{\hbar^2}{2m}\nabla''^2 + \frac{2e^2}{4\pi\varepsilon_0 r''}\right)\Psi = E_0\Psi \qquad (26)$$

each of which has an eigenvalue energy from Eq. (12) of $-13.6Z^2 = 54.4\,\text{eV}$, to give for the zeroth-order ground state energy

$$E_0 = -108.8\,\text{eV} \qquad (27)$$

The first-order correction to the energy E_1 is obtained by evaluating the integral of the interelectronic term

$$E_1 = \int |\Psi_0(r', r'')|^2 \frac{e^2}{4\pi\varepsilon_0|\mathbf{r}' - \mathbf{r}''|} d^3r' d^3r'' = \frac{5}{4}\frac{e^2}{4\pi\varepsilon_0 a_0} = \left(\frac{5}{2}\right) 13.6\,\text{eV} = 34.0\,\text{eV} \tag{28}$$

to give for the total energy to first order

$$E = E_0 + E_1 = -108.8\,\text{eV} + 34.0\,\text{eV} = -74.8\,\text{eV} \tag{29}$$

which is close to the experimental value of $-78.6\,\text{eV}$.

The overall wave function must be antisymmetric under electron exchange; in other words, it must change sign under the interchange of the two electrons

$$\Psi(r's', r''s'') = -\Psi(r''s'', r's') \tag{30}$$

Since the ground state space wave function (25) with $n' = 1$, $L' = 0$ for the first electron and $n'' = 1$, $L'' = 0$ for the second electron has the (un-normalized) form

$$\Psi_0(r', r'') = e^{-2(r'+r'')/a_0} = \phi_{1s}(r')\phi_{1s}(r'') \tag{31}$$

it is symmetric under this interchange of the space coordinates r' and r''. This means that the spin part of its wave function must be antisymmetric, and hence be a spin singlet

$$\Psi_{\text{spin}} = (\alpha'\beta'' - \beta'\alpha'')/\sqrt{2} \tag{32}$$

where α denotes spin-up, and *beta* indicates spin-down.

The next highest energy state has $n' = 1$, $n'' = 2$ corresponding to the possible states $n' = 1$, $L' = 0$ for the first electron and either $n'' = 2$, $L'' = 0$ or $n'' = 2$, $L'' = 1$ for the second electron. The space wave functions for these cases can be written in either symmetric or antisymmetric form by choosing, respectively, plus or minus signs in the expression

$$\Psi(r', r'') = [\phi_{1s}(r')\phi_{2s}(r'') \pm \phi_{1s}(r'')\phi_{2s}(r')]/\sqrt{2} \tag{33}$$

As a result, the spin part of the wave function can be either an antisymmetric singlet (32), or a symmetric triplet corresponding to

$$\Psi_{\text{spin}} = \begin{cases} \alpha'\alpha'' & M_s = -1 \\ (\alpha'\beta'' + \beta'\alpha'')/\sqrt{2} & M_s = 0 \\ \beta'\beta'' & M_s = +1 \end{cases} \tag{34}$$

chosen to make the overall wave function antisymmetric

Another approach to calculating the energy of the helium atom is to apply the variational method whereby the wave function is written as a function of one or more parameters ξ_i, and then the expectation value of the total hamiltonian $\langle\Psi(\xi_i)|\mathcal{H}|\Psi(\xi_i)\rangle = E(\xi_i)$ is minimized with respect to the parameters. According to the variational theorem the energy $E(\xi_i)$ will always lie above the true energy E_{true}, but will approach E_{true} for more judicious choices of the parameters. To take into account the partial shielding of each electron from the entire nuclear charge $2e$ by the other electron, the nuclear charge number Z in the wave function (25) can be replaced by an effective charge $Z_{\text{eff}} = Z - \sigma$, where for helium $Z = 2$, and σ is a parameter. A calculation of the expectation value $\langle\Psi(\sigma)|\mathcal{H}|\Psi(\sigma)\rangle$ of the complete Hamiltonian (24) gives (Merzbacher, Wiley, New York, 1961, p. 435)

$$E(\sigma) = -[Z^2 - \left(\frac{5}{8}\right)Z + \left(\frac{5}{8}\right)\sigma - \sigma^2]\frac{e^2}{4\pi\varepsilon_0 a_0} \tag{35}$$

which has a minimum for $\sigma = 5/16$, and gives the energy

$$E\left(\frac{5}{16}\right) = -\left[Z - \left(\frac{5}{16}\right)\right]^2 \frac{e^2}{4\pi\varepsilon_0 a_0} = -(729 \times 13.6/128) = -77.4\,\text{eV} \tag{36}$$

which is closer to the experimental value $-78.6\,\text{eV}$ than the first-order perturbation result of Eq. (29). Trial wave functions with additional parameters have given energies closer to experiment. This variational method is applicable to larger atoms.

5 ENERGIES AND SIZES OF ATOMS

The periodic table presented in Fig. 21-4 is built up in accordance with the modified hydrogen atom energy level scheme of Fig. 21-5. In this modified scheme the d and f levels move upward to join groups with higher principal quantum numbers n. The periodic table is constructed by gradually occupying the modified levels one by one, and rare gases are obtained at the magic numbers $n = 2$, 10, 18, 36, 54, and 86 when all the levels in a group are full. The ordinary elements result from adding electrons to s-states ($L = 0$) and p-states ($L = 1$). The three transition series arise from populating d-levels ($L = 2$) and the two long periods at the bottom of the periodic table containing the rare earth and the actinide atoms correspond to filling f-levels ($L = 3$).

The sizes of atoms tend to increase with the atomic number, but they also vary with the column in a systematic manner, as shown in Table 21-1. The values in this table were determined from a knowledge of the densities and the structures of solids containing only one type of atom. We see that most atoms have crystal diameters between 4 and 8 Bohr radii, or between 0.2 and

New notation — Previous IUPAC form — CAS version

1 Group IA	2 IIA	3 IIIA IIIB	4 IVA IVB	5 VA VB	6 VIA VIB	7 VIIA VIIB	8	9 VIIIA VIII	10	11 IB	12 IIB	13 IIIB IIIA	14 IVB IVA	15 VB VA	16 VIB VIA	17 VIIB VIIA	18 VIIIA	Orbit
1 H +1 −1 1.00794 1																	2 He 0 4.0020602 2	K
3 Li +1 6.941 2-1	4 Be +2 9.012182 2-2											5 B +3 10.811 2-3	6 C +2 +4 −4 12.011 2-4	7 N +1 +2 +3 +4 +5 −1 −2 −3 14.00674 2-5	8 O −2 15.9994 2-6	9 F −1 18.9984032 2-7	10 Ne 0 20.1797 2-8	K-L
11 Na +1 22.989768 2-8-1	12 Mg +2 24.3050 2-8-2											13 Al +3 26.981539 2-8-3	14 Si +2 +4 −4 28.0855 2-8-4	15 P +3 +5 −3 30.97362 2-8-5	16 S +4 +6 −2 32.066 2-8-6	17 Cl +1 +5 +7 −1 35.4527 2-8-7	18 Ar 0 39.948 2-8-8	K-L-M
19 K +1 39.0983 -8-8-1	20 Ca +2 40.078 -8-8-2	21 Sc +3 44.955910 -8-9-2	22 Ti +2 +3 +4 47.88 -8-10-2	23 V +2 +3 +4 +5 50.9415 -8-11-2	24 Cr +2 +3 +6 51.9961 -8-13-1	25 Mn +2 +3 +4 +7 54.93085 -8-13-2	26 Fe +2 +3 55.847 -8-14-2	27 Co +2 +3 58.93320 -8-15-2	28 Ni +2 +3 58.69 -8-16-2	29 Cu +1 +2 63.546 -8-18-1	30 Zn +2 65.39 -8-18-2	31 Ga +3 69.723 -8-18-3	32 Ge +2 +4 72.61 -8-18-4	33 As +3 +5 −3 74.92159 -8-18-5	34 Se +4 +6 −2 78.96 -8-18-6	35 Br +1 +5 −1 79.904 -8-18-7	36 Kr 0 83.80 -8-18-8	-L-M-N
37 Rb +1 85.4678 -18-8-1	38 Sr +2 87.62 -18-8-2	39 Y +3 88.90585 -18-9-2	40 Zr +4 91.224 -18-10-2	41 Nb +3 +5 92.90638 -18-12-1	42 Mo +6 95.94 -18-13-1	43 Tc +4 +6 +7 (98) -18-13-2	44 Ru +3 101.07 -18-15-1	45 Rh +3 102.90550 -18-16-1	46 Pd +2 +4 106.42 -18-18-0	47 Ag +1 107.8682 -18-18-1	48 Cd +2 112.411 -18-18-2	49 In +3 114.82 -18-18-3	50 Sn +2 +4 118.710 -18-18-4	51 Sb +3 +5 −3 121.75 -18-18-5	52 Te +4 +6 −2 127.60 -18-18-6	53 I +1 +5 +7 −1 126.90447 -18-18-7	54 Xe 0 131.29 -18-18-8	-M-N-O
55 Cs +1 132.90543 -18-8-1	56 Ba +2 137.327 -18-8-2	57* La +3 138.9055 -18-9-2	72 Hf +4 178.49 -32-10-2	73 Ta +5 180.9479 -32-11-2	74 W +6 183.85 -32-12-2	75 Re +4 +6 +7 186.207 -32-13-2	76 Os +3 +4 190.2 -32-14-2	77 Ir +3 +4 192.22 -32-15-2	78 Pt +2 +4 195.08 -32-16-2	79 Au +1 +3 196.96654 -32-18-1	80 Hg +1 +2 200.59 -32-18-2	81 Tl +1 +3 204.3833 -32-18-3	82 Pb +2 +4 207.2 -32-18-4	83 Bi +3 +5 208.98037 -32-18-5	84 Po +2 +4 (209) -32-18-6	85 At (210) -32-18-7	86 Rn 0 (222) -32-18-8	-N-O-P
87 Fr +1 (223) -18-8-1	88 Ra +2 226.025 -18-8-2	89** Ac +3 227.028 -18-9-2	104 Unq +4 (261) -32-10-2	105 Unp (262) -32-11-2	106 Unh (263) -32-12-2	107 Uns (262) -32-13-2												O P Q

KEY TO CHART

Atomic Number →	50 +2	← Oxidation States
Symbol →	Sn +4	
1987 Atomic Weight →	118.71	
	18 18 4	← Electron Configuration

															Orbit
*Lanthanides	58 Ce +3 +4 140.115 -20-8-2	59 Pr +3 140.90765 -21-8-2	60 Nd +3 144.24 -22-8-2	61 Pm +3 (145) -23-8-2	62 Sm +2 +3 150.36 -24-8-2	63 Eu +2 +3 151.965 -25-8-2	64 Gd +3 157.25 -25-9-2	65 Tb +3 158.92534 -27-8-2	66 Dy +3 162.50 -28-8-2	67 Ho +3 164.93032 -29-8-2	68 Er +3 167.26 -30-8-2	69 Tm +3 168.93421 -31-8-2	70 Yb +2 +3 173.04 -32-8-2	71 Lu +3 174.967 -32-9-2	N O P
**Actinides	90 Th +4 232.0381 -18-10-2	91 Pa +5 +4 231.03588 -20-9-2	92 U +3 +4 +5 +6 238.0289 -21-9-2	93 Np +3 +4 +5 +6 237.048 -22-9-2	94 Pu +3 +4 +5 +6 (244) -24-8-2	95 Am +3 +4 +5 +6 (243) -25-8-2	96 Cm +3 (247) -25-9-2	97 Bk +3 +4 (247) -27-8-2	98 Cf +3 (251) -28-8-2	99 Es +3 (252) -29-8-2	100 Fm +3 (257) -30-8-2	101 Md +2 +3 (258) -31-8-2	102 No +2 +3 (259) -32-8-2	103 Lr +3 (260) -32-9-2	O P Q

Fig. 21-4. Periodic table of the elements. (From *Chemical and Engineering News* **63**, 1985, p. 27; see also *Handbook of Chemistry and Physics*, CRC Press, front cover.)

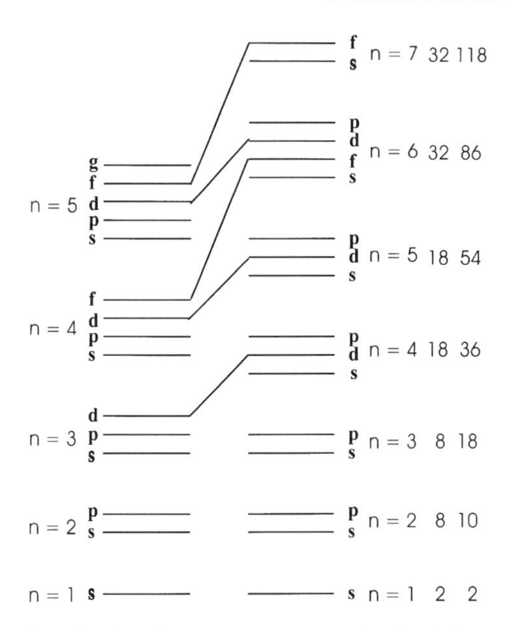

Fig. 21-5. Upward shift of hydrogen atom energy levels (left) to the positions that they occupy for heavy elements (right) showing the order in which they are filled in constructing the periodic table. The numbers in the last two columns on the right give, respectively, the number of electrons in the various n levels, and the accumulated number up to the filling of each n level group. The latter numbers 2, 10, 18, 36, ..., for full levels are the magic numbers for extra stable isotopes.

0.4 nm, as determined by the density of atomic solids. The alkali and alkaline earth metals, as well as the noble elements, are large, and the elements in the middle columns labeled III, IV, V, and VI are smaller. The transition series elements shown at the bottom of the table are more uniform in size, with less variation with Z and column.

When atoms form compounds they are attracted to each other and move closer together than expected from their crystal radii. In ionic compounds the positive ions, called cations, lose their valence electrons and hence become much smaller in size. This can be seen by comparing the crystal radii of Table 21-1 with the ionic radii of Table 21-2 for the same elements. For example, the sodium atom has a crystal radius of 3.66 Å, while the Na^+ ionic radius is only 0.97 Å. Negative ions, called anions, which gain electrons to form an ionic bond, undergo a much lesser reduction in size during ionic bonding. For example, selenium has a crystal radius of 2.32 Å, whereas the Se^{2-} cation has an ionic radius of 1.91 Å. Some atoms have several valence

TABLE 21-1. Crystal radii, or nearest neighbor distances in ångström units between atoms in monatomic crystals of the elements arranged according to their positions in the periodic table. The upper part of the table gives the ordinary elements and the lower part of the table gives the three transition series of elements.

I	II	III	IV	V	VI	VII	VIII
Li 3.02	Be 2.22	B	C 1.54	N	O 1.44	F 3.16	Ne
Na 3.66	Mg 3.20	Al 2.86	Si 2.35	P	S 2.02	Cl 3.76	Ar
K 4.53	Ca 3.95	Ga 2.44	Ge 2.45	As 3.16	Se 2.32	Br	Kr 4.00
Rb 4.84	Sr 4.30	In 3.25	Sn 2.81	Sb 2.91	Te 2.86	I 3.54	Xe 4.34
Cs 5.24	Ba 4.35	Tl 3.46	Pb 3.50	Bi 3.07	Po 3.34	At	Rn

1	2	3	4	5	6	7	8	9	10
Sc 3.25	Ti 2.89	V 2.62	Cr 2.50	Mn 2.24	Fe 2.48	Co 2.50	Ni 2.49	Cu 2.56	Zn 2.66
Y 3.55	Zr 3.17	Nb 2.86	Mo 2.72	Tc 2.71	Ru 2.65	Rh 2.69	Pd 2.75	Ag 2.89	Cd 2.98
La 3.73	Hf 3.13	Ta 2.86	W 2.74	Re 2.74	Os 2.68	Ir 2.71	Pt 2.77	Au 2.88	Hg 3.01

Source: The data are from Kittel, *Solid State Physics*, 5th Ed., Chapter 1, Table 4.

states with monotonically decreasing radii, such as manganese which forms the anions Mn^{2+} (0.80 Å), Mn^{3+} (0.66 Å), Mn^{4+} (0.60 Å), and rarely Mn^{7+} (0.46 Å).

The innermost electrons in atoms, those in the K-shell, have principal quantum number $n = 1$, and the radius r_1 of such an electron orbit may be estimated from the expression

$$r_1 \approx a_0/Z \tag{37}$$

obtained by setting $\rho = 2$ in Eq. (5) [see wave function (7)]. This gives $r_1 = 0.00064\,\text{nm}$ for the largest stable element Bi ($Z = 83$), two orders of magnitude less than the hydrogen atom radius, and only two orders of magnitude greater than the radius of the Bi nucleus. If we set $E_1 = \hbar\omega$ in Eq. (12)

TABLE 21-2. Crystal ionic radii of the elements

Element	Charge	Atomic number	Radius in A	Element	Charge	Atomic number	Radius in A	Element	Charge	Atomic number	Radius in A
Ac	+3	89	1.18	Hf	+4	72	0.78	Ra	+2	88	1.43
Ag	+1	47	1.26	Hg	+1	80	1.27	Rb	+1	37	1.47
	+2		0.89		+2		1.10	Re	+4	75	0.72
Al	+3	13	0.51	Ho	+3	67	0.894		+7		0.56
Am	+3	95	1.07	I	−1	53	2.20	Rh	+3	45	0.68
	+4		0.92		+5		0.62	Ru	+4	44	0.67
Ar	+1	18	1.54		+7		0.50	S	−2	16	1.84
As	−3	33	2.22	In	+3	49	0.81		+2		2.19
	+3		0.58	Ir	+4	77	0.68		+4		0.37
	+5		0.46	K	+1	19	1.33		+6		0.30
At	+7	85	0.62	La	+1	57	1.39	Sb	−3	51	2.45
Au	+1	79	1.37		+3		1.061		+3		0.76
	+3		0.85	Li	+1	3	0.68		+5		0.62
B	+1	5	0.35	Lu	+3	71	0.85	Sc	+3	21	0.732
	+3		0.23	Mg	+1	12	0.82	Se	−2	34	1.91
Ba	+1	56	1.53		+2		0.66		−1		2.32
	+2		1.34	Mn	+2		0.80		+1		0.66
Be	+1	4	0.44		+3		0.66		+4		0.50
	+2		0.35		+4	42	0.60		+6		0.42
Bi	+1	83	0.98		+7		0.46	Si	−4	14	2.71
	+3		0.96	Mo	+1		0.93		−1		3.84
	+5		0.74		+4	7	0.70		+1		0.65
Br	−1	35	1.96		+6		0.62		+4		0.42
	+5		0.47	N	−3		1.71	Sm	+3	62	0.964
	+7		0.39		+1		0.25	Sn	−4	50	2.94
C	−4	6	2.60		+3		0.16		−1		3.70
	+4		0.16		+5	11	0.13		+2		0.93
Ca	+1	20	1.18	NH_4	+1	41	1.43		+4		0.71
	+2		0.99	Na	+1		0.97	Sr	+2	38	1.12
Cd	+1	48	1.14	Nb	+1		1.00	Ta	+5	73	0.68
	+2		0.97		+4	60	0.74	Tb	+3	65	0.923
Ce	+1	58	1.27		+5	10	0.69		+4		0.84
	+3		1.034	Nd	+3	28	0.995	Tc	+7	43	0.979
	+4		0.92	Ne	+1	93	1.12	Te	−2	52	2.11
Cl	−1	17	1.81	Ni	+2		0.69		−1		2.50
	+5		0.34	Np	+3		1.10		+1		0.82
	+7		0.27		+4	8	0.95		+4		0.70
Co	+2	27	0.72		+7		0.71		+6		0.56
	+3		0.63	O	−2		1.32	Th	+4	90	1.02
Cr	+1	24	0.81		−1		1.76	Ti	+1	22	0.96
	+2		0.89		+1	76	0.22		+2		0.94
	+3		0.63		+6		0.09		+3		0.76
	+6		0.52	Os	+4	15	0.88	Ti	+4		0.68
Cs	+1	55	1.67		+6		0.69	Tl	+1	81	1.47
Cu	+1	29	0.96	P	−3		2.12		+3		0.95
	+2		0.72		+3	91	0.44	Tm	+3	69	0.87
Dy	+3	66	0.908		+5		0.35	U	+4	92	0.97
Er	+3	68	0.881	Pa	+3		1.13		+6		0.80
Eu	+3	63	0.950		+4	82	0.98	V	+2	23	0.88
	+2		1.09		+5		0.89		+3		0.74
F	−1	9	1.33	Pb	+2	46	1.20		+4		0.63
	+7		0.08		+4		0.84		+5		0.59
Fe	+2	26	0.74	Pd	+2	61	0.80	W	+4	74	0.70
	+3		0.64		+4	84	0.65		+6		0.62
Fr	+1	87	1.80	Pm	+3	59	0.979	Y	+3	39	0.893
Ga	+1	31	0.81	Po	+6		0.67	Yb	+2	70	0.93
	+3		0.62	Pr	+3	78	1.013		+3		0.858
Gd	+3	64	0.938		+4		0.90	Zn	+1	30	0.88
Ge	−4	32	2.72	Pt	+2	94	0.80		+2		0.74
	+2		0.73		+4		0.65	Zr	+1	40	1.09
	+4		0.53	Pu	+3		1.08		+4		0.79
H	−1	1	1.54		+4	1	0.93				

Source: CRC Handbook of Chemistry and Physics, 70th Ed., CRC Press, Boca Raton, Florida, 1990, p. F187.

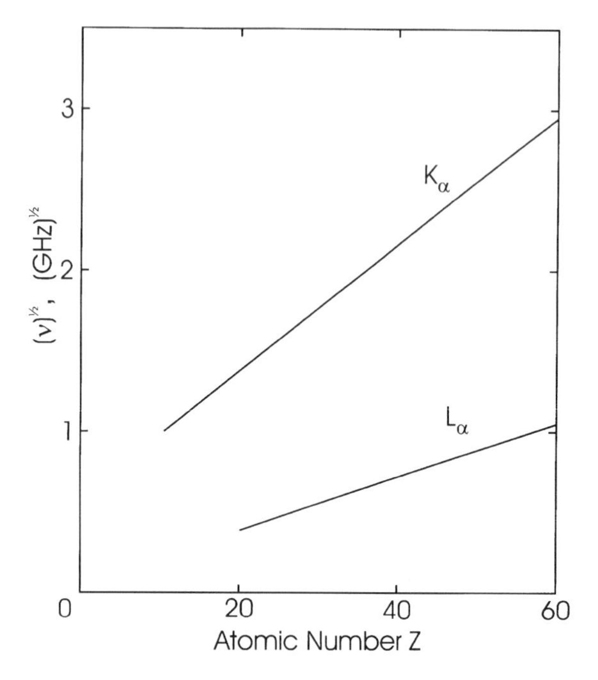

Fig. 21-6. Moseley plot showing the linear dependence of the square root of the frequencies of the K_α and L_α X-ray lines on the atomic number for the elements with Z from about 15 to 60.

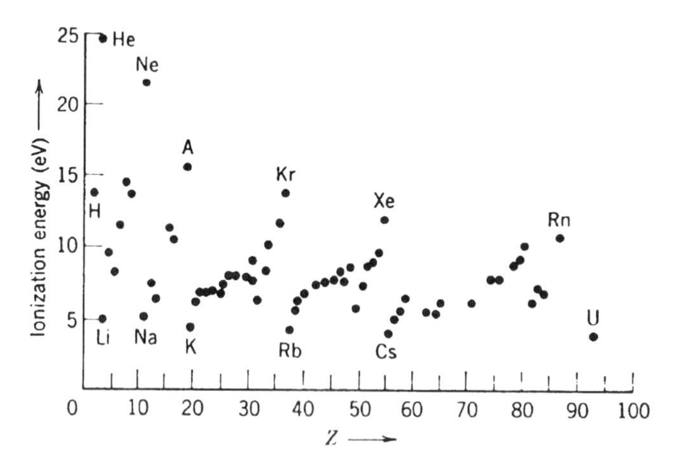

Fig. 21-7. Ionization energies of the outer electron in the elements. (From Eisberg and Resnick, *Quantum Physics*, Wiley, New York, 1974, p. 364.)

we see that the frequency ω is proportional to Z^2, and it is found experimentally that the X-ray lines of the K and L shells ($n = 0, 1$) obey Moseley's law

$$\omega^{1/2} = a(Z - b) \tag{38}$$

where $b \approx 1$ for the K-shell. Figure 21-6 shows how well Moseley's law is obeyed for many of the elements, and it is widely used for the identification of the elements in a material.

The energy of the outer electron of a neutral atom can be obtained from the ionization potential. We see from Fig. 21-7 that this is about 5 eV for the alkali elements, and it is highest for the noble metals, ranging between 24 eV for He and 11 eV for Rn. The energy for this outermost electron varies in a regular manner with the column in the periodic table, as is clear from the figure.

6 HARTREE–FOCK METHOD

Various methods have been developed for calculating the energies of heavy atoms, ranging from relatively simple approximation techniques to highly sophisticated number crunching procedures that can consume many days of super computer time. The variational method that was introduced above in the helium case has been used for much larger atoms, often involving several parameters σ_i to be minimized.

A much more sophisticated approach, called the Hartree–Fock method, is based on solving the Hartree–Fock equations (M. Tinkham, *Group Theory and Quantum Mechanics*, McGraw Hill, New York, 1964, p. 168).

$$\left[-\frac{\hbar^2\nabla_i^2}{2m} - \frac{Ze^2}{4\pi\varepsilon_0 r_1}\right]\phi_i(1) + \left[\sum_{j\neq i}\int \phi_j^*(2)\left(\frac{e^2}{4\pi\varepsilon_0 r_{12}}\right)\phi_j(2)d\tau_2\right]\phi_i(1)$$
$$-\left[\sum_{j\neq i}\frac{\int \phi_i^*(1)\phi_j^*(2)(e^2/4\pi\varepsilon_0 r_{12})\phi_j(1)\phi_i(2)d\tau_2}{\phi_i^*(1)\phi_i(1)}\right]\phi_i(1) = E_i\phi_i(1) \tag{39}$$

The first integral is called the coulomb or direct integral, and the second is the exchange integral. There are Z equations in this set, one for each electron in the atom. To start the calculation a set of trial wave functions $\phi_i(r_k)$ is used to compute the integrals, and then the equations are solved for the energies and wave functions. This provides a new set of wave functions that are used to calculate the integrals again, and then the equations are solved anew. The procedure is repeated until two successive iterations produce a negligible change in the energies and wave functions. Many hours of computer time can be consumed carrying out these calculations.

CHAPTER 22

PERTURBATION THEORY

1 INTRODUCTION

When we first take the subject of quantum mechanics we are introduced to problems that have closed form solutions, such as various one-dimensional constant potential cases, the square well, the harmonic oscillator, the hydrogen atom, etc. In the the real world it is more common to encounter problems that either do not have closed form solutions, or that have closed form solutions that are too complicated to be of use. It often happens that a real problem is close to, or is a modification of, a problem with a known solution, and when this is the case we can often employ the methods of perturbation theory to find approximate solutions. The present chapter examines ways to accomplish this.

2 POWER SERIES EXPANSION

Sometimes a quantum mechanical eigenvalue problem involves the solution of a quadratic or a cubic equation, and a power series expansion can be made to provide an approximate, more manageable, result. As an example, consider the two-spin system ($S = \frac{1}{2}, I = \frac{1}{2}$) in a magnetic field B that was examined in Chapter 18, Section 4. It has the Hamiltonian

$$\mathcal{H} = g\mu_B S_z B + T\mathbf{S} \cdot \mathbf{I} \tag{1}$$

corresponding to the Zeeman effect of a hydrogen atom, where for simplicity we neglect the nuclear Zeeman term $g_N \mu_N I_z$. In Chapter 18 the following four energies (Eq. (18)) were obtained by direct diagonalization of the Hamiltonian matrix:

$$\begin{aligned} E_1 &= -\tfrac{1}{2} g\mu_B B + \tfrac{1}{4} T \\ E_2 &= -\tfrac{1}{4} T - \tfrac{1}{2}[(g\mu_B B)^2 + T^2]^{1/2} \\ E_3 &= -\tfrac{1}{4} T + \tfrac{1}{2}[(g\mu_B B)^2 + T^2]^{1/2} \\ E_4 &= +\tfrac{1}{2} g\mu_B B + \tfrac{1}{4} T \end{aligned} \tag{2}$$

where the levels are numbered differently here than they were in Chapter 18. We can obtain approximate expressions for E_2 and E_3 in the limit $T \ll g\mu_B B$ by expanding the square root in a power series

$$(1 + x)^n \approx 1 + nx + n(n - 1)x^2/2! + \ldots \tag{3}$$

where $n = \frac{1}{2}$, and retaining the first two terms in the expansion we have

$$E_2 = -\tfrac{1}{2} g\mu_B B - \tfrac{1}{4} T - \tfrac{1}{4} T^2/g\mu_B B \tag{4}$$

$$E_3 = +\tfrac{1}{2} g\mu_B B - \tfrac{1}{4} T + \tfrac{1}{4} T^2/g\mu_B B \tag{5}$$

If the small second-order terms $\pm T^2/4g\mu_B B$ are neglected, then the result corresponds to writing the Hamiltonian in the following approximate form:

$$\mathcal{H} = g\mu_B BM_S + TM_S M_I \tag{6}$$

that is often done in electron spin resonance studies.

3 TIME INDEPENDENT PERTURBATIONS

The basic formulae for time independent perturbation theory are derived in standard quantum mechanics texts, so we merely quote them here. We assume that there is a main Hamiltonian $\mathcal{H}_0$ whose energies E_{0i} are known, and a perturbed term $\mathcal{H}'$ which is much smaller in magnitude. It is assumed that the wave functions $|i\rangle$ of E_{0i} of $\mathcal{H}_0$ are also known, and that the off-diagonal matrix elements of the main Hamilton vanish, $\langle i|\mathcal{H}_0|j\rangle = 0$ for $i \neq j$, since $|i\rangle$ and $|j\rangle$ are eigenfunctions of $\mathcal{H}_0$. The perturbation method is based on calculating matrix elements $\langle i|\mathcal{H}'|j\rangle$ of the perturbed Hamiltonian, and using them to approximate the energies E_j and wave functions Ψ_j of the total Hamiltonian $\mathcal{H}$

$$\mathcal{H} = \mathcal{H}_0 + \mathcal{H}' \tag{7}$$

Many quantum texts write down general expressions for the energies and wave functions, but in practice formulae for energies beyond second order are rarely used, so we will confine our attention to the zeroth-, first-, and second-order cases.

The zeroth-order E_{0i} and first-order E_{1i} energies of the ith energy state are given by

$$E_{0i} = \langle i|\mathcal{H}_0|i\rangle \tag{8a}$$

$$E_{1i} = \langle i|\mathcal{H}'|i\rangle \tag{8b}$$

and the energy E_i up to second order is obtained by adding the second-order summation term

$$E_i = E_{0i} + \langle i|\mathcal{H}'|i\rangle + \sum_j{}' \frac{\langle i|\mathcal{H}'|j\rangle\langle j|\mathcal{H}'|i\rangle}{E_{0i} - E_{0j}} \tag{9}$$

The prime on the summation symbol $\sum'$ indicates that the sum is taken over all j values with the $j = i$ term excluded. We note from this expression that for the case of the ground state the denominators of the second-order term are all negative, which means that a second-order perturbation correction always lowers the energy of the ground state. Higher energy states can be either raised or lowered in second order. The first-order wave functions Ψ'_{1i} are obtained from the same matrix elements as the energies

$$\Psi'_{1i} = \Psi_{0i} + \sum_j{}' \frac{\langle i|\mathcal{H}'|j\rangle}{E_{0i} - E_{0j}} \Psi_{0j} \tag{10}$$

Higher order terms can be written down, but these are ordinarily not used in practice. We should note that these wave functions Ψ'_{1i} are not normalized, and the prime on Ψ'_{1i} indicates this lack of normalization. This, however, is not so critical because Ψ'_{1i} and Ψ_{0i} are very close to each other, and the terms in the summation are all small.

It is clear that first-order corrections to the energy are computed using zero-order wave functions. If we compare the form of Eqs. (9) and (10), then we see that the second-order energies can be considered as arising from first-order wave functions. In like manner, second-order wave functions provide third-order energy corrections, and so forth.

4 HYDROGEN ATOM ZEEMAN EFFECT

As an example of how to apply time independent perturbation theory we examine the $S = \frac{1}{2}, I = \frac{1}{2}$ hydrogen atom Zeeman Hamiltonian (1) in the high field approximation where the main and perturbed Hamiltonians are, respectively

$$\mathcal{H}_0 = g\mu_B S_z B \tag{11}$$

$$\mathcal{H}' = T\mathbf{S} \cdot \mathbf{I} \tag{12}$$

$$= T[S_x I_x + S_y I_y + S_z I_z] \tag{13}$$

The matrix elements were worked out in Chapter 18, Section 4, and are given there. We have for the zero-order energies of the four levels

$$\begin{aligned} E_{01} &= E_{02} = -\tfrac{1}{2} g\mu_B B \\ E_{03} &= E_{04} = +\tfrac{1}{2} g\mu_B B \end{aligned} \tag{14}$$

and the matrix for $\mathcal{H}'$ from Eq. (17) of Chapter 18 is

$$\langle i|\mathcal{H}'|j\rangle = \begin{pmatrix} \frac{1}{4}T & 0 & 0 & 0 \\ 0 & -\frac{1}{4}T & \frac{1}{2}T & 0 \\ 0 & \frac{1}{2}T & -\frac{1}{4}T & 0 \\ 0 & 0 & 0 & \frac{1}{4}T \end{pmatrix} \tag{15}$$

The diagonal matrix elements $\langle i|\mathcal{H}'|i\rangle$ are the first-order corrections to the energy, and the square of the off-diagonal element $|\langle 2|\mathcal{H}'|3\rangle|^2$ divided by the zero-order energy difference $E_{03} - E_{02} = g\mu_B B$ from Eq. (14) provides the second-order energy correction to Eq. (9). The resulting energies

Energy level	Zeroth order	First order	Second order	Quantum number M_S	M_I
E_1	$= -\frac{1}{2}g\mu_B B$	$+\frac{1}{4}T$	—	$-\frac{1}{2}$	$-\frac{1}{2}$
E_2	$= -\frac{1}{2}g\mu_B B$	$-\frac{1}{4}T$	$-\frac{1}{4}T^2/g\mu_B B$	$-\frac{1}{2}$	$+\frac{1}{2}$
E_3	$= +\frac{1}{2}g\mu_B B$	$-\frac{1}{4}T$	$+\frac{1}{2}T^2/g\mu_B B$	$+\frac{1}{2}$	$-\frac{1}{2}$
E_4	$= +\frac{1}{2}g\mu_B B$	$+\frac{1}{4}T$	—	$+\frac{1}{2}$	$+\frac{1}{2}$

(16)

are plotted in Fig. 22-1. Note that the perturbation results (16) for the energies E_2 and E_3 agree with the power series expansion expressions of Eqs. (4) and (5). We see from the figure that the ground state E_2 is lowered by the second-order correction to the energy.

Since two of the energy states, $i = 1$ and $i - 4$, have no off-diagonal matrix elements

$$\langle i|\mathcal{H}|j\rangle = E_{0i}\delta_{ij} \quad i = 1, 4 \tag{17}$$

their wave functions are known immediately, and we write them down in the $|M_S M_I\rangle$ basis

$$\begin{aligned} \Psi_1 &= |-\tfrac{1}{2} - \tfrac{1}{2}\rangle \\ \Psi_4 &= |\tfrac{1}{2}\tfrac{1}{2}\rangle \end{aligned} \tag{18}$$

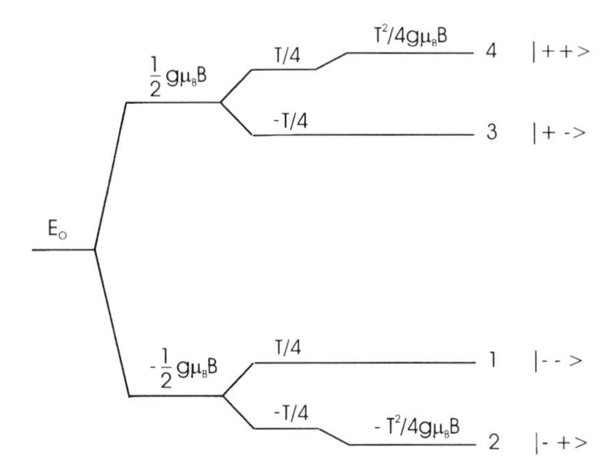

Fig. 22-1. Energy levels for the two spin $S = \frac{1}{2}, I = \frac{1}{2}$ system showing, from left to right: (a) unsplit line, (b) Zeeman splitting, (c) first-order hyperfine splitting, and (d) second-order hyperfine splitting. The levels are labeled on the right with their $|m_S m_I\rangle$ states.

The $\mathcal{H}'$ matrix has only two off-diagonal terms, both of the same form, so we define the coefficient β'

$$\beta' = \frac{\langle 3|\mathcal{H}'|2\rangle}{E_{03} - E_{02}} = \frac{\frac{1}{2}T}{g\mu_B B} \tag{19}$$

to give the orthogonal, but unnormalized, first-order wave functions

$$\begin{aligned} \Psi_2' &= \Psi_{02} - \beta'\Psi_{03} \\ \Psi_3' &= \Psi_{03} + \beta'\Psi_{02} \end{aligned} \tag{20}$$

These can be put in normalized form by writing

$$\begin{aligned} \alpha &= \frac{1}{(1 + \beta'^2)^{1/2}} \\ \beta &= \frac{\beta'}{(1 + \beta'^2)^{1/2}} \end{aligned} \tag{21}$$

and we have

$$\begin{aligned}\Psi_2 &= \alpha|-\tfrac{1}{2}\tfrac{1}{2}\rangle - \beta|\tfrac{1}{2}-\tfrac{1}{2}\rangle \\ \Psi_3 &= \alpha|\tfrac{1}{2}-\tfrac{1}{2}\rangle + \beta|-\tfrac{1}{2}\tfrac{1}{2}\rangle\end{aligned} \tag{22}$$

where $\alpha \approx 1$ and $\beta \ll 1$. All four wave functions (18) and (22) are now normalized and orthogonal.

5 DEGENERATE PERTURBATION THEORY

When two main hamiltonian energies are the same, i.e., degenerate

$$E_{0i} = E_{0j} \tag{23}$$

and there is a non-zero off-diagonal matrix element connecting them

$$\langle i|\mathcal{H}|j\rangle \neq 0 \tag{24}$$

then second-order perturbation theory cannot be applied because the energy denominator of Eq. (9) vanishes between these levels. It is necessary to diagonalize the 2×2 submatrix

$$\begin{pmatrix} \langle i|\mathcal{H}|i\rangle & \langle i|\mathcal{H}|j\rangle \\ \langle j|\mathcal{H}|i\rangle & \langle j|\mathcal{H}|j\rangle \end{pmatrix} \tag{25}$$

to raise the degeneracy before proceeding to apply perturbation theory. After diagonalization the new wave functions can be calculated and used to form a new Hamiltonian matrix that is no longer degenerate. The type of degeneracy described here is a common occurrence.

We have been discussing a totally degenerate case. If the degeneracy is not exact, then degenerate perturbation theory should be employed whenever the criterion

$$|\langle i|\mathcal{H}|j\rangle|^2 \ll |\langle i|\mathcal{H}|i\rangle - \langle j|\mathcal{H}|j\rangle| \tag{26}$$

fails to be satisfied for a pair of off-diagonal matrix elements. This ensures that every second-order term in the summation of Eq. (9) will be much less than the main energy term E_{0i}.

6 TIME DEPENDENT PERTURBATIONS

Until now we have been treating the case in which both the known Hamiltonian $\mathcal{H}_0$ and the perturbation Hamiltonian $\mathcal{H}'$ are independent of time. We now examine what happens when $\mathcal{H}_0$ remains independent of time, but the perturbation $\mathcal{H}'(\mathrm{t})$ depends on the time

$$\mathcal{H}(\mathbf{r}, t) = \mathcal{H}_0(\mathbf{r}) + \mathcal{H}'(\mathbf{r}, t) \tag{27}$$

The time dependent Schrödinger equation

$$\mathcal{H}\Psi = i\hbar \frac{\partial \Psi}{\partial t} \tag{28}$$

has the solution

$$\Psi(\mathbf{r}, t) = \sum_n a_n(t)\phi_n(\mathbf{r})\,\mathrm{e}^{-iE_n t/\hbar} \tag{29}$$

where the coefficients $a_n(t)$ depend only on the time. If we substitute this into the time dependent Schrödinger equation (28), and make use of the unperturbed Schrödinger equation $\mathcal{H}_0\phi_n = E_{0n}\phi_n$ to cancel two of the terms, then we obtain

$$i\hbar \sum_n \mathrm{e}^{-iE_n t/\hbar}\,\phi_n \frac{da_n}{dt} = \sum_n \mathrm{e}^{-iE_n t/\hbar} a_n \mathcal{H}'\phi_n \tag{30}$$

Multiplying both sides on the left by the unperturbed wave function ϕ_m^* and integrating over space provides the time dependence of $a_m(t)$

$$\frac{da_m}{dt} = -(i/\hbar) \sum_n \mathrm{e}^{-i\omega_{mn}t} a_n(t)\mathcal{H}'_{mn}(t) \tag{31}$$

where the Bohr angular frequency ω_{mn} is defined by

$$\omega_{mn} = (E_m - E_n)/\hbar \tag{32}$$

the matrix element $\mathcal{H}'_{mn}(t)$

$$\mathcal{H}'_{mn}(t) = \int \phi_m^* \mathcal{H}'(t)\phi_n \, d\tau \tag{33}$$

can depend on the time, and use was made of the orthonormality of the unperturbed wavefunctions $\phi_n(\mathbf{r})$.

Ordinarily, at the beginning, $t = 0$, the system will be in a single initial quantum state, i, which means

$$a_j(0) = \begin{cases} 1 & j = i \\ 0 & j \neq i \end{cases} \tag{34}$$

and then Eq. (31) becomes

$$\frac{da_i}{dt} \approx -(i/\hbar)a_i(t)\mathcal{H}'_{ii}(t) \qquad \text{initial state } i \tag{35a}$$

$$\frac{da_n}{dt} \approx -(i/\hbar)\,\mathrm{e}^{-i\omega_{ni}t}\mathcal{H}'_{ni}(t) \quad \text{other state } n \neq i \tag{35b}$$

where we retained $a_i(t)$ in the initial state equation (35a), and set $a_i(t) = 1$ in the $n \neq i$ equation (35b). These are the basic equations for first-order time dependent perturbation theory. They are valid for times short enough so that the magnitude of the initial state amplitude, $|a_i(t)|$, remains close to 1. As time proceeds, of course, $a_i(t)$ must decrease as other states become populated, but first-order theory assumes that the system is still close to its initial state. A typical time evolution of the initial state $a_i(t)$ and of two other amplitudes $a_j(t)$ for times $t \ll \hbar/\mathcal{H}'_{ij}$ is shown in Fig. 22-2.

A particularly simple case is a perturbation $\mathcal{H}'$ which is turned on at time $t = 0$, but is otherwise independent of time. We confine our attention to times short enough so that the system still stays close to its initial state, i.e., $a_i(t)$ remains close to 1 in magnitude. For this situation Eq. (36a) integrated from $t = 0$ to a later time t gives

$$a_i(t) = \mathrm{e}^{-i\mathcal{H}_{ii}t/\hbar} \tag{36}$$

so that the initial state amplitude $a_i(t)$ fluctuates harmonically at the frequency

$$\omega_i = \mathcal{H}_{ii}/\hbar \tag{37}$$

and Eq. (36) applies as long as the time is short enough so that the first-order expressions (35) remain a good approximation.

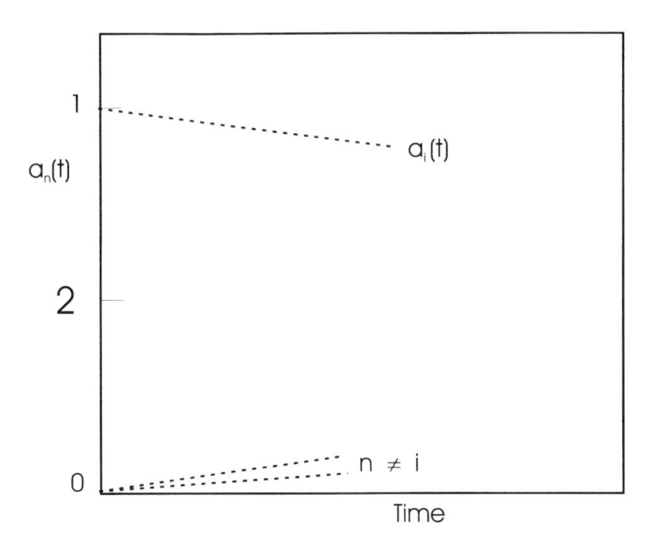

Fig. 22-2. Time evolution of the initial coefficient $a_i(t)$ and two other coefficients $a_n(t)$ for $n \neq i$ in time dependent perturbation theory for times $t \ll \hbar/\mathcal{H}'_{ii}$.

The amplitude coefficients $a_n(t)$ of the states being populated by the perturbation $\mathcal{H}'$ are found by integrating Eq. (35b) over the same interval of time

$$a_n(t) = \frac{\mathcal{H}'_{ni}(1 - e^{-i\omega_{ni}t})}{\omega_{ni}\hbar} \tag{38}$$

and this has the magnitude

$$|a_n(t)| = \frac{2|\mathcal{H}'_{ni}|\,|\sin\frac{1}{2}\omega_{ni}t|}{|E_n - E_i|} \tag{39}$$

where use was made of Eq. (32). For times short enough so that $\omega_{ni}t \ll 1$, we have

$$a_n(t) = -i(\mathcal{H}'_{ni}/\hbar)t \tag{40}$$

Thus, $a_n(t)$ initially rises linearly with time, as shown in Fig. 22-2, then later when it is no longer true that $\omega_{ni}t \ll 1$ the factor $a_n(t)$ fluctuates with the period $\tau = 2/\omega_{ni}$ and amplitude $[2\mathcal{H}'_{ni}/\hbar\,\omega_{ni}]$, subject to the standard assumption

$$\mathcal{H}'_{ni} \ll |E_n - E_i| \tag{41}$$

of time independent perturbation theory.

7 TRANSITION PROBABILITY

The probability of finding the system in the state n at time t is given by the absolute value of the square of the amplitude (39)

$$|a_n(t)|^2 = \frac{4|\mathcal{H}'_{ni}|^2 \sin^2 \frac{1}{2}\omega_{ni}t}{\hbar^2\omega_{ni}^2} \tag{42}$$

The transition probability per unit time w of inducing a transition to some excited state among a group of closely lying states is obtained by summing over this range of excited states

$$w = t^{-1}\sum |a_n(t)|^2 \tag{43}$$

and for many closely lying levels this summation can be conveniently carried out as an integration over the energy

$$w = t^{-1}\int |a_n(t)|^2 \rho(n)\, dE_n \tag{44}$$

where $\rho(n)$ is the density of states. Ordinarily $\mathcal{H}'_{ni}$ and $\rho(n)$ are slowly varying functions of the energy, so they can be taken out of the integral to give

$$w = \frac{4|\mathcal{H}'_{ni}|^2 \rho(n)}{t\hbar}\int \frac{\sin^2 \frac{1}{2}\omega_{ni}t}{\omega_{ni}^2}\, d\omega_{ni} \tag{45}$$

The integrand is a sharply peaked function, as shown in Fig. 22-3, so the range of integration can be extended from $-\infty$ to $+\infty$. Making use of the definite integral

$$\int_{-\infty}^{\infty} x^{-2}\sin^2 x\, dx = \pi \tag{46}$$

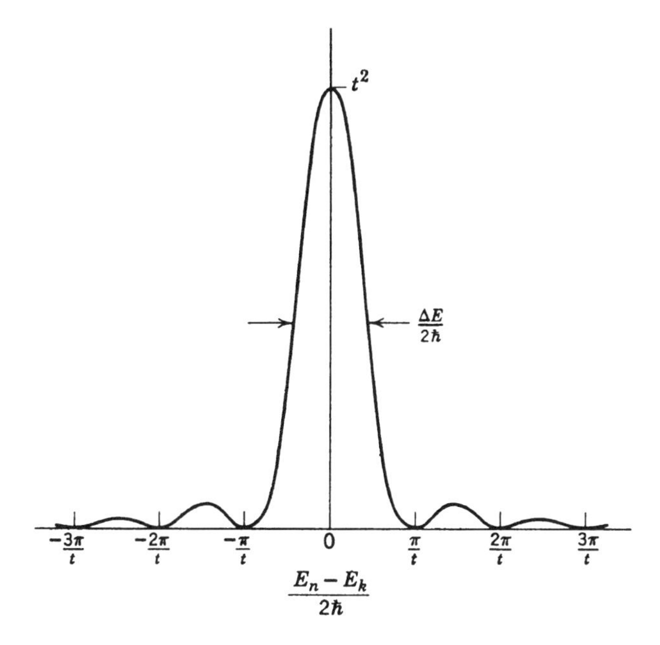

Fig. 22-3. Sharply peaked function in the integrand of the integral appearing in the expression (45) for the transition probability w. (From R. M. Eisberg, *Fundamentals of Modern Physics*, Wiley, New York, 1961, Fig. 9-8.)

we obtain the following expression of w:

$$w = \frac{2\pi}{\hbar} \rho(n) |\mathcal{H}'_{ni}|^2 \tag{47}$$

which is independent of time. This is the probability that the perturbation will induce a transition from the initial ground state E_i to an excited state E_n within a unit time, assuming that the perturbation is weak enough or the time is short enough so that the ground state still remains appreciably populated.

8 SCATTERING

In scattering problems we frequently start with an incoming plane wave e^{1kz}, and we assume that the outgoing wavefunction $\psi(r, \theta, \phi)$

$$\psi(r, \theta, \phi) = e^{ikz} + \chi(\mathbf{r}) \tag{48}$$

is the sum of the unscattered incoming wave plus an outgoing spherical wave $\chi(\mathbf{r})$ with the angular dependence $f(\theta, \phi)$

$$\psi(r, \theta, \phi) = e^{ikz} + r^{-1} f(\theta, \phi)\, e^{ikr} \tag{49}$$

For simplicity we assume axial symmetry, so $f(\theta, \phi) = f(\theta)$. In many problems it is assumed that the potential $U(\mathbf{r})$ that produces the scattering is a perturbation, so the scattered wave $\chi(r)$ is a small addition to the unperturbed incoming wave e^{1kz}.

The differential cross-section $\sigma(\theta)$

$$\sigma(\theta) = |f(\theta)|^2 \tag{50}$$

is often expanded in partial waves with phase shifts δ_L

$$\sigma(\theta) = (1/k^2) |\sum (2L+1)\, e^{i\delta_L} \sin \delta_L P_L(\cos\theta)|^2 \tag{51}$$

and carrying out an integration gives the total cross-section as a summation over L partial waves

$$\sigma = 2\pi \int_0^{\pi} \sigma(\theta) \sin\theta\, d\theta = (4\pi/k^2) \sum (2L+1) \sin^2 \delta_L \tag{52}$$

Thus the phase shift δ_L is a measure of the extent to which the Lth Legendre polynomial $P_L(\cos\Theta)$ contributes to the scattering, and this can be determined experimentally.

CHAPTER 23

FLUIDS AND SOLIDS

1 INTRODUCTION

In this chapter we are concerned with the three states of matter, namely gases, liquids, and solids. We begin with an examination of gases and liquids, which collectively are called fluids. The discussion of solids focuses on the structure of materials. The next chapter will treat the conduction of electric current through solids. Additional properties of solids have been discussed in other chapters, such as lattice vibrational modes (Chapter 4), specific heat and thermal conductivity (Chapter 8), dielectric and magnetic properties (Chapters 10 and 11, respectively), and optical properties (Chapters 13 and 14).

2 GASES

In the gaseous state there is no order, neither short range nor long range. The molecules move around randomly, and either do not interact, or they interact weakly via what is called the van der Waals force. The simplest gas law, in which the molecules do not interact, is the ideal gas law whereby the product of the pressure P and volume V is proportional to the absolute temperature T

$$PV = NRT \tag{1}$$

where N is the number of moles, and the gas constant R equals Avogadro's number $N_A = 6.022 \times 10^{23}$ mol^{-1} times Boltzmann's constant k_B

$$R = N_A k_B \tag{2}$$

This law (1) can also be written in terms of the molar density $\rho = N/V$

$$P = \rho RT \tag{3}$$

A more realistic law assumes that the molecules of the gas induce electric dipoles in each other, dipoles arising from a shift of the negative electron clouds relative to the positive nuclei. The induced dipole moments interact via an attractive force that falls off with the distance as l/r^6, a quantity that scales as $1/V^2$ for a gas. The overall effect is to reduce the pressure in the gas by the factor a/V^2, where a is a constant characteristic of the particular gas. In addition the volume available for the movement of gaseous molecules must be reduced by the actual volume b occupied by the molecules themselves, which gives for the pressure P

$$P = \frac{RT}{V - b} - \frac{a}{V^2} \tag{4}$$

where we assume $N = 1$. This equation of state, shown plotted in Fig. 23-1 for several temperatures, is easily rearranged to the usual form

$$\left\{P + \frac{a}{V^2}\right\}(V - b) = RT \tag{5}$$

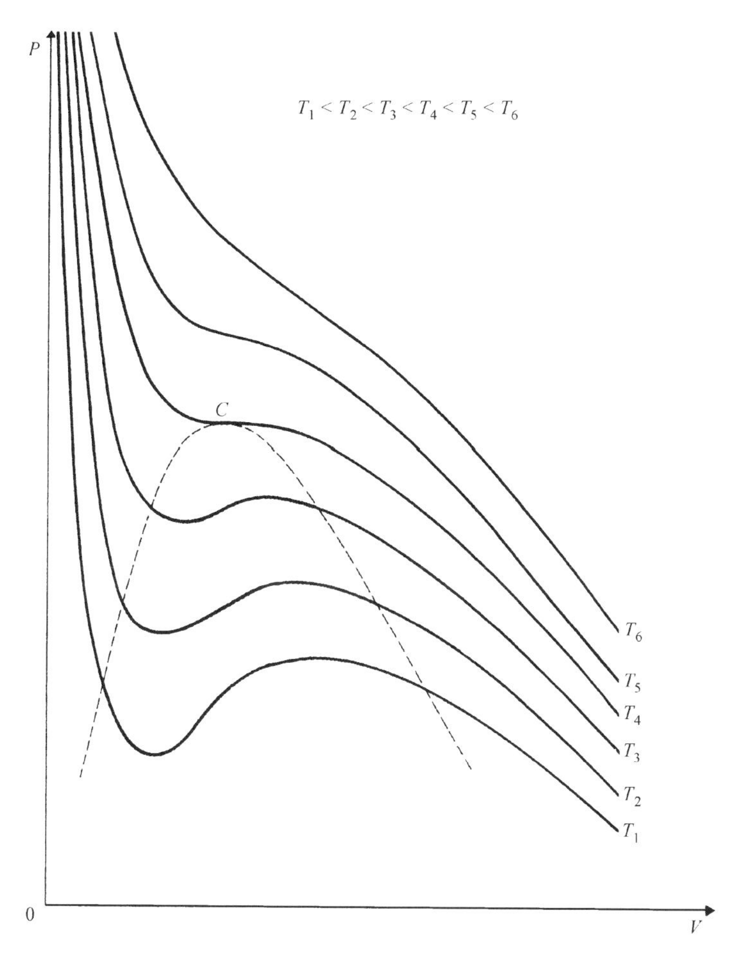

Fig. 23-1. Pressure–volume diagram of a van der Waals gas showing the critical isotherm T_4 that separates the gas from the liquid and vapor phases, with the critical point C indicated. (From F. Reif, *Statistical Thermal Physics*, McGraw-Hill, New York, 1965, p. 307.)

called the van der Waals equation, which provides a good representation of the qualitative features of real gases, including the gas to liquid phase transition. The van der Waals constants a and b have been determined for many gases. Equation (5) reduces to the ideal gas law $PV = RT$ for the limits $PV^2 \gg a$ and $V \gg b$ which occur at high temperatures. Many other, more realistic, gas laws have been proposed over the years, but they all have limited ranges of validity.

3 PHASE TRANSITIONS

The isotherms plotted in Fig. 23-1 for temperatures below the critical temperature T_c show an abrupt change in volume at the transition between the liquid and the vapor state. Above the critical point, defined by the expressions

$$\left(\frac{\partial P}{\partial V}\right) = 0 \qquad \left(\frac{\partial^2 P}{\partial V^2}\right) = 0 \tag{6}$$

the liquid state cannot exist. For a van der Waals gas we have for the critical point

$$P_c = a/27b^2 \qquad V_c = 3b \qquad T_c = 8a/27bR \tag{7}$$

together with the ratio

$$RT_c/P_cV_c = 8/3 = 2.67 \tag{8}$$

and for most real gases we find values of RT_c/P_cV_c close to the range from 1.2 to 1.3.

The P versus T diagram of Fig. 23-2 plots lines of constant volume, called isochors, and gives the sublimation line, the fusion line, and the vaporization line that separate the phases. The triple point, where all three phases meet, is characteristic of each material, and occurs at a particular temperature and pressure. Sublimation, or the direct passage between the solid and gas phase, occurs below the triple point on the figure.

The melting line on Fig. 23-2 is drawn with a negative slope corresponding to a material like water which expands on freezing, so ice floats in liquid water. Most materials have a positive slope so that the solid phase sinks in the liquid phase. The sign of the slope dP/dT is given by the equation of

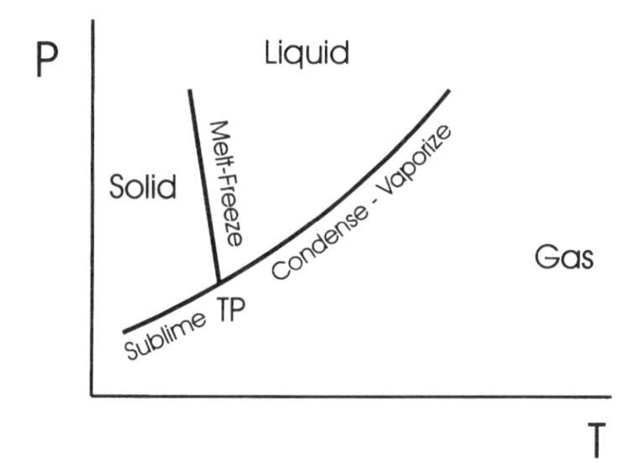

Fig. 23-2. Pressure–volume diagram of a material showing the solid, liquid, and gas phases. The triple point (TP) where all three phases are in equilibrium is indicated. The melting–freezing line is drawn with a negative slope for a material like water in which the solid phase floats in the liquid.

state for the melting line, called the Clausius–Clapeyron equation (Chapter 8, Section 8)

$$\frac{dP}{dT} = \frac{L}{T(V_L - V_S)} \tag{9}$$

where L is the latent heat, and in the case of water $V_L < V_S$. The sublimation and vaporization curves always have a positive slope.

4 LIQUID STATE AND FLUCTUATIONS

A liquid is a material which has short-range order but lacks long-range order. The configuration around an individual molecule is similar to that around others in terms of intermolecular distances and molecular orientations. Nearest neighbor molecules tend to be situated in a particular way relative to each other, but more widely separated ones tend to be randomly oriented relative to each other. For example, when an ionic compound like $Cr(NO_3)_3$ is dissolved in water the Cr^{3+} ions tend to have an octahedron of H_2O dipoles around them with negatively charged $O^{-\delta}$ pointing toward Cr^{3+}, an arrangement corresponding to good short range order in the neighborhood of each Cr^{3+} ion. There is no long-range order because the various CrO_6 octahedra undergo random tumbling motions relative to each other and become unrelated in orientation.

The reason for the lack of long-range order is the continuous thermal agitation of the molecules relative to each other arising from what is called Brownian motion. According to the equipartition theorem each quadratic degree of freedom has a thermal energy of $\frac{1}{2}k_B T$, so a molecule of mass m in a liquid has the following average x-direction kinetic energy:

$$\langle \tfrac{1}{2}mv_x^2 \rangle = \tfrac{1}{2}k_B T \tag{10}$$

and similarly for the other two directions, to give for the average kinetic energy

$$\langle \tfrac{1}{2}mv^2 \rangle = \tfrac{3}{2}k_B T \tag{11}$$

where $\langle v_x^2 \rangle = \langle v_y^2 \rangle = \langle v_z^2 \rangle = \frac{1}{3}\langle v^2 \rangle$.

A simple approach to Brownian motion involves the Langevin equation

$$m\frac{dv}{dt} = F'(t) \tag{12}$$

where $F'(t)$ is a force that describes the interaction of a mass m in the liquid with the many degrees of freedom of the system. $F'(t)$ depends on the positions of the nearby atoms that are in constant motion, and hence it fluctuates in a rapid and highly irregular manner. The rate of fluctuation is characterized by a correlation time τ_c which is of the order of a mean molecular separation divided by a mean molecular velocity. For typical low viscosity liquids $\tau_c \approx 10^{-13}$ s.

The fluctuating force $F'(t)$ can be separated into a rapid randomly fluctuating part $F_r(t)$ which has an average value of zero

$$\langle F'(t) \rangle = 0 \tag{13}$$

and a slowly varying part $-\alpha v$ which is proportional to the particle's velocity v and retards its motion. The frictional coefficient α depends on the size, the shape, the velocity, and perhaps the orientation of the moving mass. The Langevin equation then assumes the form

$$m\frac{dv}{dt} = -\alpha v + F_r(t) \tag{14}$$

If an external force is also acting, then it can be added to the right-hand side. The frictional constant α for a small spherical particle of radius a undergoing streamline motion is given by Stokes' law

$$\alpha = 6\pi\eta a \tag{15}$$

where η is the viscosity. When turbulence is present the frictional drag force depends on a higher power of the velocity.

To gain some insight into the significance of α consider applying an external electric field $\mathcal{E}$ to the liquid containing a suspended particle of mass m and charge e. The Langevin equation for this case is

$$m\frac{dv}{dt} = e\mathcal{E} - \alpha v + F_{\mathrm{r}}(t) \tag{16}$$

After an initial period of acceleration there is, aside from fluctuations, an overall steady state motion at the terminal speed v_0

$$e\mathcal{E} = \alpha v_0 \tag{17}$$

If we define the mobility μ as the steady state speed per unit electric field

$$\mu = v_0/\mathcal{E} \tag{18}$$

then we have for the frictional constant α

$$\alpha = e/\mu \tag{19}$$

Mobilities of various ions in aqueous solutions have been tabulated. They are important in the technique of electrophoresis that is used to separate and perhaps identify macro molecules.

A similar result is obtained for a particle falling in a fluid under the force of gravity mg. Equation (16) applies with mg replacing $e\mathcal{E}$, and we obtain the terminal speed

$$v_0 = mg/\alpha = mg/6\pi\eta a \tag{20}$$

where Stokes' law (15) is used for the case of a spherical particle. In a centrifuge rotating at the angular velocity ω the external force is centrifugal, $mr\omega^2$, to give for the terminal radial speed

$$v_0 = mr\omega^2/\alpha = mr\omega^2/6\pi\eta a \tag{21}$$

5 DIFFUSION

When the concentration n of molecules in a solution varies from place to place the molecules will slowly diffuse toward equalizing the concentration. If there is a uniform concentration gradient $\partial n/\partial z$ in the z direction, then there will be a flux J_z of molecules along z given by

$$J_z = -D\frac{\partial n}{\partial z} \tag{22}$$

where D is the diffusion constant, also called the coefficient of self-diffusion, with the units m^2/s. The diffusion process in the z direction is governed by the diffusion equation

$$\frac{\partial n}{\partial t} = D\frac{\partial^2 n}{\partial z^2} \tag{23}$$

Diffusion is often an activated process with an activation energy E_A

$$D = D_0 \exp[-E_A/k_B T] \tag{24}$$

where D_0 is called the prefactor.

When an electric field E is applied to a solution of ions of charge q the ions experience a force qE and flow in the E direction. This sets up a concentration gradient (22) and a flux of ions J_z in the opposite direction which tends to balance the flux established by the applied field. The diffusion constant for this case is given by the Einstein–Nernst relation

$$D = k_B T\mu/q \tag{25}$$

for ions of charge q and mobility μ.

Much of the random molecular motion in liquids is of the rotational type and the probability $\psi_{(\theta,\phi,t)}$ that the axis of a molecule points in the direction θ, ϕ evolves in time as

$$\frac{\partial\psi}{\partial t} = \frac{D_{\rm rot}}{a^2}\left\{\frac{1}{\sin\theta}\frac{\partial}{\partial\theta}\sin\theta\frac{\partial\psi}{\partial\theta} + \frac{1}{\sin^2\theta}\frac{\partial^2\psi}{\partial\phi^2}\right\} \tag{26}$$

where a is the molecular radius. The correlation time τ_c for this process is inversely proportional to the diffusion constant

$$\tau_c = a^2/2D_{rot} \tag{27}$$

and it is often referred to as the Debye correlation time.

Diffusion on the atomic or molecular scale in low viscosity liquids is very rapid; for example, the rotational correlation time τ_c in water at 20°C is 3.5×10^{-12} s. However, diffusion over macroscopic distances, such as 1 cm, is a very slow process in undisturbed liquids, with time scales of perhaps hours or days, and mild stirring greatly enhances the speed of the process. Diffusion can occur in solids but the time scales are much longer than in liquids. Glasses are essentially frozen solids which have liquid-like arrangements of molecules but solid-like time scales for molecular diffusion.

6 STRUCTURE OF SOLIDS

Crystalline solids have long-range order as well as short-range order. If two adjacent molecules have a particular orientation relative to each other and relative to crystallographic axes, then two like molecules 10 or 100 lattice spacings away should have the same orientations. The atoms remain at regular sites in the lattice, although thermal energy causes them to undergo vibrational motion centered at this site, and some molecular groups such as methyl ($-CH_3$) can rotate in solids.

There are seven crystal systems for arranging atoms in a crystal lattice, and these are classified according to the lengths a, b, c and angles α, β, γ of the conventional unit cell, as shown in Table 23-1 and Fig. 23-3. Three of them have all of their axes at 90° relative to each other: the cubic system, in which all three lengths are the same; the tetragonal system, in which two of the three lengths are the same; and the orthorhombic system, in which all three lengths are different ($a \neq b \neq c$). Triclinic is the lowest possible symmetry with all lengths and all angles different. Monoclinic is second lowest in symmetry with all lengths different ($a \neq b \neq c$) and two angles of 90°. Trigonal has all three lengths and all three angles the same, but not 90°. Hexagonal is a special high symmetry case with two lengths the same and two 90° angles.

All conventional unit cells, shown in Fig. 23-3, have an atom on each vertex, there are eight vertices, and each vertex atom is shared with eight other unit cells, so a vertex atom counts as one atom per unit cell. A primitive unit cell (P) has atoms at vertex positions only. A body centered

TABLE 23-1. Characteristics of the seven crystal systems. The 14 lattice types are classified as primitive (P, 1 atom per unit cell), body centered (I, 2 atoms per cell), face centered (F, 4 atoms per cell), and side centered or top and bottom centered (B, C, 2 atoms per cell).

System	Axes	Angles	Lattice types
Cubic	$a = b = c$	$\alpha = \beta = \gamma = 90^\circ$	P, I, F
Tetragonal	$a = b \neq c$	$\alpha = \beta = \gamma = 90^\circ$	P, I
Orthohombic	$a \neq b \neq c$	$\alpha = \beta = \gamma = 90^\circ$	P, I, F, C
Hexagonal	$a = b \neq c$	$\alpha = \beta = 90^\circ,\ \gamma = 120^\circ$	P
Trigonal	$a = b = c$	$\alpha = \beta = \gamma < 120, \neq 90^\circ$	P (also called R)
Monoclinic	$a \neq b \neq c$	$\alpha = \beta = 90^\circ \neq \gamma$	P, B
Triclinic	$a \neq b \neq c$	$\alpha \neq \beta \neq \gamma$	P

unit cell (I) has an additional atom in the center of the cell, corresponding to two atoms per cell. A face centered cell (F) has atoms at vertices and the centers of the faces. Each face is shared by two unit cells, and there are six faces so there are four atoms per unit cell. Finally, there is the top and bottom centered case (C) and the equivalent side centered one (B) with two atoms per unit cell. Some crystal systems have two or more of these special crystal types, as shown in Fig. 23-3 and Table 23-1. The 14 crystal types are called Bravais lattices. Some characteristics of the three cubic Bravais lattices are given in Table 23-2. The packing fraction given in the table is the fraction of the cell volume occupied by the atoms.

7 CLOSE PACKING

A particularly important arrangement of atoms in solids is what is called close packing. The most efficient way to cover a flat surface with circles, all of the same size, is the hexagonal arrangement sketched in Fig. 23-4. This closest packing configuration covers 90.7% ($50\pi/\sqrt{3}\%$) of the surface, and provides the largest number of circles per unit area. The closest packing arrangement in three dimensions consists in stacking two-dimensional hexagonal layers one on top of the other so that the atoms in one layer fit between the atoms of the layers below and above. If the third layer is directly above the first, the fourth directly above the second, etc., corresponding to an A B A B A . . . layering scheme, then the crystal has a hexagonal close packed (HCP) structure. If, on the other hand, the third layer is put on in the third possible way so that it is not above the first or the

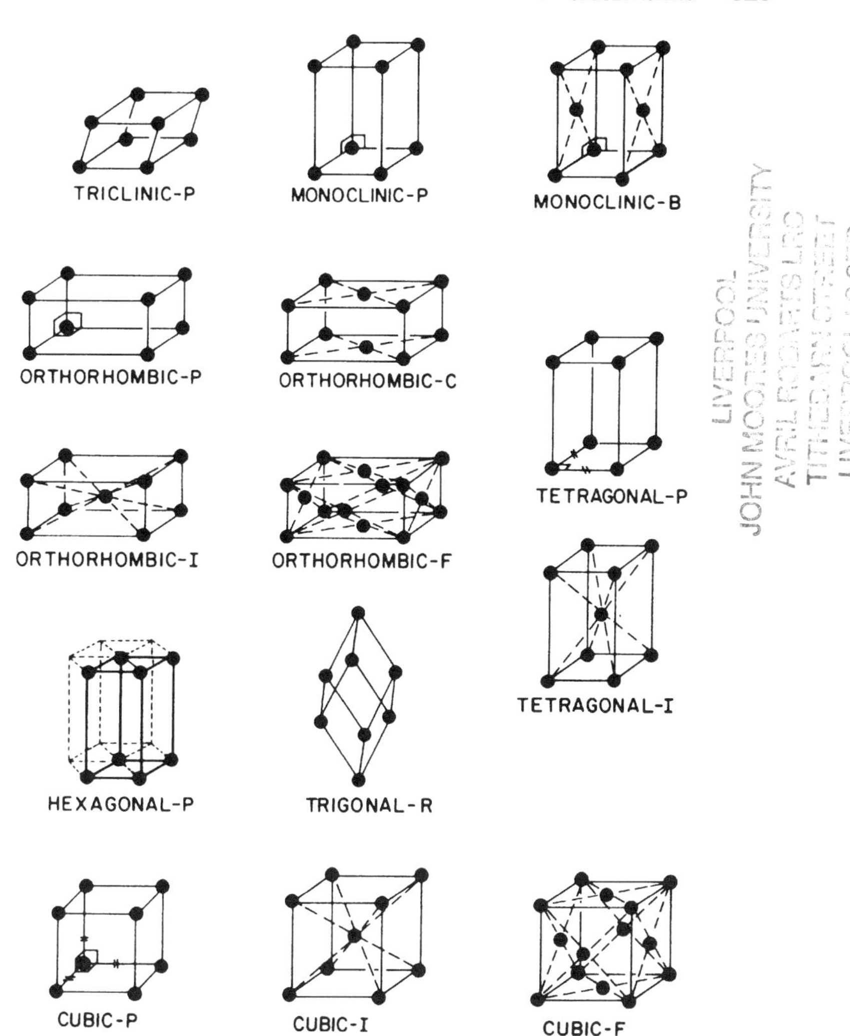

Fig. 23-3. Conventional unit cells of the 14 Bravais or space lattices. (From G. Burns, *Solid State Physics*, Academic Press, New York, 1985, p. 24.)

second, corresponding to an A B C A B C A . . . layering scheme, the result is a face centered cubic (FCC) structure. It is not easy to visualize that this layering scheme is equivalent to FCC. The FCC and HCP arrangements of spheres occupy 74.0% ($50\sqrt{2}\pi/3\%$) of the volume, corresponding to the packing fraction of Table 23-2.

TABLE 23-2. Characteristics of conventional cubic unit cells of lattice constant a displayed in Fig. 23-3

Characteristic	Primitive (simple)	Body centered	Face centered
Volume	a^3	a^3	a^3
Points of lattice	1	2	4
Nearest neighbors	6	8	12
Nearest neighbor distance	a	$\frac{1}{2}\sqrt{3}a = 0.866a$	$a/\sqrt{2} = 0.707a$
Second nearest neighbors	12	6	6
Second neighbor distance	$\sqrt{2}a = 1.414a$	a	a
Packing fraction	$\pi/6 = 0.524$	$\sqrt{3}\pi/8 = 0.680$	$\sqrt{2}\pi/6 = 0.740$

Wyckoff (Vol. 1, Chapter 2) lists 32 elements with HCP structures, 30 that are FCC, and 24 that have body centered cubic (BCC) structures. In addition there are a few elements with lower symmetry structures, and some such as carbon and tin have more than one structure at room temperature. In fact close packing arrangements occur quite commonly among compounds as well as among atoms alone. For example, consider the compounds formed from the three atoms Al, Mg, and O. The oxygen ions O^{2-} are large and the metallic ions Mg^{2+} and Al^{3+} are small, with the following ionic radii:

$$\begin{array}{ll} Al^{3+} & 0.051\ \text{nm} \\ Mg^{2+} & 0.066\ \text{nm} \\ O^{2-} & 0.132\ \text{nm} \end{array} \tag{28}$$

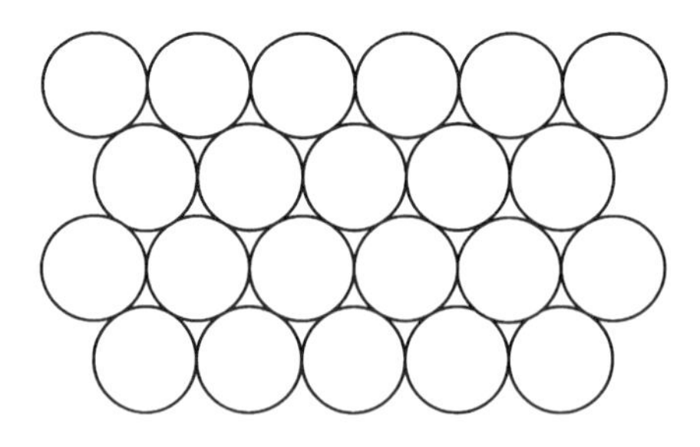

Fig. 23-4. Close packing arrangement of circles on a planar surface.

The crystal periclase, MgO, has a perfect arrangement of FCC oxygens with their octahedral sites occupied by magnesium ions which are also FCC. The crystal corundum, Al_2O_3, has a slightly distorted arrangement of HCP oxygens with the smaller aluminums lying in two-thirds of the octahedral sites between the oxygens. The crystal spinel, $MgAl_2O_4$, has an almost perfect FCC arrangement of oxygen ions with the Mg ions occupying one-eighth of the smaller tetrahedral sites with four oxygen nearest neighbors, and the Al ions occupying one-half of the larger octahedral sites with six oxygen nearest neighbors.

8 DIRECT AND RECIPROCAL LATTICE

To determine a crystal structure we need to know the basis vectors **a**, **b**, and **c** that define the unit cell, and then we need the coordinates u, v, w of all the atoms in that cell. For example, the ith atom has the position

$$\mathbf{r}_i = u_i\mathbf{a} + v_i\mathbf{b} + w_i\mathbf{c} \tag{29}$$

Knowing the basis vectors means knowing their lengths and knowing their angles or orientations in space. They define the direct or coordinate space lattice.

Associated with every direct lattice there is a reciprocal lattice with the basis vectors

$$\mathbf{A} = \frac{2\pi\mathbf{b}\times\mathbf{c}}{\mathbf{a}\cdot(\mathbf{b}\times\mathbf{c})} \qquad \mathbf{B} = \frac{2\pi\mathbf{c}\times\mathbf{a}}{\mathbf{a}\cdot(\mathbf{b}\times\mathbf{c})} \qquad \mathbf{C} = \frac{2\pi\mathbf{a}\times\mathbf{b}}{\mathbf{a}\cdot(\mathbf{b}\times\mathbf{c})} \tag{30}$$

These basis vectors have the following orthonormality properties:

$$\begin{gathered} \mathbf{a}\cdot\mathbf{A} = \mathbf{b}\cdot\mathbf{B} = \mathbf{c}\cdot\mathbf{C} = 2\pi \\ \mathbf{a}\cdot\mathbf{B} = \mathbf{a}\cdot\mathbf{C} = \mathbf{b}\cdot\mathbf{A} = \mathbf{b}\cdot\mathbf{C} = \mathbf{c}\cdot\mathbf{A} = \mathbf{c}\cdot\mathbf{B} = 0 \end{gathered} \tag{31}$$

We are interested in reciprocal lattice vectors **G**

$$\mathbf{G} = h\mathbf{A} + k\mathbf{B} + \ell\mathbf{C} \tag{32}$$

for which the coefficients h, k, ℓ are integers. These integers define a crystallographic plane, and **G** has the physical significance that the magnitude of its reciprocal provides the distance d between the $hk\ell$ planes

$$d_{hk\ell} = 2\pi/|\mathbf{G}| \tag{33}$$

The integers h, k, ℓ are called Miller indices, and their reciprocals are proportional to the intercepts a_M, b_M, c_M of the corresponding $(hk\ell)$ lattice plane with the a, b, and c axes, respectively, as shown in Fig. 23-5. In other words, the ratios $a_M{:}b_M{:}c_M$ are the same as the ratios $h^{-1}{:}k^{-1}{:}\ell^{-1}$, and the Miller indices h, k, ℓ are the smallest set of integers that satisfy these ratios.

9 CRYSTAL STRUCTURE DETERMINATION

Crystal structures are determined by reflecting X-ray photons, electrons, or neutrons off the atomic planes of crystal lattices, and thereby obtaining the spacings and orientations of these planes. The incoming ray of known wave vector $\mathbf{k}$ reflects off the crystal to produce the outgoing wave vector $\mathbf{k}'$ which is measured to provide the difference wave vector $\mathbf{\Delta}k$

$$\mathbf{k} + \Delta\mathbf{k} = \mathbf{k}' \tag{34}$$

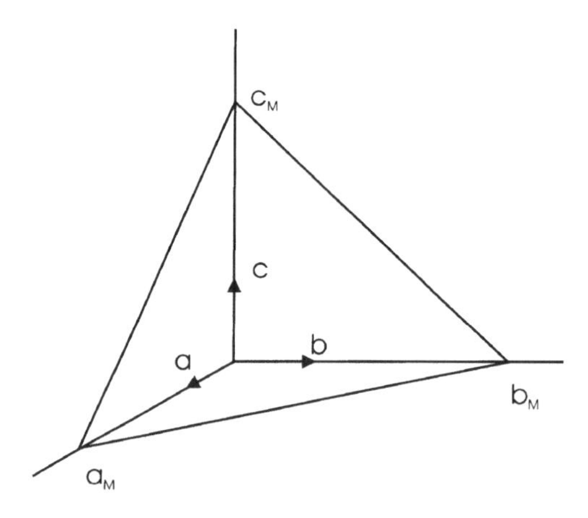

Fig. 23-5. Crystallographic plane $(hk\ell)$ and its intercepts a_M, b_M, and c_M along the three coordinate axes a, b, c respectively. The Miller indices h, k, ℓ of the plane are the smallest set of integers proportional, respectively, to the receiprocals $1/a_M$, $1/b_M$, and $1/c_M$.

The outgoing wave $\mathbf{k}'$ is high in intensity when the difference wave vector is equal to a reciprocal lattice vector, $\mathbf{\Delta k} = \mathbf{G}$, with integer coefficients, as defined by Eq. (32). Taking the scalar product of both sides gives

$$(\mathbf{k} + \mathbf{G})^2 = k'^2 \tag{35}$$

Since the reflection involves elastic scattering with no change in wavelength the magnitudes $k = k' = 2\pi/\lambda$ are equal, and we have Bragg's law in reciprocal lattice notation

$$2\mathbf{k} \cdot \mathbf{G} + G^2 = 0 \tag{36}$$

With the observation from Fig. 23-6 that $\pi/2 - \theta$ is the angle between the vectors $\mathbf{k}$ and $\mathbf{G}$, and the aid of Eq. (33) Bragg's law may be written in a more familiar way

$$2d \sin\theta = n\lambda \tag{37}$$

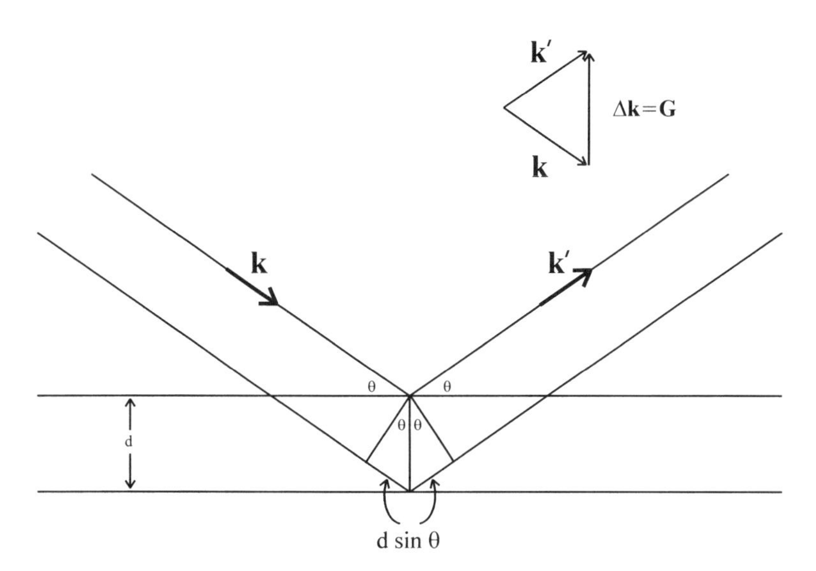

Fig. 23-6. An X-ray beam incident along the **k** direction shown reflected off two parallel planes separated by the distance d to produce the outgoing beam with wave vector $\mathbf{k}'$, where $|\mathbf{k}| = |\mathbf{k}'|$.

where the minus sign is dropped because **G** and $-\mathbf{G}$ are equally valid reciprocal lattice vectors. Figure 23-6 shows the geometry justifying this relation.

When X-ray photons form the incoming wave, the expression for the energy

$$E = \hbar\omega = hc/\lambda \tag{38}$$

gives the following relation between the wavelength in nm and the energy in keV:

$$\lambda = 1.24/E \quad \text{X-rays} \tag{39}$$

For electron and neutron beams the incoming energy is kinetic

$$E = \tfrac{1}{2}mv^2 = \hbar^2k^2/2m = h^2/2m\lambda^2 \tag{40}$$

which gives the wavelength (nm) versus energy (measured in electron volts eV) relationships

$$\lambda = 1.2/\sqrt{E} \qquad \text{electrons} \tag{41a}$$

$$\lambda = 0.028/\sqrt{E} \quad \text{neutrons} \tag{41b}$$

10 WIGNER–SEITZ CELL

It is easy to see that the 14 conventional unit cells displayed in Fig. 23-3 can fill space. They, however, suffer from two liabilities. All of them have atoms on the vertices instead of only in the middle, and the seven non-primitive ones have two or four atoms per cell. For all 14 Bravais lattices there is a systematic way to generate a primitive, space filling unit cell called a Wigner–Seitz cell which contains only one atom located in the center, i.e., $u_i = v_i = w_i = \frac{1}{2}$ in Eq. (29). We illustrate how to generate such a cell in two dimensions and then extend the process to three dimensions.

Consider one atom of the oblique two-dimensional lattice sketched in Fig. 23-7(a). A straight line is drawn from this particular atom to each near neighbor, and a perpendicular line is constructed at the center of and perpendicular to each such line. The resulting lines enclose the atom in what is usually an irregular hexagon which fills space. This hexagon is a primitive

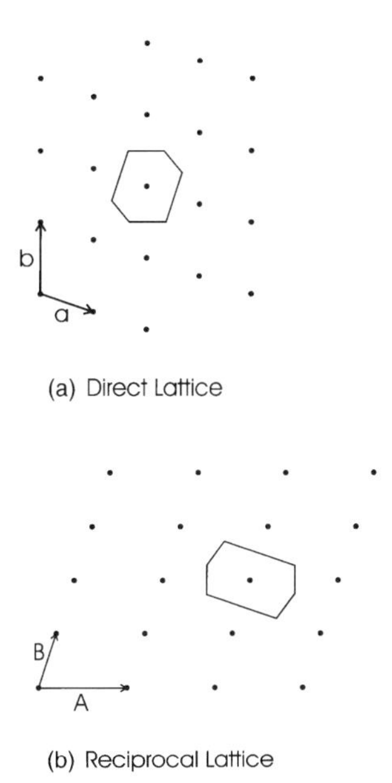

Fig. 23-7. Part (a) of the figure shows the construction of a Wigner–Seitz unit cell in an oblique two-dimensional coordinate space lattice with basis vectors **a** and **b**. Part (b) shows the construction of a Brillouin zone in the associated reciprocal lattice with basis vectors **A** and **B**. The coordinate space direct lattice and the k-space reciprocal lattice are duals of each other, related through Eqs. (30) and (31).

unit cell of the two-dimensional oblique lattice. The analogous construction for the associated two-dimensional oblique reciprocal lattice is shown in Fig. 23-7(b), where the two pairs of basis vectors satisfy expressions (31): $\mathbf{a} \cdot \mathbf{A} = \mathbf{b} \cdot \mathbf{B} = 2\pi$ and $\mathbf{a} \cdot \mathbf{B} = \mathbf{b} \cdot \mathbf{A} = 0$.

A similar procedure is followed in three dimensions. A straight line is drawn from a particular atom to each near neighbor, and a perpendicular plane is constructed at the center of and perpendicular to each such line. The resulting planes enclose the atom in a space filling polyhedron which constitutes a primitive unit cell of the three-dimensional lattice. The Wigner–Seitz cell of a simple cubic lattice is also a cube, but with the atom in the center at the $[\frac{1}{2}\frac{1}{2}\frac{1}{2}]$ position, instead of on the vertices. Figure 23-8(a) shows the tetrakaidekahedron, really a truncated regular octahedron, which constitutes the BCC Wigner-Seitz cell, and Fig. 23-8(b) provides a sketch of the rhombic dodecahedron FCC Wigner–Seitz cell.

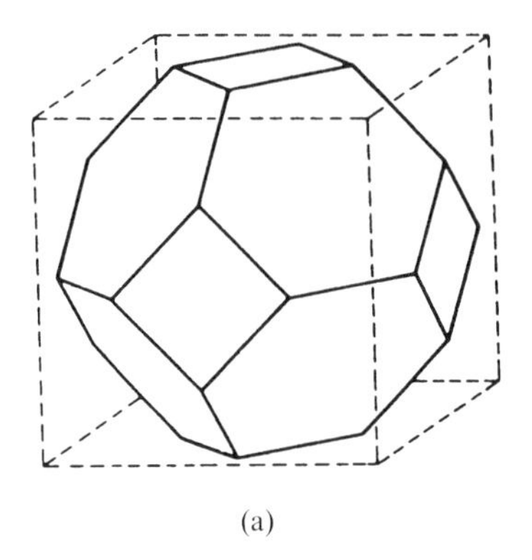

(a)

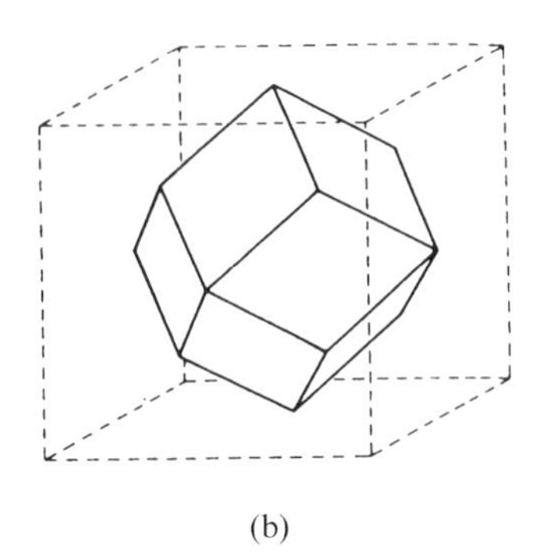

(b)

Fig. 23-8. Wigner–Seitz cell in a three-dimensional coordinate space (a) BCC lattice, and (b) FCC lattice. These figures (a) and (b) also constitute Brillouin zones associated, respectively, with FCC and BCC coordinate space lattices. (From G. Burns, *Solid State Physics*, Academic Press, New York, 1985, pp. 270, 271.)

The primitive unit cell generated in the reciprocal lattice using the Wigner–Seitz construction is called a Brillouin zone. Figure 23-7(b) shows the reciprocal lattice and Brillouin zone for the two-dimensional oblique lattice of Fig. 23-7(a). In three dimensions the reciprocal lattice of a FCC direct lattice is BCC, and the reciprocal lattice of a BCC direct lattice is FCC. This means that Fig. 23-8(a) is a picture of the Wigner–Seitz cell of a BCC direct lattice, and it is also the picture of the Brillouin zone corresponding to a FCC direct lattice. An analogous statement could be made of Fig. 23-8(b).

11 PHONONS AND OTHER PARTICLES

In Chapters 4 and 8 we discussed lattice vibrations of solids. These vibrations can be localized in individual molecules or groups of atoms, or they can involve many atoms vibrating coherently with each other. Localized lattice vibrations called phonons can move through solids, and be reflected and refracted. In an ionic solid phonons involving positive and negative charges vibrating in phase have low frequencies and are called acoustic phonons, and when the positive and negative charges vibrate out of phase the phonons have high frequencies and are called optical. Phonons can also involve longitudinal and transverse vibrations of atoms. The locations of earthquakes can be determined by timing the arrival of the longitudinal, transverse, and surface phonons called seismic waves which travel at different speeds.

Transverse optical phonons and electromagnetic waves can interact, and a quantum of this coupled photon–phonon field is called a polariton.

An electrical conductor has a background of positive ions with delocalized electrons moving around through the positive background. This constitutes a plasma, an equal concentration of positive and negative charges with at least one type being mobile. The electrons of this plasma can oscillate, and a plasma oscillation or charge density wave is a collective excitation of the gas of conduction electrons. There is a characteristic plasma frequency ω_p given by

$$\omega_p^2 = ne^2/\varepsilon m \tag{42}$$

where n is the electron concentration. Incident electromagnetic waves will be reflected if their frequency is less that ω_p, and they will be transmitted for $\omega > \omega_p$. This explains the reflectivity of common metals in the visible region, and their transparency in the ultraviolet. A plasmon is a quantum of plasma oscillation that can exhibit particle type properties.

The electron moving against the positive ion background will tend to polarize nearby atoms and strain the lattice in its vicinity. The electron plus its induced lattice polarization strain field is mobile, and constitutes a polaron.

A coupled electron and hole, i.e., an electron coupled to a positive charge due to a missing electron, constitutes an entity called a Mott–Wannier exciton. This electrically neutral 'particle' has hydrogen atom like energies E_n, but in semiconductors they are much lower in magnitude because of the large dielectric constant ε and small effective mass m^*:

$$E_n = -\frac{m^*/m_0}{(\varepsilon/\varepsilon_0)^2}\frac{E_0}{n^2} \tag{43}$$

where m_0 is the mass of a free electron and ε_0 is the dielectric constant of free space. The hydrogen atom ground state energy is $E_0 = -13.6\,\text{eV}$ and the principal quantum number n takes on the values $n = 1, 2, 3, 4 \ldots$. For silicon and germanium the ground state ($n = 1$) exciton energies are -14.7 and -4.15 milli electron volts (meV), respectively. A hydrogen atom has a Bohr radius $a_0 = 0.51\,\text{Å}$, and the corresponding exciton radius a_e given by

$$a_e = \frac{\varepsilon/\varepsilon_0}{m^*/m_0}a_0 \tag{44}$$

is much larger than a_0. The Mott–Wannier exciton can move through the lattice transporting its excitation energy from place to place.

There is another type of exciton called a Frenkel exciton corresponding to an excited electronic state of a single atom which can move around the lattice by hopping from one atom to another due to the coupling between neighbors. This exciton motion can take the form of propagating waves. The energy states of Frenkel excitons are generally much larger than those of the Mott–Wannier type.

12 SUPERCONDUCTIVITY

A superconductor is a material that has two characteristic properties, it has zero electrical resistance, and it excludes magnetic flux from its interior. A superconducting wire carries electrical current without any dissipation of energy, a property that can be potentially important for the long-range transportation of electricity, for magnetic energy storage, and for the construction of lossless electronic devices such as computers.

A material that superconducts does so below a characteristic temperature T_c called the superconducting transition temperature. Prior to 1986 the highest known T_c was 23.2 K achieved in 1973 by the A15 type compound Nb_3Ge. The discovery of high temperature superconductivity in 1986 by J. G. Bednorz and K. H. Müller opened up a new era, and a decade later the transition temperature is up to 133 K for the compound $HgBa_2Ca_2Cu_3O_8$, and 164 K has been reached at a pressure of 30 GPa. Conventional superconductors such as NbTi must be operated at liquid helium temperature, 4.2 K, which is very costly. High temperature superconductors, on the other hand, can operate at the temperature of liquid nitrogen, 77 K, which is much less expensive.

A type I superconductor is a perfect diamagnet since it has a susceptibility $\chi = -1$, and from the expression

$$B = \mu H = \mu_0(H + M) = \mu_0 H(1 + \chi) \tag{45}$$

this makes the magnetization $M = \chi H$ negative so it cancels the internal H field, and as a result the magnetic B field vanishes inside the material. High temperature superconductors are type II, which means that they have zero electrical resistance, but their susceptibility, although negative, is not as negative as -1, so they only partially expel magnetic flux. The flux that is present is in the form of vortices or tubes of normal material each of which contains one quantum of flux $\Phi_0 = h/2e = 2.068 \times 10^{-15}$ tesla m^2 which is called a fluxon. The expulsion of flux causes the superconductor to repel magnets and any magnetic fields in its vicinity. This is the property responsible for the levitation of magnets by superconductors. Some day this may provide us with frictionless transportation by trains levitated above their tracks.

The carriers of supercurrent are paired electrons called Cooper pairs, and ordinarily the superelectron energy state is separated from the higher energy normal state conduction band by an energy gap E_g. The two characteristic length parameters of a superconductor are the coherence length ξ and the penetration length λ_L given by

$$\xi = \frac{2\hbar V_F}{\pi E_g} \qquad \lambda_L = \left(\frac{m_e}{\mu_0 n_s e^2}\right)^{\frac{1}{2}}$$

where V_F is the Fermi velocity and n_s is the density of superelectrons. At the interface between a normal metal and a superconductor n_s builds up from zero at the boundary to its value far inside over a length scale determined by ξ. A magnetic field outside a Type I superconductor penetrates a distance λ_L into its interior. A superconductor is Type I for $(\lambda_L/\xi) < 1/\sqrt{2}$ and Type II for $(\lambda_L/\xi) > 1/\sqrt{2}$.

Conventional electromagnets are difficult to build for magnetic fields above about 2 tesla, but high temperature superconductors can, in principle, attain fields two orders of magnitude higher. The main technological problems holding up the achievement of this goal are the necessity of improving the ductility of the wire material, and the need to increase the amount of electric current that the superconducting wires can carry without going normal.

CHAPTER 24

CONDUCTION IN SOLIDS

1 INTRODUCTION

In atoms the energy levels are narrow, and when atoms form compounds the interaction between the atoms broadens the energy levels of the outer electrons into bands in the manner shown in Fig. 24-1. The X-ray emission from

Fig. 24-1. Energy levels of an atom (left) broadened into energy bands (right) when the atoms form a solid.

the innermost K and L shells ($n = 1, 2$) of atoms in solids provides sharp lines since the inner shells of adjacent atoms in solids are shielded from each other, and so are not broadened into bands. The electrical conduction of solids is determined by the configuration of the outer broadened electron bands. If the outermost occupied band is full it constitutes a valence band, and cannot carry current. If, in addition, the next higher band, called the conduction band lies far above the top of the valence band, i.e., $E_g \gg k_B T$, so that it cannot be thermally populated, then the compound is an insulator. If the conduction band is half full of delocalized electrons, then the material is a conductor. If the energy gap E_g, shown on the figure, is comparable with $k_B T$, then some electrons will be promoted or raised from the full valence band to the empty conduction band and the material is an intrinsic semiconductor. If the material is doped to enhance this process of electron promotion, then the material is a more efficient semiconductor.

As an example, consider the ionic compound NaCl in which the sodium atom gives its valence electron to chlorine

$$\mathrm{Na} \Rightarrow \mathrm{Na}^+ + \mathrm{e}^- \tag{1a}$$

$$\mathrm{Cl} + \mathrm{e}^- \Rightarrow \mathrm{Cl}^- \tag{1b}$$

so both Na^+ and Cl^- have filled electron shells, the outer energy band called the valence band is full, the empty conduction band lies far above, and the material is an insulator. Solid sodium, on the other hand, has a background lattice of Na^+ ions whose electrons fill the inner shells and the valence band, and in addition there are delocalized electrons from the ionization process of

Eq. (1a) which half fill the conduction band, thereby making solid sodium a good conductors.

The present chapter begins by summarizing some of the theoretical approaches to electron conduction, and then discusses the frequency and temperature dependences of electrical conductivity. A brief mention will be made about the relationship between electrical and thermal conductivity. Then we explain the Fermi surface and energy bands in k-space, and apply these ideas to conductors and semiconductors. Finally we will conclude by showing how the Hall effect can provide important information on the conduction process.

2 ELECTRON TRANSPORT THEORIES

The electrical conductivity of a metal may be described most simply in terms of the constituent atoms losing their valence electrons to form a background lattice of positive ions called cations through which the delocalized conduction electrons move. The number density n of conduction electrons in a metallic element of density ρ, atomic mass number A, and valence Z is given by

$$n = N_A Z\rho/A \tag{2}$$

where N_A is Avogadro's number. The typical values of n for electrons in metallic conductors ($8.47 \times 10^{28}/\mathrm{m}^3$ for Cu) are more than a thousand times greater than the number density of molecules in a gas at room temperature and atmospheric pressure ($2.7 \times 10^{25}/\mathrm{m}^3$).

The simplest approximation to explain electrical conductivity is the Drude model, which makes several assumptions about conduction electrons:

1. They do not interact with the cations (free electron approximation) except when one collides elastically with a cation, on the average $1/\tau$ times per second, abruptly and randomly changing the direction of its velocity $\mathbf{v}$ (relaxation time approximation).
2. They maintain thermal equilibrium through collisions, using Maxwell–Boltzmann statistics (classical statistics approximation).
3. They do not interact with each other (independent electron approximation).

This model predicts many general features of electrical conduction, but it fails to account for other such as tunneling and band gaps.

More satisfactory explanations of electron transport relax or discard one or more of these assumptions. Ordinarily one abandons the free electron approximation by having the electrons move in a periodic potential arising from the background lattice of positive ions. The classical statistics assumption is generally replaced by recognizing that the electrons obey Fermi–Dirac statistics with the distribution of velocities $f_0(\mathbf{v})$ given by

$$f_0(\mathbf{v}) = \frac{1}{\exp[(\frac{1}{2}mv^2 - \mu)/k_B T] + 1} \tag{3}$$

where μ, called the chemical potential, is the energy involved in removing an electron from this gas. Non-interacting conduction electrons treated by Fermi-Dirac statistics are said to constitute a Fermi gas.

The relaxation time approximation assumes that the velocity distribution function $f(\mathbf{v}, t)$ is time dependent, and when it is disturbed from equilibration then collisions bring it back to its equilibrium state f_0 with a time constant τ in accordance with the expression

$$\frac{df}{dt} = \frac{f(t) - f_0}{\tau} \tag{4}$$

Ordinarily the relaxation time τ is assumed to be independent of the velocity, resulting in a simple exponential return ($e^{-t/\tau}$) to equilibrium. In systems of interest, $f(\mathbf{v}, t)$ always remains close to equilibrium (3).

3 ELECTRICAL CONDUCTIVITY

When a potential difference exists between two points along a conducting wire, a uniform electric field E is established which exerts the force $F = -eE$ that accelerates the electrons

$$-eE = m(dv/dt) \tag{5}$$

and during a time t of the order of the collision time τ they attain the velocity

$$v = -(eE/m)\tau \tag{6}$$

The electron motion consists of successive periods of acceleration interrupted by collisions, and on the average each collision reduces the electron velocity to zero before the start of the next acceleration.

We can write an expression for the current density J

$$J = ne\langle v\rangle = (ne^2\tau/m)E \tag{7}$$

by assuming that the average velocity $\langle v\rangle$ of the electrons is given by Eq. (6). The dc electrical conductivity σ_0 is defined by Ohm's Law

$$j = \sigma_0 E \tag{8}$$

and the resistivity ρ_0 is the reciprocal of the conductivity

$$\sigma_0 = 1/\rho_0 = ne^2\tau/m \tag{9}$$

Metals typically have room temperature resistivities between 1.5 and 20 μΩcm, with Cu, Ag, and Au between 1.5 and 2. High temperature superconductors in the normal state fall between 300 and 10,000 μΩcm. Semiconductor resistivities have values from 10^4 to 10^{15} μΩcm, and insulators range from 10^{20} to 10^{28} μΩcm.

4 AC ELECTRICAL CONDUCTIVITY

When a harmonically varying electric field $E = E_0 e^{-i\omega t}$ acts on the conduction electrons they are periodically accelerated in the forward and backward directions as E reverses sign every cycle. They also undergo random collisions with the average time τ between them. The collisions, which interrupt their regular oscillations, may be taken into account by adding the frictional damping term p/τ to Eq. (5).

$$d\mathbf{p}/dt + \mathbf{p}/\tau = -e\mathbf{E} \tag{10}$$

where $p = mv$ is the momentum. The momentum has the same harmonic time variation, $p = mv_0 e^{-i\omega t}$, and if we substitute this into Eq. (10) and solve for the velocity v_0 we obtain

$$v_0 = \frac{-eE_0}{m}\frac{\tau}{1 - i\omega\tau} \tag{11}$$

Comparing this with Eqs. (7) and (9) with v_0 playing the role of $\langle \mathrm{v} \rangle$ we obtain the ac conductivity $\sigma(\omega)$

$$\sigma(\omega) = \frac{\sigma_0}{1 - i\omega\tau} \tag{12}$$

This reduces to the dc case $\sigma(\omega) = \sigma_0$ when the frequency ω is zero.

In the low frequency limit $\omega\tau \ll 1$ many collisions occur during each cycle of the E field, the average electron motion follows the oscillations, and $\sigma(\omega) \sim \sigma_0$. On the other hand, in the high frequency limit $\omega\tau \gg 1$ then E oscillates much more rapidly than collisions occur and the electrical conductivity becomes predominantly imaginary, corresponding to a reactive impedance. For very high frequencies the collision rate becomes immaterial and the electron gas behaves like a plasma, i.e., an electrically neutral ionized gas in which the negative charges are mobile electrons and the positive charges are fixed in position. Electromagnetic wave phenomena can be described in terms of the frequency dependent dielectric constant $\varepsilon(\omega)$

$$\varepsilon(\omega) = \varepsilon_0 \left(1 - \frac{\omega_\mathrm{p}^2}{\omega^2} \right) \tag{13}$$

where ω_p is the plasma frequency

$$\omega_\mathrm{p} = (ne^2/\varepsilon_0 m)^{1/2} \tag{14}$$

Thus ω_p has the physical significance of being the characteristic frequency of the conduction electron plasma below which the dielectric constant is negative so electromagnetic waves cannot propagate, and above which ε is positive and propagation occurs. As a result metals are opaque for $\omega < \omega_\mathrm{p}$ and transparent for $\omega > \omega_\mathrm{p}$.

5 RESISTIVITY

Electrons moving through a metallic conductor are not only scattered by phonons but also by lattice defects, impurity atoms, and other imperfections in the otherwise perfect lattice. These impurities produce a temperature independent contribution which puts an upper limit on the overall electrical conductivity of the metal.

According to Matthiessen's rule the conductivities arising from the impurity and phonon contributions add as reciprocals, which is the same as saying that their respective individual resistivities, ρ_0 and ρ_{ph}, respectively, add directly to give for the total resistivity $\rho(T)$

$$\rho(T) = \rho_0 + \rho_{ph}(T) \tag{15}$$

The phonon term $\rho_{ph}(T)$ is proportional to the temperature T at high temperatures and to T^5 via the Bloch law at low temperatures. Near and above room temperature the impurity contribution is negligible, so the resistivity of metallic elements is roughly proportional to the temperature

$$\rho(T) \approx \rho(300K)[T/300] \qquad 250\text{K} < T \tag{16}$$

At low temperatures, far below the Debye temperature, the Bloch T^5 law applies to give

$$\rho(T) = \rho_0 + AT^5 \qquad T \ll \Theta_D \tag{17}$$

We see from Eqs. (16) and (17) that metals have a positive temperature coefficient of resistivity so they become better conductors at low temperature. In contrast to this the resistivity of a semiconductor has a negative temperature coefficient so it increases with decreasing temperature. As is explained in Section 9, this occurs because lowering the temperature of an n-type material causes thermally excited electrons in the conduction band to return to their positions on donor atoms, and lowering T for a p-type causes the excited electrons on acceptor sites to return to the valence band. Both processes reduce the number of mobile charge carriers, and the result is an increase in the resistivity of a semiconductor as the temperature is lowered.

6 THERMAL CONDUCTIVITY

When a temperature gradient exists in a metal the motion of the conduction electrons involves the transport of heat in the form of kinetic energy from the hotter to the cooler regions. In good conductors such as copper and silver this transport involves the same phonon collision processes that are responsible for the transport of electric charge. The ratio $k_{th}/\sigma T$ involving the thermal (K_{th}, J cm^{-1} s^{-1} K^{-1}) and electrical (σ, Ω^{-1} cm^{-1}) conductivities for most common metallic elements in the range from 273 K to 373 K is between twice and thrice the value predicted by the Law of Wiedermann and Franz

$$K_{\text{th}}/\sigma T = \tfrac{3}{2}(k_{\text{B}}/e)^2 \quad (18)$$

$$= 1.11 \times 10^{-8} \text{W}\Omega/\text{K}^2 \quad (19)$$

where the universal constant $\frac{3}{2}(k_{\text{B}}/e)^2$ is called the Lorenz number.

7 FERMI SURFACE

Conduction electrons obey Fermi–Dirac (FD) statistics, and the FD distribution function (3) written in terms of the energy E

$$f(E) = \frac{1}{\exp[(E-\mu)/k_{\text{B}}T] + 1} \quad (20)$$

is plotted in Chapter 9, Fig. 9-3(a) for $T = 0$ and in Fig. 9-3(b) for $0 < T \ll T_{\text{F}}$. The chemical potential μ corresponds through the expression

$$\mu \approx E_{\text{F}} = k_{\text{B}}T_{\text{F}} \quad (21)$$

to a Fermi temperature T_{F} that is typically in the neighborhood of 10^5 K. This means that the distribution function $f(\mathbf{v})$ is 1 for energies below E_{F}, is 0 above E_{F}, and assumes intermediate values only in a region $k_{\text{B}}T$ wide near E_{F}, as shown in Fig. 9-2(b).

The electron kinetic energy can be written in several ways

$$E_{\text{K}} = \tfrac{1}{2}mv^2 = p^2/2m = \hbar^2k^2/2m = (\hbar^2/2m)(k_x^2 + k_y^2 + k_z^2) \quad (22)$$

where $p = \hbar k$, and the energy is quantized in reciprocal or k space. For a rectangular solid of dimensions L_x, L_y, and L_z each cartesian component of $\mathbf{k}$ can assume the discrete values $2\pi m_x/L_x$ in the x direction, and likewise for the y and z directions, where m_x is an integer between 1 and L_x/a, and similarly for m_y and m_z.

The one-dimensional case corresponding to the energy

$$E_k = \hbar^2k_x^2/2m = E_0(k_xa/\pi)^2 \quad (23)$$

where E_0 is the energy at the edge of the first Brillouin zone ($k_x = \pm\pi/a$)

$$E_0 = (\pi^2\hbar^2/2ma^2) \quad (24)$$

is sketched in Fig. 24-2. It is of interest to express the energy in terms of k_x values that lie within the first Brillouin zone

$$-\pi/a \leq k_x \leq +\pi/a \tag{25}$$

and to do so we write

$$E_k = E_0(2n_x + k_x a/\pi)^2 \tag{26}$$

where

$$n_x, n_y = 0, \pm 1, \pm 2, \ldots \tag{27}$$

for the various energies.

At absolute zero the k-space levels of Eq. (23) are doubly occupied by electrons of opposite spin up to the Fermi energy E_F

$$E_F = \hbar^2 k_F^2/2m \tag{28}$$

where we have returned to the three-dimensional case. Partial occupancy occurs in the narrow region of width $k_B T$ at E_F shown in Fig. 9-3(b). For simplicity we assume a cubic shape in coordinate space, so

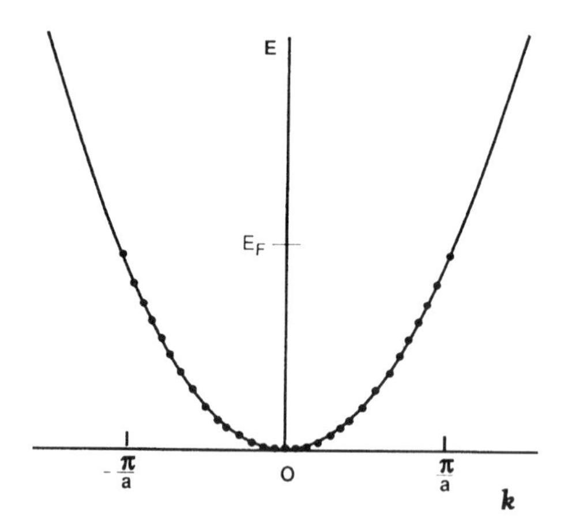

Fig. 24-2. One-dimensional free electron energy band shown occupied up to the Fermi energy E_F which is at the first Brillouin zone boundaries $k = \pm\pi/a$. (From C. P. Poole, Jr et al., *Superconductivity*, Academic Press, New York, 1995, p. 9.)

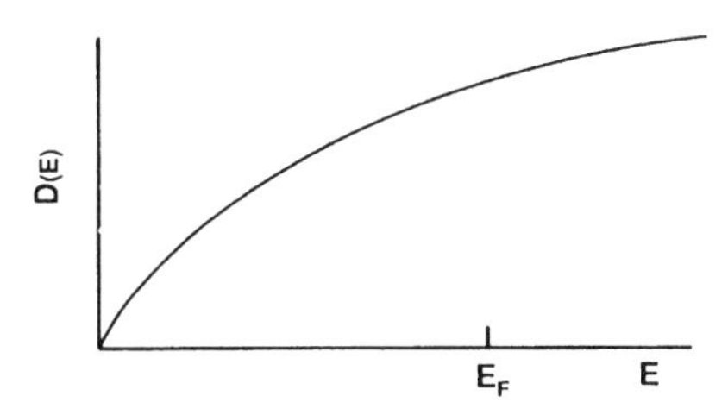

Fig. 24-3. Density of states of a free electron energy band in three dimensions. (From C. P. Poole, Jr et al., *Superconductivity*, Academic Press, New York, 1995, p. 10.)

$L_x = L_y = L_z = L$, and a spherical Fermi surface in reciprocal space, to give for the total number of electrons N

$$N = 2\frac{\text{occupied } k\text{-space volume}}{k\text{-space volume per electron}} = 2\frac{4\pi k_{\mathrm{F}}^3/3}{(2\pi/L)^3} \tag{29}$$

The electron density $n = N/V = N/L^3$ at the energy $E = E_{\mathrm{F}}$ is

$$n = k_{\mathrm{F}}^3/3\pi^2 = \frac{1}{3\pi^2}(2mE_{\mathrm{F}}/\hbar^2)^{3/2} \tag{30}$$

and the density of states $D(E)$ per unit volume obtained from evaluating the derivative dn/dE of this expression (with E_{F} replaced by E) is

$$D(E) = \frac{d}{dE}n(E) = \frac{1}{2\pi^2}(2m/\hbar^2)^{2/3}E^{1/2} = D(E_{\mathrm{F}})[E/E_{\mathrm{F}}]^{1/2} \tag{31}$$

and this is shown sketched in Fig. 24-3. With the aid of Eqs. (30 and (31), respectively, the density to states at the Fermi level can be written in two equivalent ways

$$D(E_{\mathrm{F}}) = \begin{cases} 3n/2k_{\mathrm{B}}T_{\mathrm{F}} \\ mk_{\mathrm{F}}/\hbar^2\pi^2 \end{cases} \tag{32}$$

for this isotropic case in which the energy is independent of direction in k-space so the Fermi surface is spherical. In many actual conductors, including the high temperature superconductors in their normal states above T_c, this is not the case, and $D(E)$ is a more complicated expression.

It is convenient to express the electron density n and the total electron energy E_T in terms of integrals over the density of states

$$n = \int D(E)\, f(E)\, dE \tag{33}$$

$$E_T = \int D(E)\, f(E)\, E\, dE \tag{34}$$

The product $D(E)f(E)$ that appears in these integrands is shown plotted versus energy in Fig. 24-4(a) for $T = 0$ and in Fig. 24-4(b) for $0 < T \ll T_F$.

The free electron kinetic energy of Eq. (22) is obtained from the plane wave solution $\phi = e^{-i\mathbf{k}\cdot\mathbf{r}}$ of the Schrödinger equation

$$-(\hbar^2/2m)\nabla^2\phi(\mathbf{r}) + V(\mathbf{r})\phi(\mathbf{r}) = E\phi(\mathbf{r}) \tag{35}$$

with the potential $V(\mathbf{r})$ set equal to zero. When a potential is included in the Schrödinger equation the free electron energy parabola in the one-dimen-

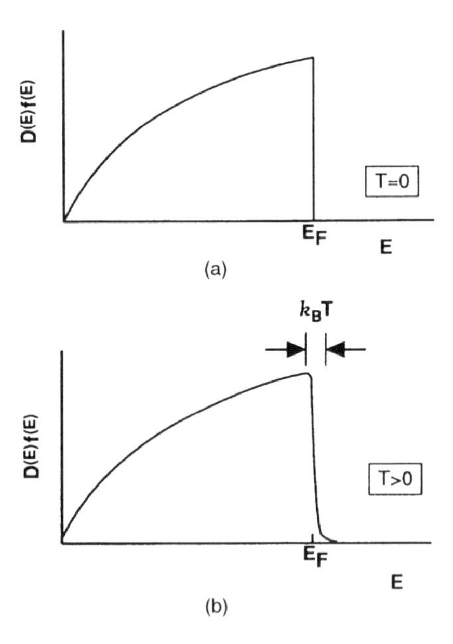

Fig. 24-4. Energy dependence of the occupation of a free electron energy band by electrons (a) at 0 K, and (b) for $T > 0$ K. (From C. P. Poole, Jr et al., *Superconductivity*, Academic Press, New York, 1995, p. 10.)

sional case of Fig. 24-2 develops energy gaps, as indicated in Figs. 24-5 and 24-8(a). These gaps appear at the boundaries $k = \pm\pi/a$ of the unit cell in k-space, called the first Brillouin zone, and of successively higher Brillouin zones, as shown. The energy levels are closer near the gap, which means that the density of states $D(E)$ is larger there, as shown in Figs. 24-6 and 24-7. For weak potentials, $|V| \ll E_F$, the density of states is close to its free electron form away from the gap, as indicated on the figures. The number of points in k-space remains the same, i.e., is conserved, when the gap forms; it is the density $D(E)$ that changes.

If the kinetic energy near an energy gap is written in the free electron form

$$E_K = \hbar^2 k^2/2m^* \tag{36}$$

then the first derivative of E_K with respect to k gives the electron velocity $v = \hbar k/m^*$ near the gap

$$v = \frac{1}{\hbar}\left(\frac{dE_K}{dk}\right)_{E_F} \tag{37}$$

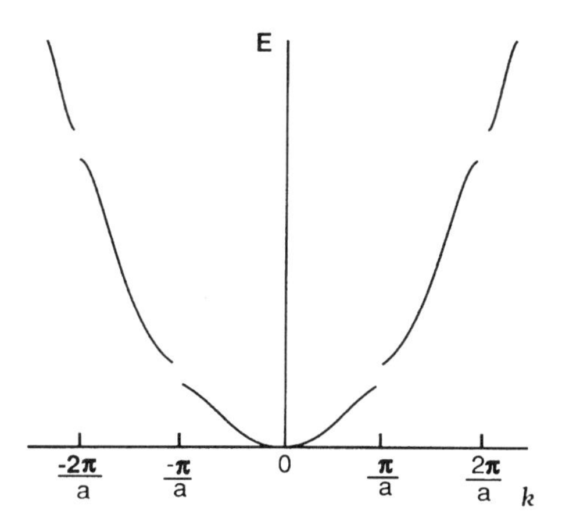

Fig. 24-5. One-dimensional free electron energy band perturbed by the presence of a weak periodic potential. The gaps open up at the zone boundaries $k = \pm n\pi/a$, where $n = 1, 2, 3, \ldots$. (From C. P. Poole, Jr et al., *Superconductivity*, Academic Press, New York, 1995, p. 11.)

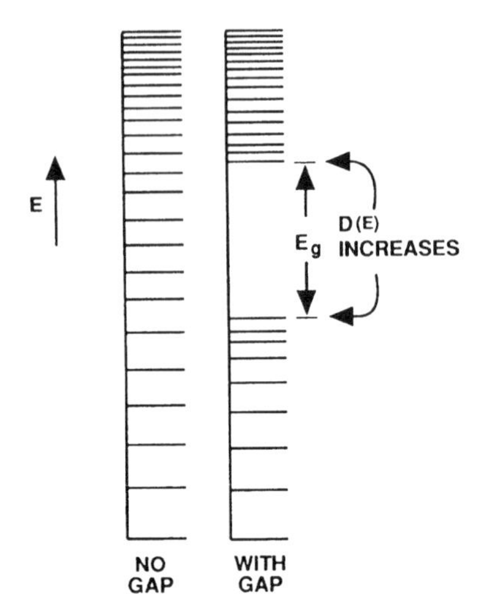

Fig. 24-6. Spacing of free electron energy levels in the absence of a gap (left) and in the presence of a gap (right) of the type shown in Fig. 24-5. (From C. P. Poole, Jr et al., *Superconductivity*, Academic Press, New York, 1995, p. 11.)

and the second derivative provides the effective mass $m^*(k)$

$$\frac{1}{m^*} = \frac{1}{\hbar^2}\left(\frac{d^2 E_K}{dk^2}\right)_{E_F} \tag{38}$$

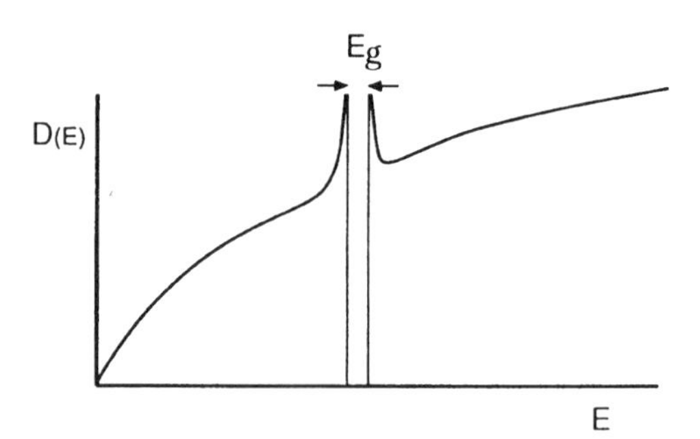

Fig. 24-7. Energy dependence of the density of states $D(E)$ in the presence of a gap corresponding to the case of Fig. 24-6. (From C. P. Poole, Jr et al., *Superconductivity*, Academic Press, New York, 1995, p. 11.)

The velocity v is plotted in Fig. 24-8(b) taking into account the bending of the free electron parabola near the gap as shown in Fig. 24-8(a), and the differentiation can be carried out if one knows the shapes of the energy bands near the Fermi level. The density of states $D(E_F)$ also deviates from

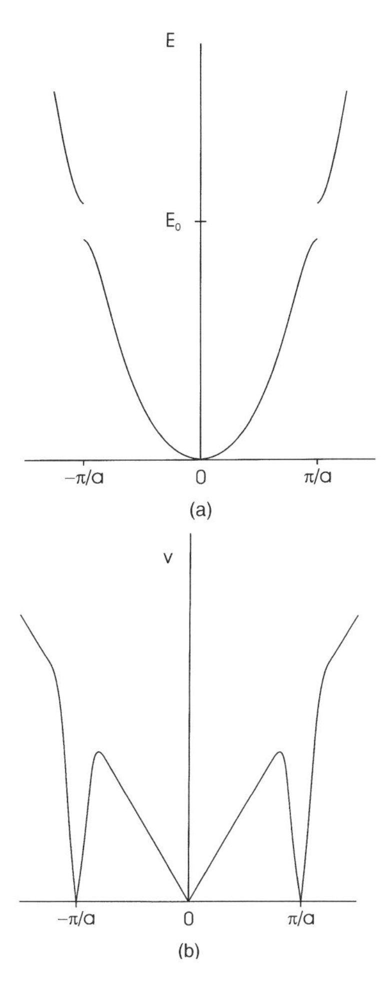

Fig. 24-8. Plots of (a) energy E, and (b) electron velocity v versus k for the model of a one-dimensional energy parabola that distorts to become horizontal at a gap which opens up at the edges ($\pm\pi/a$) of the Brillouin zone. Equation (37) was used to calculate v from part (a) to make the plot presented in part (b).

the free electron value near the gap, being proportional to the effective mass m^*

$$D(E_F) = m^* k_F / \hbar^2 \pi^2 \tag{39}$$

as may be inferred from Eq. (32). We can see from a plot of the slopes of Fig. 24-8(b) that for this idealized case the effective mass jumps from a strongly negative to a strongly positive value at the gap, and it is close to the 'parabolic energy' value far from the gap.

8 ENERGY BANDS IN TWO DIMENSIONS

The one-dimensional plots of the energy versus k shown in Figs. 24-2 and 24-5 are simple and easy to understand. Plots of this type become very complex and hard to visualize in three dimensions. To acquire some experience with these complexities we provide an analysis of the energy bands of a two-dimensional simple square lattice of size $L \times L$ in which the spacing between the atoms is a and $L/a = 10$. Figure 24-9 shows the lattice in k-space with the overall dimensions of $2\pi/a$ and the distance between points of $2\pi/L$, and also shows the k_x, k_y values at various positions of the k-space lattice. the unit cell in k-space has the dimensions $(2\pi/L)^2$, as shown. We are interested in calculating the energy along the paths from the point Γ in the center to M at the corner, then to X in the center of a side, and finally back to Γ, as indicated in Fig. 24-10.

The energy is all kinetic, given by

$$E_K = E_0(a/\pi)(k_x^2 + k_y^2) \tag{40}$$

where E_0 was defined in Eq. (24). We wish to express the energy in terms of k_x and k_y values that lie within the first Brillouin zone

$$-\pi/a \leq k_x, k_y \leq +\pi/a \tag{41}$$

corresponding to the limits from Fig. 24-11 using the format of Eq. (26), and to do so we write

$$E_K = E_0[(2n_x + k_x a/\pi)^2 + (2n_y + k_y a/\pi)^2] \tag{42}$$

where

$$n_x, n_y = 0, \pm 1, \pm 2, \ldots \tag{43}$$

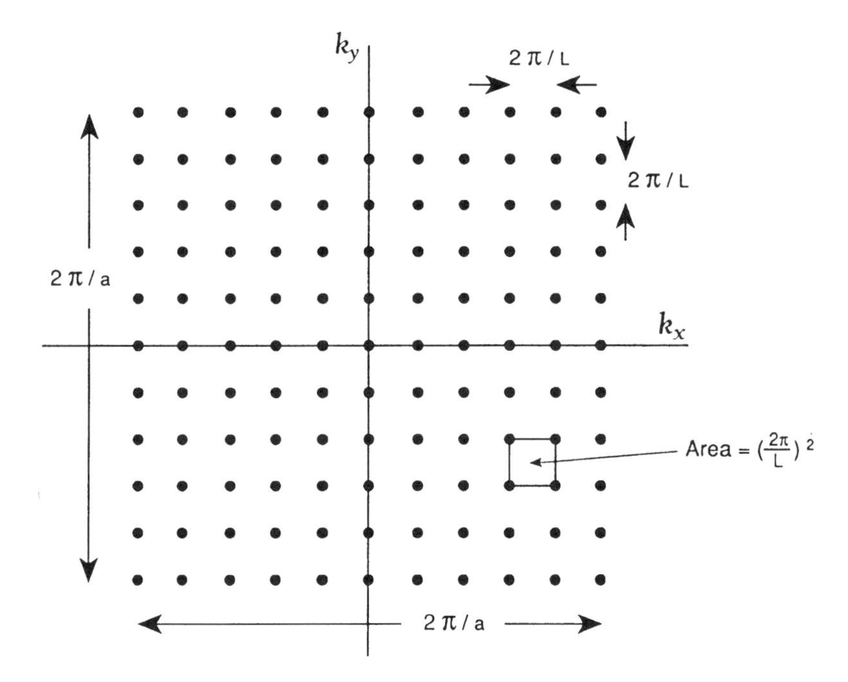

Fig. 24-9. Points of a rectangular lattice in two-dimensional k-space drawn for $L = 10a$. (From C. P. Poole, Jr et al., *Superconductivity*, Academic Press, New York, 1995, p. 212.)

for the various energies. Equation (41) is the energy of the (0,0) bands that have $n_x = n_y = 0$. Along the path $\Gamma \Rightarrow X$ we have $k_y = 0$, to give for Eq. (42)

$$E_K = E_0[(2n_x + k_x a/\pi)^2 + (2n_y)^2] \quad \Gamma \Rightarrow X \tag{44a}$$

along the path $\Gamma \Rightarrow M$ we have $k_x = k_y$, to give

$$E_K = E_0[(2n_x + k_x a/\pi)^2 + (2n_y + k_x a/\pi)^2] \quad \Gamma \Rightarrow M \tag{44b}$$

and along the path $X \Rightarrow M$ we have $k_x = \pi/a$, to give

$$E_K = E_0[(2n_x + 1)^2 + (2n_y + k_y a/\pi)^2] \quad X \Rightarrow M \tag{44c}$$

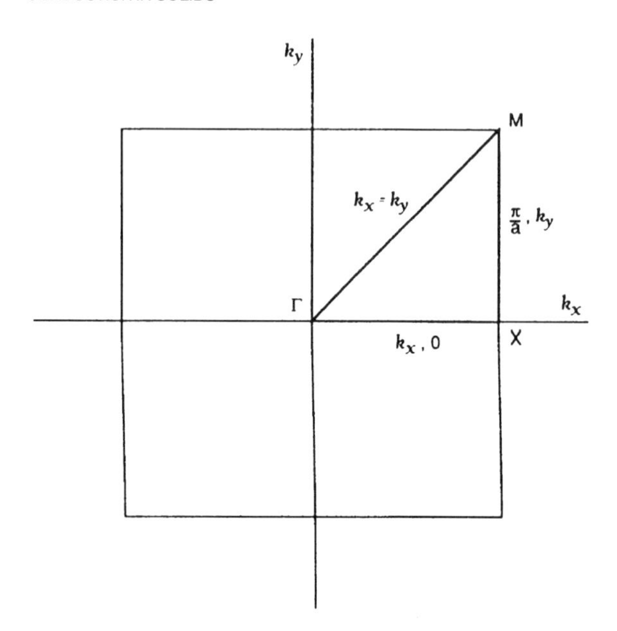

Fig. 24-10. Special symmetry points Γ for the center (0,0), M for the corner $(\pi/a, \pi/a)$, and X for the midpoint of the side $(\pi/a, 0)$ of the Brillouin zone of Fig. 24-9. (From C. P. Poole, Jr et al., *Superconductivity*, Academic Press, New York, 1995, p. 212.)

The energies obtained from these equations (44) are plotted in Fig. 24-12, each labeled with its (n_x, n_y) value. It is left as an exercise for the reader to identify the bands labeled α, β, γ, and δ at the top of the figure. We see from the figure that the energy bands come together at the special points Γ, M, and X.

In a particular material the energy bands of Fig. 24-12 will be filled up to the Fermi level, and be empty above E_F. For example, if $E_F = 1.5E_0$, then the (0,0) band will be filled except in the neighborhood of the point M, and the $(-1, 0)$ band will be full up to the energy $E = E_F$ in the $\Gamma \Rightarrow X$ and $X \Rightarrow M$ directions. The remaining bands will be empty.

If the effect of a potential is included, then the energy bands of Fig. 24-12 will have gaps in the energy at the edges of the Brillouin zone, and these gaps will appear at the points Γ, M, and X where energy bands meet from different directions. This is an analogy with the one-dimensional case depicted in Fig. 24-5.

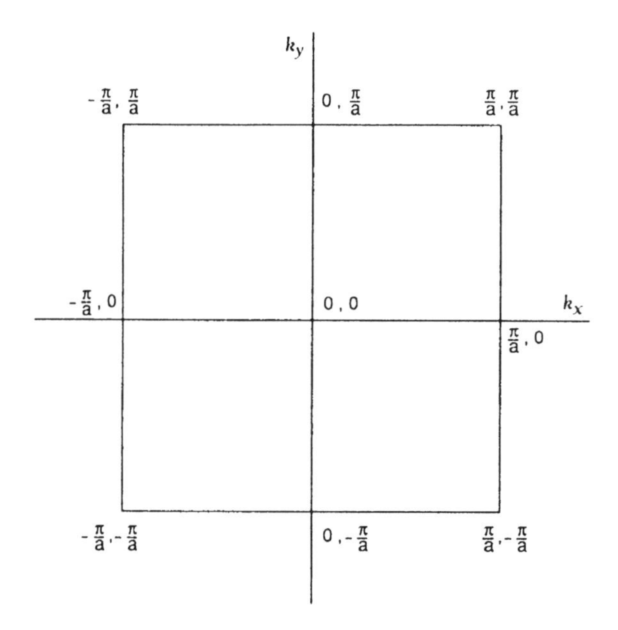

Fig. 24-11. The $(k_x k_y)$ values of the special symmetry points of the Brillouin zone of Fig. 24-10. (From C. P. Poole, Jr et al., *Superconductivity*, Academic Press, New York, 1995, p. 211.)

9 SEMICONDUCTORS

We mentioned in the previous section that the valence band of an insulator is full of electrons, and the upper lying conduction band is empty. The Boltzmann factor $\exp(-\Delta E_g/k_B T)$ for typical insulators is so small that the number of electrons excited to the conduction band is negligible. An intrinsic semiconductor has a small enough gap ΔE_g so that some electrons are thermally excited to the conduction band where they can carry electric current. In addition, when an electron goes from a full valence band to the conduction band it leaves behind the absence of an electron which acts like the presence of a positive electron called a hole in an otherwise full valence band. The hole is capable of moving around because an electron in a neighboring chemical bond can jump to occupy the site of the missing electron, thereby transferring the hole to the new bond. The holes can also act as though they are delocalized throughout space. The delocalized electrons in

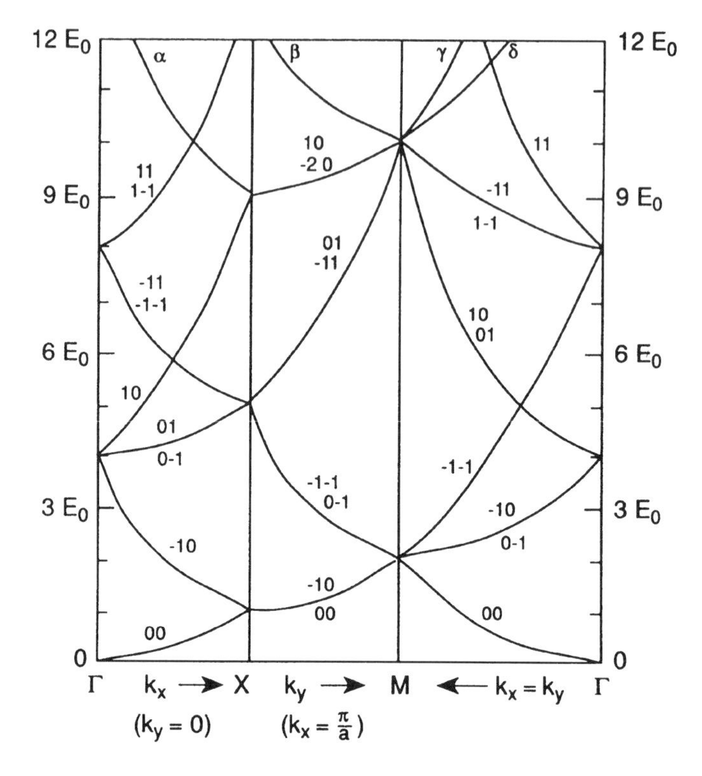

Fig. 24-12. Energy bands for the free electron approximation on a square lattice plotted for the three principal directions $\Gamma \Rightarrow X$, $X \Rightarrow M$, and $M \Rightarrow \Gamma$. The reader should try to identify the bands α, β, γ, and δ at the top of the figure. (From C. P. Poole, Jr et al., *Superconductivity*, Academic Press, New York, 1995, p. 213.)

the conduction band and the delocalized holes in the valence band both carry electric current, but the number of charge carriers is not large, so the measured conductivity is small. Such a semiconductor is referred to as intrinsic because the number of holes in the valence band equals the number of electrons in the conduction band.

Consider the semiconductor silicon or germanium. The elements Si and Ge, which are in column IV of the periodic table of Fig. 21-4 in Chapter 21, and have four valence electrons each which they use to form chemical bonds with their four nearest neighbors. If the Si is doped with a group V element such as P, As, or Sb, each of which has five valence electrons, then when the As atom in a Si site uses four of its valence electrons to form the four bonds tetrahedrally coordinated to nearest neighbor Si atoms, it has one extra valence electron left. The As atom energy levels containing the extra electron are located in the energy gap close to the conduction band, a distance ΔE_D below, as shown in Fig. 24-13. Since $\Delta E_D/k_BT \ll 1$ the Boltzmann factor $\exp(-\Delta E_D/k_BT)$ is close to 1, and most of the donor electrons from the As are excited to the conduction band where they contribute to the conductivity. This is called n-type conductivity because it arises from electrons in the conduction band.

If the dopant in Si or Ge is a group III atom such as B, Ga, or In which has only three valence electrons instead of four, then it will take an electron from the valence band to complete its fourth bond. This produces a hole in the valence band which is mobile and can carry current. Energetically we represent acceptor atoms like Ga as having energies ΔE_A slightly above the top of the valence band, as shown in Fig. 24-13, where they attract electrons from the valence band. There is a high probability that electrons will jump to these acceptor levels because $\Delta E_A/k_BT \ll 1$ in the corresponding Boltzmann factor $\exp(-\Delta E_A/k_BT) \sim 1$. This type of conductivity is referred to as p-type since the current is carried by positive charges or holes left behind in the valence band by the electrons that jumped to the acceptor levels.

Semiconductors can be selectively doped with donors or acceptors to provide the desired level of conductivity. There can also be some contribution from the intrinsic mechanism mentioned earlier, but ordinarily the n-type or p-type mechanism will dominate. Some typical energy gaps and impurity energy differences are shown in Table 24-1, and this gives engineers a wide choice of characteristics for device design.

The situation in actual semiconductors is more complicated than we have presented because the energy bands shift with the direction in the crystal,

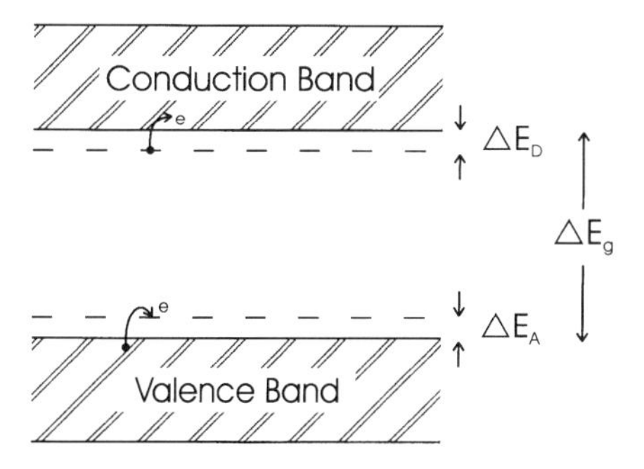

Fig. 24-13. Valence and conduction bands of a semiconductor showing the energy gap ΔE_g, and the acceptor (E_A) and donor (E_D) levels.

TABLE 24-1 Energy gaps in semiconductors

System	Energy in eV		Type of semiconductivity
	Si	Ge	
ΔE_G	1.12	0.67	Intrinsic
ΔE_D, P	0.044	0.012	n-type
ΔE_D, Sb	0.039	0.010	n-type
ΔE_A, B	0.046	0.010	p-type
ΔE_A, In	0.16	0.011	p-type

and the energies listed in the table vary with the direction. The overall picture, however, does not change appreciably.

10 HALL EFFECT

The Hall effect employs crossed electric and magnetic fields to obtain information on the sign and the mobility of charge carriers. The experimental arrangement illustrated in Fig. 24-14 shows a magnetic field B_0 applied in the z direction perpendicular to a slab, and a battery that establishes an electric field E_y along the y direction which causes the current $I = JA$ to flow, where $J = nev$ is the current density. The Lorentz force

$$\mathbf{F} = q\mathbf{v} \times \mathbf{B}_0 \tag{45}$$

of the magnetic field on each moving charge q is in the positive x direction for both positive and negative charge carriers, as is clear from Figs. 24-15(a) and 24-15(b), respectively. This causes a charge separation to build up on the sides of the plate which produces an electric field E_x perpendicular to the current (y) and magnetic field (z) directions. The induced electric field is in the negative x direction for positive q, and in the positive x direction for negative q, as shown in Figs. 24-15(c) and 24-15(d), respectively. After the charge separation has built up the electric force $q\mathbf{E}_x$ balances the magnetic force $q\mathbf{v} \times \mathbf{B}_0$ to give

$$q\mathbf{E}_x = q\mathbf{v}\mathbf{B}_0 \tag{46}$$

and the charge carriers q proceed along the wire undeflected.

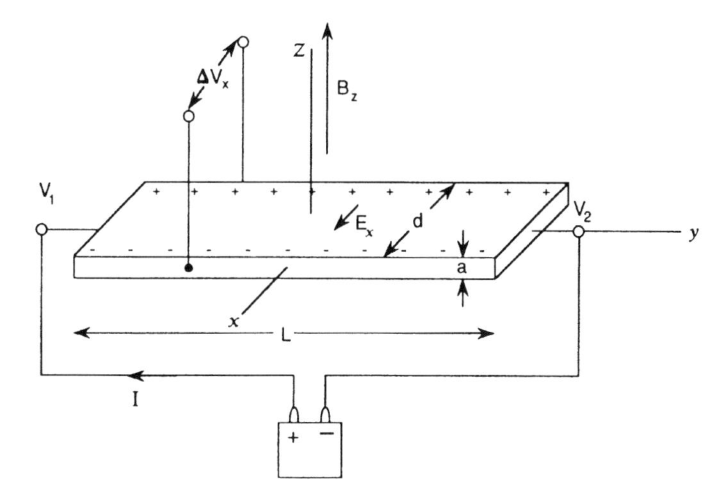

Fig. 24-14. Experimental arrangement for Hall effect measurements showing an electrical current I passing through a flat plate of width d and thickness a in a uniform transverse magnetic field B_z. The voltage drop $V_2 - V_1$ along the plate, the voltage difference ΔV_x across the plate, and the electric field E_x across the plate are indicated. The figure is drawn for negative charge carriers (electrons). (From C. P. Poole, Jr et al., *Superconductivity*, Academic Press, New York, 1995, p. 18.)

The Hall coefficient R_H is defined as the ratio

$$R_H = E_x / J_y B_z \tag{47}$$

of three measurable quantities, and substituting Eq. (7) $J = nev$ and Eq. (46) into (47) we obtain for holes ($q = e$) and electrons ($q = -e$), respectively

$$R_H = 1/ne \quad \text{(holes)} \tag{48a}$$

$$R_H = -1/ne \quad \text{(electrons)} \tag{48b}$$

where the sign of R_H is determined by the sign of the charges. Thus, the Hall effect distinguishes electrons from holes, and when all of the charge carriers are the same, this experiment provides the charge density n. When both positive and negative charge carriers are present there is a partial (or total) cancellation of their Hall effects.

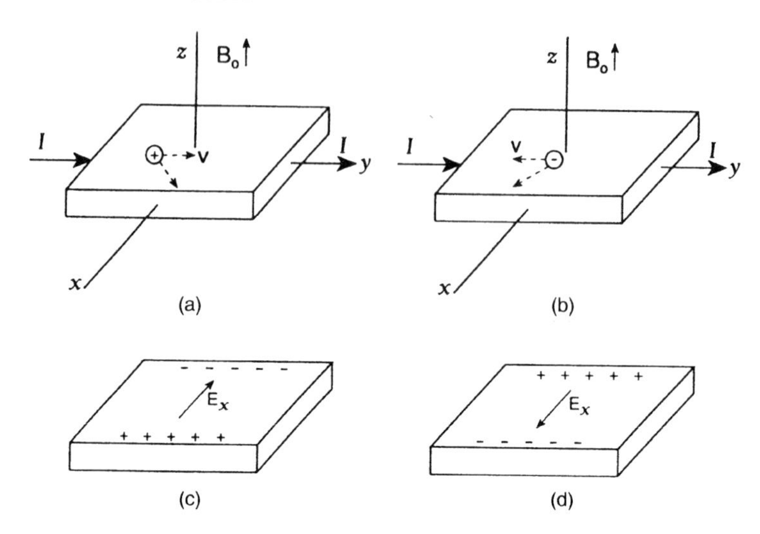

Fig. 24-15. Charge carrier motion (upper figures) and transverse electric field (lower figures) for the Hall effect experimental arrangement of Fig. 24-14. Positive charge carriers deflect as indicated in (a) and produce the transverse electric field E_x shown in (c). The corresponding deflection and resulting electric field for negative charge carriers are sketched in (b) and (d) respectively. (From C. P. Poole, Jr et al., *Superconductivity*, Academic Press, New York, 1995, p. 19.)

CHAPTER 25

NUCLEI

1 INTRODUCTION

In Chapter 21 we discussed atoms, and we elaborated on the details of the electron distribution. The nucleus simply provided the force that held the electrons in the atom, but its internal structure was of no concern. In the present chapter we examine this internal structure that arises from the protons and neutrons, collectively called nucleons. In the next chapter we discuss the internal structure of nucleons that arises from the presence of quarks, and we describe the other elementary particles formed by them.

2 FUNDAMENTAL FORCES

This section surveys the characteristics of the four fundamental forces, and of the various particles that are involved with ordinary matter, namely atoms, the electrons and nuclei of atoms, the nucleons of nuclei, and the quarks of nucleons.

The gravitational force is a long range ($1/r^2$ distance dependence) force which is far too weak to influence the properties of atoms or nuclei. Its strength relative to the strong force is 10^{-39} for distances less than 1 fm. The weak force, with a strength of 10^{-6} relative to the strong force at an energy of 1 GeV, is involved in the conversion of protons and neutrons into each other. It is not important in determining the characteristics of nuclei, and will be discussed in the next chapter.

The electric force between two charged particles q and q'

$$F_{\mathrm{EM}} = qq'/4\pi\varepsilon_0 r^2 \tag{1}$$

constitutes Coulomb's law. It is also a long-range force, with the same distance dependence as the gravitational force, and its coupling strength relative to the strong force is down by the dimensionless fine structure constant α

$$\alpha = e^2/4\pi\varepsilon_0 \hbar c \approx 1/137 \tag{2}$$

The Coulomb attraction between the negative electrons and the positive nucleus is the main force holding atoms together. The repulsive Coulomb force between each pair of protons in a nucleus counteracts the strong force which binds nucleons together, and it makes nuclei in the second half of the periodic table less and less stable with increasing Z.

The strengths and distance dependences of the four fundamental forces are compared in Fig. 25-1. Two protons in contact experience a strong force of attraction of relative strength 1, an electromagnetic force of repulsion of relative strength 1/137, and a gravitational force of attraction of negligible relative strength, namely reduced by the factor 10^{-39}.

The strong force does not act on electrons, and it is not influenced by the charge of a particle. It has about the same strength between two protons,

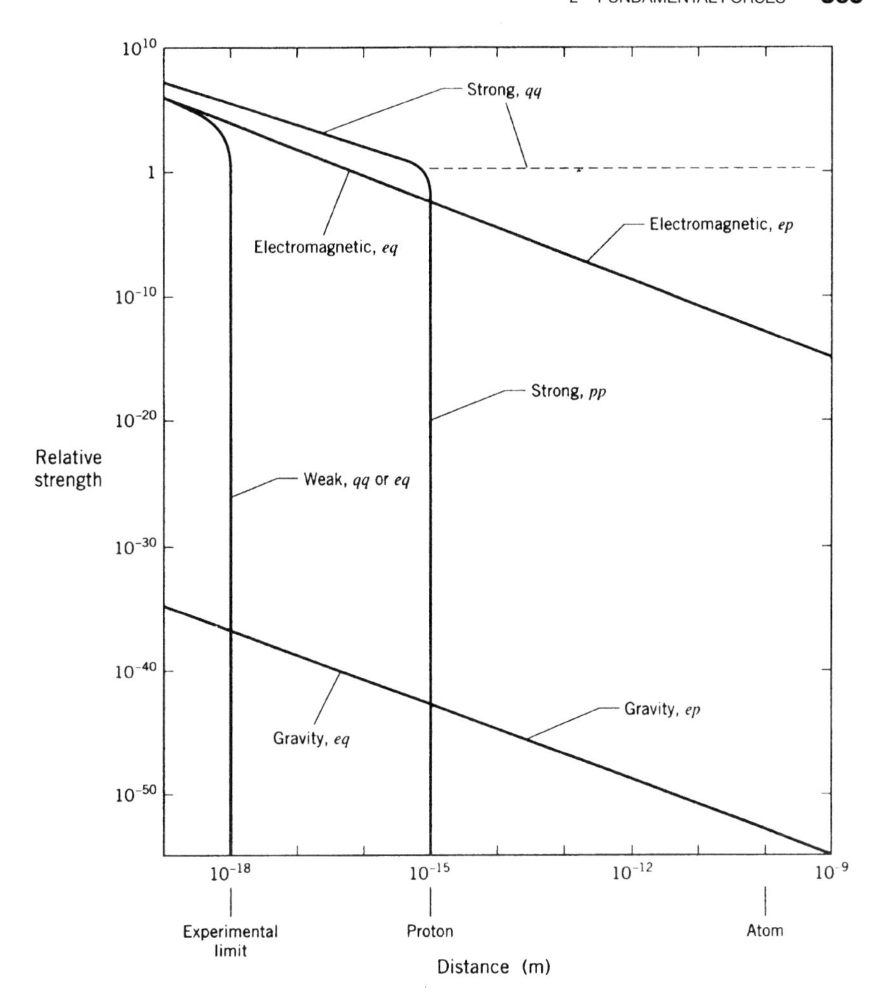

Fig. 25-1. Distance dependence of the relative strengths of the four fundamental forces between electrons (e), protons (p), and quarks (q). The horizontal dashed line represents the force of attraction between two separated quarks. (From J. W. Rohlf, *Modern Physics from α to Z^0*, Wiley, New York, 1994, p. 27.)

between two neutrons, and between a proton and a neutron. Light neuclei tend to have isotopes with equal numbers of protons and neutrons as a result of this equality of interaction strength because the Coulomb interaction is not yet strong enough to have an effect. In heavy nuclei the increase in charge makes the Coulomb repulsive forces between protons more and more important, and the nucleus compensates for this by increasing the ratio of neutrons to protons because the neutrons themselves do not participate in the Coulomb interaction, and their presence weakens it for protons by moving them farther apart. Eventually the number of protons becomes so large that the Coulomb interaction begins to dominate, causing most of the actinides to be radioactive, and it prevents larger nuclei from existing.

3 SIZE OF THE NUCLEUS

Atoms are all more or less the same size; having radii between 0.1 and 0.3 nm, i.e., between 2 and 6 Bohr radii. Hydrogen is an exception with a radius $a_0 = 0.053$ nm. This is proven by plotting the atom density of various solids as a function of the atomic number Z of the constituent atoms. The density of electrons in atoms increases by perhaps a fractor of 100 in going from the light to the heavy elements.

In nuclei the situation is the opposite; all nuclei tend to have the same density, and the radius R of a nucleus in femtometers (fm, 10^{-15} m) is given by

$$R = 1.2A^{1/3} \tag{3}$$

where A is the atomic mass number, the number of nucleons in the nucleus. The density ρ_N of a nucleus is

$$\rho_N = A/(4\pi R^3/3) = 0.14 \text{ nucleons}/(\text{fm})^3 \tag{4}$$

More specifically, the density is constant out to a radius of about $0.7R$, it decreases more or less linearly from 0.8 to $1.2R$, and reaches zero at about $1.3R$. The constituent nucleons, i.e., protons and neutrons, each have radii of about 1 fm, the range of the strong force that binds nucleons together. The nucleus may be looked upon as a close packing of nucleons in the shape of spheres.

A neutron star is believed to approximate a close packed neutron structure, so it has a density comparable with that of Eq. (4), and a typical radius is several kilometers. It must be small enough so that attractive gravitational interaction cannot collapse it to a black hole. This is analogous to the case of a nucleus which must be small enough so that the repulsive Coulomb interaction does not cause it to blow apart. Thus, inverse square long-range forces can dominate over strong interaction forces under some limiting conditions.

We have assumed that a nucleus has a spherically symmetric nucleon distribution. For many nuclei the charge distribution is elongated or flattened corresponding, respectively, to a prolate or oblate ellipsoidal shape. This is shown by the presence of a positive (prolate shape) or negative (oblate shape) electric quadrupole moment. The charge distribution, of course, arises from the protons and not from the neutrons. This is a classical picture. Quantum mechanically a nucleus must have a nuclear spin greater than one half, $I > \frac{1}{2}$, to have an electric quadrupole moment. A nucleus cannot have an electric dipole moment.

4 BINDING ENERGY

The binding energy of a nucleus E_B is the rest energy of the constituent nucleons minus the rest energy $M_N c^2$ of the nucleus itself

$$E_B = Zm_p c^2 + (A - Z)m_n c^2 - M_N c^2 \tag{5}$$

where

$$m_p c^2 = 938.3 \text{ MeV} \tag{6a}$$

$$m_n c^2 = 939.6 \text{ Mev} \tag{6b}$$

A helium nucleus, called an alpha particle, has $Z = 2$, $A = 4$, and the rest energy $Mc^2 = 3.7274\,\text{GeV}$, and from Eq. (5) the binding energy $E_B = 28.3\,\text{Mev}$. Of greater interest is the binding energy per nucleon, E_B/A, which has the representative values:

$$E_B/A = \begin{cases} 7.08 \text{ MeV} & ^4\text{He} \\ 8.79 \text{ MeV} & ^{56}\text{Fe} \\ 7.57 \text{ MeV} & ^{238}\text{U} \end{cases} \tag{7}$$

We see from the plot of E_B/A versus A in Fig. 25-2 that the most stable nuclei, namely those with the largest values of E_B/A, are in the range from $A = 50$ to $A = 65$, and that ^{56}Fe is particularly stable. Therefore energy can be gained by the fusion process of combining two small nuclei into a larger one or by the fission process of splitting a large nucleus into smaller ones. It is also clear from the plot that the very light nuclei such as ^{2}H and ^{3}He have especially small binding energies so fusion reactions involving them are particularly efficient in providing energy.

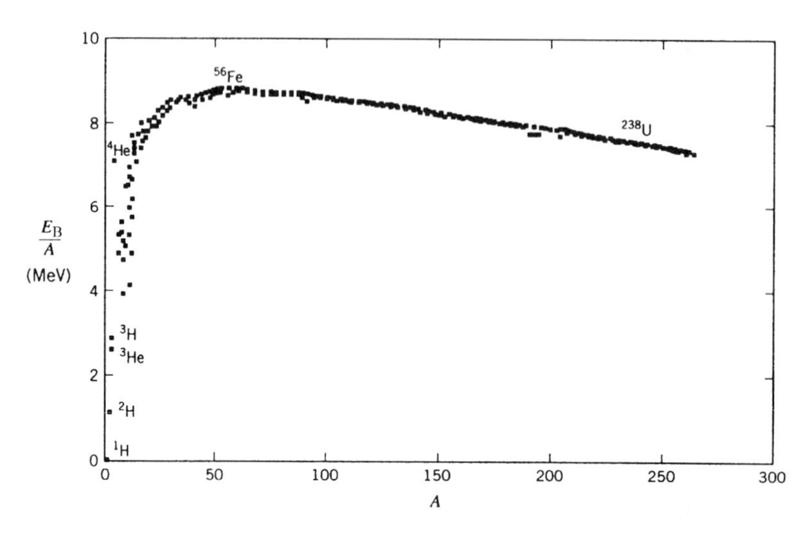

Fig. 25-2. Dependence of the binding energy per nucleon on the atomic mass A. (From J. W. Rohlf, *Modern Physics from α to Z^0*, Wiley, New York, 1994, p. 302.)

A simple model of a nucleus is a spherical liquid drop of nuclear matter with a binding energy αA proportional to the volume minus a surface energy $-\beta A^{2/3}$ representing the incomplete bonding of the nucleons at the surface. Such a model is inadequate, so additional factors must be taken into account. The most important factor that makes large nuclei less stable is the Coulomb interaction which goes as $-q^2/r$, so it adds the term $-\gamma Z^2/A^{1/3}$. There is also a tendency, from the Pauli exclusion principle, for the number of neutrons to equal the number of protons, and for the number of each to be even, which adds two more terms to the binding energy. The result is the semiempirical binding energy formula, sometimes called the Weizsacker formula, which we give in the units of MeV

$$\begin{array}{ccccc} \text{Volume} & \text{Surface} & \text{Coulomb} & \text{Pauli term} & \text{Even/odd factor} \end{array}$$

$$\frac{E_B}{A} = 15.75 - \frac{17.8}{A^{1/3}} - \frac{0.711Z^2}{A^{4/3}} - 23.7\left\{1 - \frac{2Z}{A}\right\}^2 + \begin{vmatrix} +1 \\ 0 \\ -1 \end{vmatrix} \frac{11.18}{A^{3/2}} \tag{8}$$

where in the last term $+1$ applies to even–even nuclei, 0 to even–odd and odd–even nuclei, and -1 to odd–odd nuclei. This expression fits the data of Fig. 25-2 quite well, except for the very light nuclei. For example, the three nuclei ^{7}Li, ^{56}Fe, and ^{208}Pb have the following values for the various terms of Eq. (8):

$$
\begin{array}{lcccccccccccl}
 & \text{Volume} & & \text{Surface} & & \text{Coulomb} & & \text{Pauli term} & & \text{Even/odd factor} & & \text{Total} & \text{Isotope} \\
E_B/A = \left\{ \begin{array}{l} \\ \\ \\ \end{array} \right. & \begin{array}{c} 15.75 \\ 15.75 \\ 15.75 \end{array} & \begin{array}{c} - \\ - \\ - \end{array} & \begin{array}{c} 9.31 \\ 4.65 \\ 3.00 \end{array} & \begin{array}{c} - \\ - \\ - \end{array} & \begin{array}{c} 0.48 \\ 2.24 \\ 3.88 \end{array} & \begin{array}{c} - \\ - \\ - \end{array} & \begin{array}{c} 0.48 \\ 0.12 \\ 1.06 \end{array} & \begin{array}{c} + \\ + \\ + \end{array} & \begin{array}{l} 0 \\ 0.03 \\ 0.004 \end{array} & & \begin{array}{l} = 5.48 \text{ MeV} \\ = 8.76 \text{ MeV} \\ = 7.81 \text{ MeV} \end{array} & \begin{array}{l} ^{7}\text{Li} \\ ^{56}\text{Fe} \\ ^{208}\text{Pb} \end{array}
\end{array} \quad (9)
$$

and their total E_B/A values compare well with the respective experimental values of 5.61, 8.79, and 7.88 MeV. These results confirm that as the atomic number Z increases, the surface term becomes less dominant and the Coulomb term becomes more dominant in determining the binding energy.

If the semiempirical binding energy formula (8) is minimized with respect to Z, then we obtain the expression

$$
Z = \frac{\frac{1}{2}A}{1 + 0.0075A^{2/3}} \quad (10)
$$

which fits the data for the Z versus $A - Z$ plot presented in Fig. 25-3 for stable nuclei.

5 SHELL MODEL

The semiempirical formula (8) gives a good overall fit to the experimental binding energy data, but it does not provide any explanation for nuclear spectroscopy, magnetic moments and spin, decay schemes, or the existence of particularly stable nuclei which have what are called magic numbers of neutrons or protons. These magic numbers are 2, 8, 20, 28, 50, 82, and 126, and especially stable are the double magic number nuclei $^4He_2$, $^{16}O_8$, $^{40}Ca_{20}$, $^{48}Ca_{20}$, and $^{208}Pb_{82}$. These nuclear magic numbers differ somewhat from their counterparts in atoms, namely 2, 10, 18, 36, 54, and 86, which correspond to the atomic numbers of the rare gases.

The nuclear shell model is constructed from successive energy levels labeled with a principal quantum number $n = 1, 2, 3, \ldots$, an orbital quantum number ℓ, and of course spin $s = \frac{1}{2}$. For $\ell = 0$ we have the total angular momentum $j = \frac{1}{2}$, and for $\ell > 0$ there are two values of j for each ℓ given by

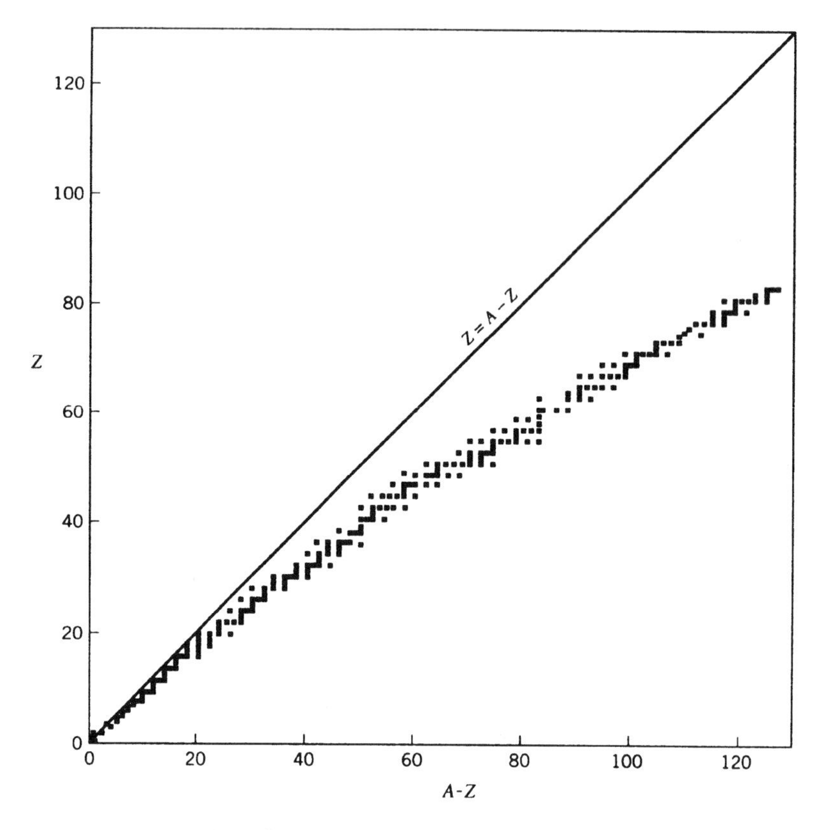

Fig. 25-3. Plot of the number of protons Z versus the number of neutrons $(A\text{–}Z)$ in the stable nuclei. (From J. W. Rohlf, *Modern Physics from α to Z^0*, Wiley, New York, 1994, p. 302.)

$$j = \ell \pm \tfrac{1}{2} \tag{11}$$

since $m_s = \pm\frac{1}{2}$. The model has the following values and nomenclature:

$$\begin{array}{rcccccccc}
\text{orbital } \ell = & 0 & 1 & 2 & 3 & 4 & 5 & 6 & \ldots \\
\text{symbol} = & \text{s} & \text{p} & \text{d} & \text{f} & \text{g} & \text{h} & \text{i} & \ldots \\
j \text{ values} = \left\{ \begin{array}{l} \\ \\ \end{array} \right. & \begin{array}{c} - \\ \frac{1}{2} \end{array} & \begin{array}{c} \frac{1}{2} \\ \frac{3}{2} \end{array} & \begin{array}{c} \frac{3}{2} \\ \frac{5}{2} \end{array} & \begin{array}{c} \frac{5}{2} \\ \frac{7}{2} \end{array} & \begin{array}{c} \frac{7}{2} \\ \frac{9}{2} \end{array} & \begin{array}{c} \frac{9}{2} \\ \frac{11}{2} \end{array} & \begin{array}{c} \frac{11}{2} \\ \frac{13}{2} \end{array} & \begin{array}{c} \ldots \\ \ldots \end{array}
\end{array} \tag{12}$$

Each ℓ-level holds $2(2\ell + 1)$ electrons, and each j-level holds $2j + 1$ electrons.

The spin–orbit interaction $\lambda\ell \cdot \mathbf{s}$ is given by

$$\lambda\ell \cdot \mathbf{s} = \tfrac{1}{2}\lambda[j(j+1) - \ell(\ell+1) - s(s+1)] \tag{13}$$

$$= \begin{cases} \tfrac{1}{2}\ell\lambda & j = \ell + \tfrac{1}{2} \\ -\tfrac{1}{2}(\ell+1)\lambda & j = \ell - \tfrac{1}{2} \end{cases} \tag{14}$$

where the spin–orbit interaction constant λ is negative for nucleons, and hence the largest j level of each pair is lower in energy. We also see that the splitting between the two j-levels is $(\ell + \tfrac{1}{2})\lambda$, which increases with ℓ in the manner shown in Fig. 25-4. The center of gravity of the splitting (13) is preserved, as may be shown by multiplying each level shift by its degeneracy $2j + 1$.

As the atomic number Z increases, the energy levels gradually fill, starting with the lowest, and the numbers in parentheses on the right-hand side of the figure are the number of nucleons $(2j + 1)$ needed to fill each level. Horizontal dashed lines are drawn on the figure where there are relatively large gaps between successive energy levels, and these gaps occur where the nuclei contain magic numbers of nucleons, indicated on the right, and are in very stable energy states.

Protons and neutrons have the same energy level scheme shown in Fig. 25-4, and for light nuclei the energies of both are close together. However, for heavier nuclei the Coulomb interaction begins to be appreciable, and the proton energies shift upward more and more as Z increases. Figure 25-3 gives the dependence of Z on A–Z. The divergence of Z from the value $\tfrac{1}{2}A$, corresponding to the solid line drawn on the figure for slope 1, is so pronounced for large Z that the double magic number isotope of lead, $^{208}Pb_{82}$, has 82 protons and 126 neutrons.

6 RADIOACTIVE DECAY

The stable isotopes cluster around the stability line in the Z versus A–Z plot of Fig. 25-3, and the unstable isotopes far from the line, emit alpha particles (^4He nuclei) or beta particles (electrons) to move closer to stability. Elements with $Z > 83$ are also unstable and evolve toward stability by emitting α and β particles.

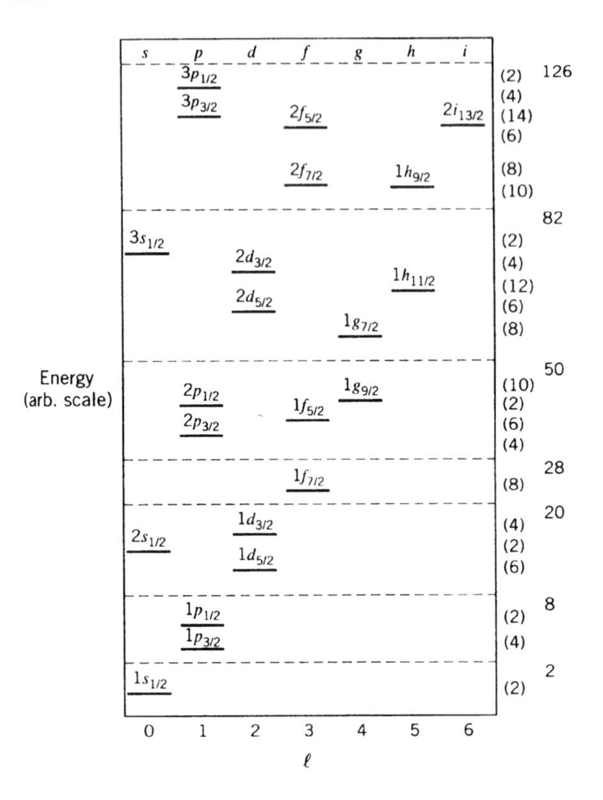

Fig. 25-4. Sketch of the energy levels of the nuclear shell model. The numbers in parentheses on the right are the number of nucleons in a subshell, and the numbers without parentheses on the far right are the number of nucleons in filled subshells up to that level. (From J. W. Rohlf, *Modern Physics from* α *to* Z^0, Wiley, New York, 1994, p. 305.)

If there are N radioactive nuclei in a material, then they will decay at a rate $-dN/dt$ which is proportional to the number $N(t)$ that is present

$$\frac{dN}{dt} = -\lambda N \tag{15}$$

and this can be integrated to give the expression

$$N(t) = N_0 e^{-\lambda t} \tag{16}$$

where the mean lifetime τ is

$$\tau = 1/\lambda \tag{17}$$

The time needed for half of the nuclei $N(t)$ of Eq. (15) to decay is called the half life $t_{1/2}$, and it is related to τ by the expression

$$t_{1/2} = \tau \ln 2 \tag{18}$$

The situation is more complicated when the daughter nucleus is itself radioactive with its own characteristic decay scheme and half life.

In α-particle decay the atomic number Z of the parent nucelus decreases by 2 and A is lowered by 4:

$$Z_{\text{Daug}} = Z_{\text{Par}} - 2 \tag{19}$$

$$A_{\text{Daug}} = A_{\text{Par}} - 4 \tag{20}$$

The energy Q that is released

$$Q = (M_{\text{Par}} - M_{\text{Daug}} - M_{\alpha})c^2 \tag{21}$$

goes into the kinetic energy of the daughter nucleus and the α-particle

$$Q = \frac{p_{\text{Daug}}^2}{2M_{\text{Daug}}} + \frac{p_{\alpha}^2}{2M_{\alpha}} \tag{22}$$

where, from the conservation of momentum, the daughter nucleus and the α-particle have equal and opposite momenta, $\mathbf{p}_{\text{Daug}} = -\mathbf{p}_{\alpha}$, if the parent nucleus decays at rest.

In β decay a neutron in the nucleus converts to a proton and an electron is emitted, so Z increases by 1 and A stays the same. An electron antineutrino $\bar{\nu}_e$ accompanies the decay

$$n \Rightarrow \text{p} + \text{e}^- + \bar{\nu}_e \tag{23}$$

and serves to conserve momentum in the process. The decay reaction (23) also occurs outside a nucleus since $m_p < m_n$ and the free neutron is unstable with a lifetime of about 17 min. Inverse beta decay or β^+ decay

$$\text{p} \Rightarrow \text{n} + \text{e}^+ + \nu_e \tag{24}$$

whereby a proton converts to a neutron and emits a positron e^+ can occur inside a nucleus, but cannot occur outside since $m_p < m_n$ and the reaction is not energetically favored. The proton, in fact, is quite stable outside a nucleus, with an estimated lifetime in excess of 10^{32} years. The reaction (24) decreases Z of the parent nucleus by 1 without changing A.

Electron capture involving the reaction

$$\mathrm{p} + \mathrm{e}^- \Rightarrow \mathrm{n} + \nu_\mathrm{e} \tag{25}$$

often takes place in nuclei that undergo β^+ decay, and it results in a decrease of Z by 1 with no change in A. Usually the innermost shell or K electron is captured, so the process is called K-capture.

When radioactive $^{60}\mathrm{Co}$ decays via the reaction

$$^{60}\mathrm{Co} \Rightarrow {}^{60}\mathrm{Ni} + \mathrm{e}^- + \bar{\nu}_\mathrm{e} + \gamma \tag{26}$$

a γ-ray is emitted because the $^{60}\mathrm{Ni}$ nucleus is created in a nuclear excited state, and it decays to the ground state by emitting the γ-ray. By a nuclear excited state we mean a nucleon in an energy level above the ground state, and when it jumps back to its ground state the γ-ray photon is emitted. Many radioactive disintegrations are accompanied by the emission of γ-rays.

There are several naturally occurring radioactive elements that have very long half lives and decay through a sequence of steps to a final stable nucleus. For example radioactive $^{238}\mathrm{U}$ with $Z = 92$, $N = 146$, $A = 238$ is 99.3% abundant and has a half life of 4.46 billion years. Its decay scheme to stable $^{206}\mathrm{Pb}$ with $Z = 82$, $N = 124$, $A = 206$ involves the emission of eight α-particles to lower A by 32, thereby lowering Z by 16, accompanied by six β^- decays to raise Z by 6 since the lowering of Z by the α-particle emission was too much. There are several paths for accomplishing this, as shown in Fig. 25-5. There are four radioactive series in nature, starting with the isotopes $^{232}\mathrm{Th}$ (100% abundant), $^{235}\mathrm{U}$ (0.72% abundant), $^{237}\mathrm{Np}$ (no natural abundance), and $^{238}\mathrm{U}$, and ending, respectively, with the stable magic number nuclei $^{208}\mathrm{Pb}$, $^{207}\mathrm{Pb}$, $^{209}\mathrm{Bi}$, and $^{206}\mathrm{Pb}$, where $^{208}\mathrm{Pb}$ is a double magic number nucleus. Bismuth ($^{209}\mathrm{Bi}$, $Z = 83$, 100% abundant) is the highest atomic number stable isotope, all higher Z nuclei being radioactive.

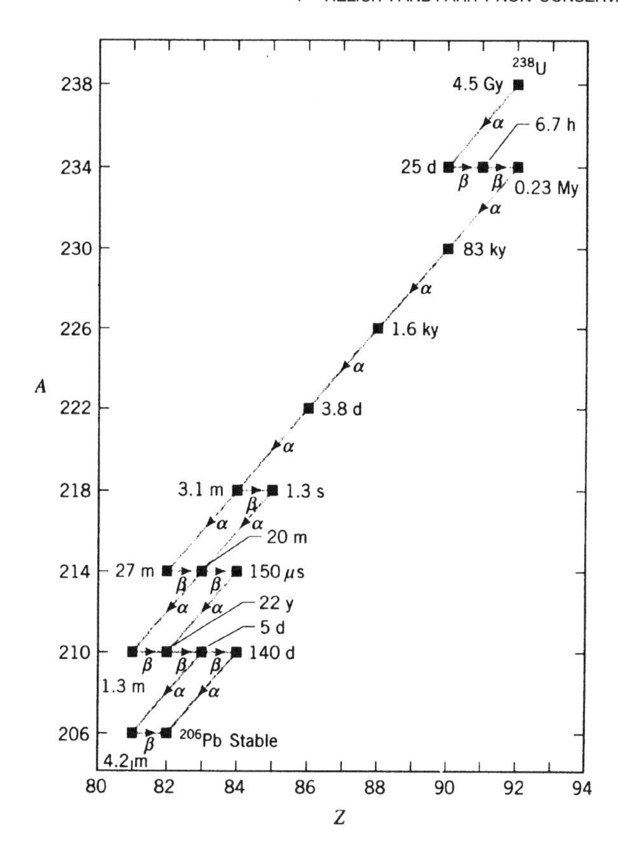

Fig. 25-5. Sequence of reactions involving the radioactive decay of ^{238}U to the double magic number isotope ^{206}Pb. (From J. W. Rohlf, *Modern Physics from α to Z^0*, Wiley, New York, 1994, p. 313.)

7 HELICITY AND PARITY NON-CONSERVATION

The helicity H of a particle is defined as the component of its intrinsic spin angular momentum $\hbar\mathbf{S}$ in the direction of its velocity $\mathbf{v}$

$$H = \frac{\mathbf{S} \cdot \mathbf{v}}{Sv} \tag{27}$$

We will refer to $\mathbf{S}$ as simply the spin. Helicity can be frame dependent. For example, consider an electron moving with positive helicity, i.e., with $\mathbf{S}$ and

v parallel. If a Lorentz transformation is carried out with a parallel velocity $v_L > v$, then the electron's velocity will be reversed in the new Lorentz frame, but **S** will be unchanged in direction, so the helicity of the electron will have changed to negative. Now consider a neutrino. It moves at the speed of light so no Lorentz transformation can change the direction of its velocity, and the helicity remains the same, always negative. Experimentally all neutrinos are found to have negative helicities, so the helicity is a fundamental characteristic of a neutrino. In contrast to this an antineutrino has a positive helicity.

The 1957 C. S. Wu et al. experimentally measured the β^- decay of ^{60}Co nuclei to ^{60}Ni nuclei through the reaction of Eq. (26) in the presence of a magnetic field which was first oriented one way and then reversed in direction. It was found that more electrons were emitted in the direction opposite to the magnetic field, i.e., opposite to the direction of the nuclear spin. If parity had been conserved, then equal numbers of electrons would have been emitted parallel and antiparallel to the field direction.

Beta decay is a weak interaction involving the weak force, and parity is not conserved for weak interactions. It is conserved, however, for electromagnetic and strong interactions. More will be said about this in the next chapter.

8 MÖSSBAUER EFFECT

The Mössbauer effect is the recoilless transition from an excited nuclear state to a nuclear ground state. An example, depicted in Fig. 25-6, is the decay of ^{57}Co by electron capture to form ^{57}Fe in a nuclear excited state, and the subsequent decay of the ^{57}Fe to another excited state by the emission of a 122 keV γ-ray, followed by the Mössbauer recoilless emission of a 14.4 keV γ-ray. If this reaction had happened in outer space, then the newly formed ^{57}Fe nucleus would have to recoil to balance momentum with the γ-ray, and this reduces the energy of the γ-ray so that it can no longer satisfy the condition $\Delta E = \hbar\omega$ for reabsorption by another ^{57}Fe nucleus. When the ^{57}Fe atom is in a crystalline solid the whole crystal can take up the momentum of the recoil so the γ-ray photon energy is unchanged. This essentially recoilless γ-emission constitutes the essence of the Mössbauer effect. It is detected when another ^{57}Fe nucleus absorbs the emitted photon.

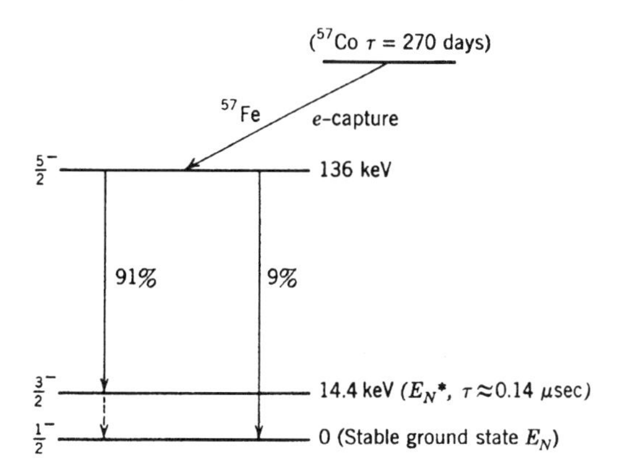

Fig. 25-6. Decay scheme of the nucleus ^{57}Co to ^{57}Fe via electron capture. The vertical arrows are γ-ray emissions, and the dashed vertical arrow on the lower left is the recoilless γ-ray Mössbauer transition. (From A. J. Freeman and R. E. Watson, in *Magnetism*, G. T. Rado and H. Suhl (eds), Academic Press, New York, Vol. IIA, 1965, p. 167.)

9 CARBON DATING AND THE CARBON CYCLE

We end this chapter by describing some nuclear reaction schemes. The first involves radioactive dating based on the creation of the radioactive ^{14}C nucleus by protons that come to earth from the sun. The protons react with nuclei in the atmosphere, transform them to other nuclei, producing neutrons in the process.

$$\text{p} + (\text{nucleus}) \Rightarrow (\text{nucleus}') + n \tag{28}$$

The neutrons in turn react with nitrogen, transforming it to radioactive carbon

$$\text{n} + {}^{14}\text{N} \Rightarrow {}^{14}\text{C} + \text{p} \tag{29}$$

and the ^{14}C decays back to ^{14}N with a half life of 5730 years.

$$^{14}\text{C} \Rightarrow {}^{14}\text{N} + \text{e}^- + \bar{\nu}_\text{e} \tag{30}$$

There is a dynamic equilibrium in the earth's atmosphere between the concentrations of ^{14}C and ^{12}C so the ratio ^{14}C/^{12}C in the air does not change with time. However, when carbon becomes bound in a material such as wood or cloth the ^{14}C present there decays slowly by reaction (30), and the ratio $^{14}C/^{12}C$ in the material decreases with time. A measurement of this ratio provides us with the age of the material.

Another nuclear reaction scheme to be considered is fusion. There are a number of ways in which this can occur, and an example is

$$d + d \Rightarrow {}^3He + n \qquad Q = 4.0\,\text{MeV} \tag{31}$$

$$^3He + {}^3He \Rightarrow \alpha + p + p \qquad Q = 2.9\,\text{Mev} \tag{32}$$

The overall ^3He conserving reaction

$$4d \Rightarrow \alpha + 2n + 2p \qquad Q = 20.9\,\text{MeV} \tag{33}$$

thus provides an energy gain of 5.2 MeV per reacting deuteron, with the ^3He acting as a catalyst.

Nuclear fusion is responsible for the radiation emitted by stars. The carbon cycle proposed for explaining the origin of this radiation begins with the reaction

$$p + {}^{12}C \Rightarrow {}^{13}N + \gamma \tag{34}$$

and ends with a reaction that forms the starting material ^{12}C again

$$p + {}^{15}N \Rightarrow {}^{12}C + \alpha \tag{35}$$

The overall reaction from the sequence of six reactions that constitute the carbon cycle is

$$4p \Rightarrow \alpha + 2e^+ + 2\nu_e + 3\gamma \quad Q = 25\,\text{MeV} \tag{36}$$

where p^+ and α^{2+} are both positively charged. Here the carbon acts as a catalyst for the burning of four protons to form an α-particle. The carbon cycle dominates above a temperature of 10^8 K, and an alternative cycle called the proton cycle which starts with a direct $p + p$ reaction dominates below this temperature. The temperature of the sun's core is $\approx 1.5 \times 10^7$ K, so solar emission arises from the proton cycle that involves the same overall reaction (36).

CHAPTER 26

ELEMENTARY PARTICLES

1 INTRODUCTION

In Chapter 21 we mentioned that atoms have radii between 0.1 and 0.3 nm and energies ranging from several electron volts for the valence electrons to about 10^5 eV for the innermost electrons of the heaviest elements. In Chapter 25 we noted that nuclei have radii between 1 and 8 fm and they have binding energies between 2 MeV for the lightest nucelus ^2H and 2 GeV for the heaviest nuclei. We mentioned in these chapters that atoms consist of electrons interacting with a positive nucleus and with each other via the electromagnetic (Coulomb) interaction, and nuclei consist of nucleons, i.e., neutrons and protons, attracting each other via the strong interaction, with the protons also undergoing mutual repulsion via the electromagnetic interaction.

This chapter begins by giving more details on the four fundamental interactions that were introduced in the previous chapter. Each interaction is mediated by one or more particles, with the photon playing this role for the electromagnetic interaction. In addition to these mediating particles there are two broad classes of elementary particles, namely the heavier hadrons that participate in strong interactions, and the lighter leptons that are fermions, i.e., particles with half integral spin ($s = \frac{1}{2}, \frac{3}{2}, \frac{5}{2}, \ldots$) with no strong interaction. The hadrons themselves are divided into two classes, the heavier group called baryons which are fermions, and the lighter group called mesons which are bosons, i.e., particles with integer spin ($s = 0, 1, 2, \ldots$). Examples of particles in these groups are: baryons (neutron n and proton p), mesons (pion π and kaon K), and leptons (electron e and muon μ). Since quarks and antiquarks are the constituents of the hadrons we begin our discussion by analyzing the nature of quarks, and then proceed to discuss baryons, mesons, and leptons in turn. The quark discussion clarifies the various quantum numbers and other properties that characterize the elementary particles, and the selection rules that govern their decay schemes and interactions.

2 CHARACTERISTICS OF THE FUNDAMENTAL FORCES

In Chapter 25 we gave a brief introduction to the four fundamental forces. Each force is characterized by a relative strength α_i, a typical lifetime τ_i, a source, and a particle or particles that mediate interactions involving the force. We discuss these characteristics for each force law.

The electromagnetic interaction involves the Coulomb force law which has the form

$$F = \frac{e^2}{4\pi\varepsilon_0 r^2} \tag{1}$$

for two source charges e separated by the distance r. The strength of this interaction

$$(\text{strength})_{\text{em}} = e^2/4\pi\varepsilon_0 \tag{2}$$

has the units energy times distance. To obtain a measure of the dimensionless coupling strength of this interaction consider the elastic scattering of two electrons represented by the Feynman diagram of Fig. 26-1

$$e + e \Rightarrow e + e \tag{3}$$

in which the momentum transfer takes place through the emission of a photon $\gamma^* = h\nu$ by one electron

$$e \Rightarrow \gamma^* + e \tag{4}$$

and its absorption by the other

$$\gamma^* + e \Rightarrow e \tag{5}$$

The intermediate particle γ^* that is transferred is not free, so it is called a virtual particle or virtual photon. We say that the force between two charges is transmitted by or mediated by massless particles called photons. The strength of this transfer interaction, with the units energy times distance, is the energy of the photon $h\nu$ times its wavelength λ, and this is a universal constant, hc

$$(\text{strength})_{\text{ph}} = (h\nu)\lambda = hc \tag{6}$$

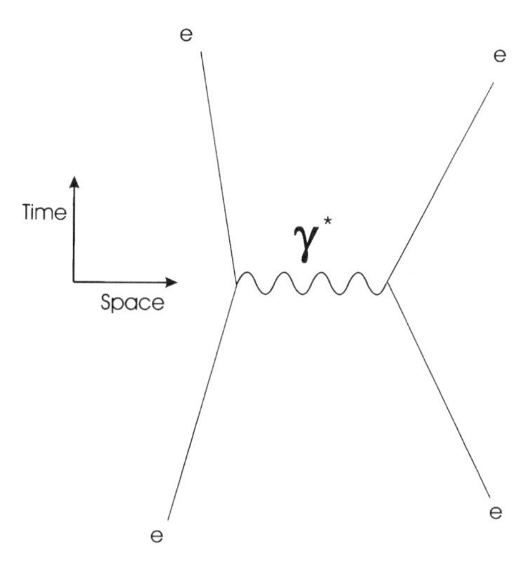

Fig. 26-1. Feynman diagram for momentum transfer between interacting electrons taking place through the exchange of a virtual photon γ^*.

The dimensionless strength α_{em} of the electromagnetic interaction is equal to the fine structure constant: $\alpha = \alpha_{em}$

$$\alpha = 2\pi \frac{(\text{strength})_{em}}{(\text{strength})_{ph}} = \frac{e^2}{2\varepsilon_0 hc} = \frac{1}{137} \tag{7}$$

which we encountered in Chapter 21 when we found that the energy of a hydrogen atom in its ground state is given by Eq. (15) in that chapter: $E_0 = \frac{1}{2}mc^2\alpha^2$.

The gravitational interaction between two masses has the same inverse square distance dependence

$$F = \frac{Gm_1m_2}{r^2} \tag{8}$$

as the Coulomb interaction (1). The source of the gravitational interaction is energy, and it is mediated by a hypothetical massless particle called a graviton. Its dimensionless strength α_g is obtained by comparing the two forces for an electron interacting with a proton

$$\alpha_g = \alpha \frac{Gm_em_p}{e^2/4\pi\varepsilon_0} = \frac{Gm_em_p}{\hbar\, c} \sim 10^{-39} \tag{9}$$

and we see that the gravitational interaction is negligible compared with the electromagnetic one when the interacting particles are electrons and/or protons.

Protons and neutrons in close proximity attract each other by a short-range force called the strong interaction force. The quantum theory of this strong interaction is called quantum chromodynamics (QCD). The strong interaction between two quarks is mediated by massless particles called gluons. The dimensionless strength of the strong interaction is one, $\alpha_s = 1$.

The weak force is involved in transforming protons and neutrons into each other via the reactions: $n \Rightarrow p + e$ and $e + p \Rightarrow n$. The weak force acts between leptons and quarks, and is transmitted by heavy particles called W^+, W^-, and Z^0.

The various forces have time constants associated with them, and these time constants are inversely proportional to the squares of the coupling strengths. The time constants represent typical particle decay times. The time constant τ_s for the strong force is the time it takes light to travel a distance equal to the diameter of a proton:

$$\tau_s = (10^{-15}\text{m})/(3 \times 10^8\,\text{m/s}) = 10^{-23}\,\text{s} \tag{10}$$

For the electromagnetic interaction we have

$$\tau_{em} = \tau_s/\alpha^2 \sim 10^{-19}\,\text{s} \tag{11}$$

and for the weak interaction

$$\tau_w = \tau_s/\alpha_w^2 \sim 10^{-11}\,\text{s} \tag{12}$$

Table 26-1 compares the characteristics of these four interactions.

3 PARTICLE CHARACTERISTICS

Three fundamental properties of a particle are its mass m, its spin s, and its charge q/e, where e is the magnitude of the electric charge of an electron and proton. When two, three, or several particles are the same type (baryon, meson, or lepton) and have the same spin and nearly the same mass, they can be considered different states of the same particle. This small group is characterized by a quantity called isospin I, with $2I + 1$ equal to the number of particles in the group. Each particle of the group is designated by a projection m_I of I, where $-I \leq m_I \leq I$ in analogy with the projection states of angular momentum. For example, the proton and neutron are two states of the baryon particle called a nucleon with isospin $I = \frac{1}{2}$, and the projections are $m_I = +\frac{1}{2}$ for the proton and $m_I = -\frac{1}{2}$ for the neutron. The particle called the pion is a meson which comes in three charge states so it has isospin $I = 1$ with the respective isospin projections $m_I = +1, 0,$ and -1 for π^+, π^0, and π^-, respectively.

TABLE 26-1. Characteristics of the four fundamental forces

Force	Source	Mediator	Relative strength α_i	Time constant τ_i (s)
Strong	quark color[a]	gluon	1	10^{-23}
Electromagnetic	electric charge	photon	1/137	10^{-19}
Weak	weak charge	W and Z_0 bosons	10^{-6}	10^{-11}
Gravity	energy	graviton	10^{-39}	

[a]Also called strong charge.

Every elementary particle is assigned a baryon quantum number B which has the values

$$\begin{aligned} B &= +1 \quad \text{baryon} \\ B &= -1 \quad \text{antibaryon} \\ B &= +\tfrac{1}{3} \quad \text{quark} \\ B &= -\tfrac{1}{3} \quad \text{antiquark} \\ B &= 0 \quad \text{other particles} \end{aligned} \tag{13}$$

for the various classes of particles. The charge q/e of an elementary particle is given by the modified Gell-Mann, Nishijima formula.

$$q/e = m_I + \tfrac{1}{2}(B + S + C) \tag{14}$$

where S and C are additional quantum numbers called strangeness and charm, respectively, which will be defined below. This equation agrees with the isotopic spin projections m_I assigned above for the nucleons and the pions because they have strangeness $S = 0$. This value for S was originally assigned because the nucleons and the pions were the ordinary particles of their larger groups, so the remaining particles were called 'strange', i.e., for them $S \neq 0$. Strangeness is an additive quantum number so for several particles we have

$$S = \sum S_i \tag{15}$$

There is a strangeness conservation law whereby for strong and electromagnetic interactions the net strangeness (15) of the system before a reaction equals the net strangeness after the reaction. This conservation law does not necessarily hold for weak interactions. An example of a weak interaction is beta-decay (β-decay) whereby a neutron inside a nucleus transforms to a proton and the nucleus emits a positive electron plus an antineutrino

$$\mathrm{p} \Rightarrow \mathrm{n} + \mathrm{e}^+ + \nu_\mathrm{e} \tag{16}$$

We are familiar with the positron e^+ as the antiparticle of the electron e^-. They have the same mass and opposite charge. Antiparticles in general have the same mass and opposite signs for the four quantities appearing in Eq. (14), namely q, m_I, B, and S. For example, the charged pions π^+ and π^-, which have the mass $m = 139.6\, m_\mathrm{e}$, are antiparticles of each other, and the neutral π^0 with the slightly smaller mass 135.0 m_e is its own antiparticle. We encounter several more examples of antiparticles later in this chapter.

4 THREE-QUARK MODEL

The three-quark model has been successful in explaining many of the properties of hadrons, both the heavier fermion types called baryons and the lighter boson types called mesons. The characteristics of the up, down, and strange quarks of this model, called u, d and s, respectively, are listed in Table 26-2. A baryon is composed of three quarks, and since baryons have baryon number $B = 1$ it follows that $B = \frac{1}{3}$ for each quark. Mesons consist of quark-antiquark pairs, where an antiquark is the antiparticle of a quark, and since $B = 0$ for a meson we conclude that $B = -\frac{1}{3}$ for an antiquark.

We see from the table that the up and down quarks have strangeness $S = 0$, much lower rest energies mc^2, and they constitute an isospin doublet state $I = \frac{1}{2}$ with $m_I = +\frac{1}{2}$ for u and $m_I = -\frac{1}{2}$ for d. The third quark of the model, called the strange quark s, has strangeness $S = -1$, a much larger rest energy, and is an isospin singlet, $I = 0$, so $m_I = 0$, as shown in Fig. 26-2. With the aid of the Gell-Mann, Nishijima formula (14) we find the charges $q/e = +\frac{2}{3}$, $-\frac{1}{3}$, and $-\frac{1}{3}$ for quarks u, d, and s, respectively. Characteristics of their antiquark counterparts $\bar{\text{u}}$, $\bar{\text{d}}$, and $\bar{\text{s}}$ are also listed in the table, as well as

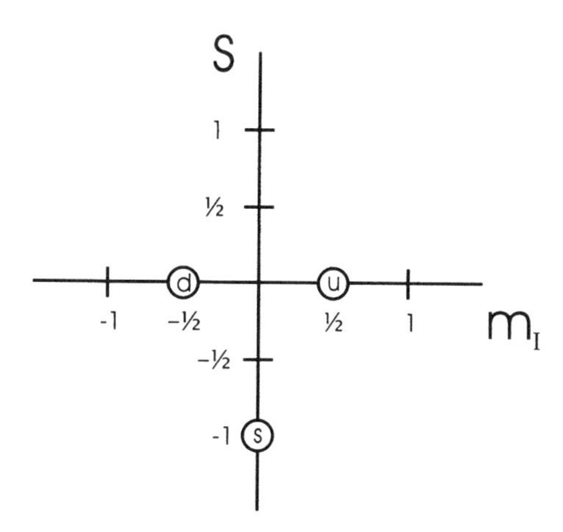

Fig. 26-2. Strangeness S versus isotopic spin projection m_I plot for the three quarks up (u), down (d), and strange (s).

TABLE 26-2. Characteristics of the first four quarks and antiquarks. They all have spin $s = \frac{1}{2}$.

	Regular quarks				Antiquarks			
Property	up, u	down, d	strange, s	charm, c	up, $\bar{u}$	down, $\bar{d}$	strange, $\bar{s}$	charm, $\bar{c}$
Isospin, I	$\frac{1}{2}$	$\frac{1}{2}$	0	0	$\frac{1}{2}$	$\frac{1}{2}$	0	0
Isospin projection, m_I	$+\frac{1}{2}$	$-\frac{1}{2}$	0	0	$-\frac{1}{2}$	$+\frac{1}{2}$	0	0
Charge, q/e	$+\frac{2}{3}$	$-\frac{1}{3}$	$-\frac{1}{3}$	$+\frac{2}{3}$	$-\frac{2}{3}$	$+\frac{1}{3}$	$+\frac{1}{3}$	$-\frac{2}{3}$
Baryon number, B	$+\frac{1}{3}$	$+\frac{1}{3}$	$+\frac{1}{3}$	$+\frac{1}{3}$	$-\frac{1}{3}$	$-\frac{1}{3}$	$-\frac{1}{3}$	$-\frac{1}{3}$
Strangeness, S	0	0	-1	0	0	0	$+1$	0
Charm, C	0	0	0	$+1$	0	0	0	-1
Parity	$+1$	$+1$	$+1$	$+1$	-1	-1	-1	-1
mc^2, MeV	5 ± 3	10 ± 5	200 ± 100	1500 ± 200	5 ± 3	10 ± 5	200 ± 100	1500 ± 200

the properties of a fourth quark, c, called charm, which will be introduced later in the chapter.

Group theory has been used to classify the hadrons into multiplets of the special unitary group of dimension 3, called SU(3). Since there are three quarks u, d, and s, and three antiquarks ū, d̄, and s̄, there are $3^2 = 9$ possibilities for mesons, and their properties are listed in Table 26-3. The SU(3) mathematical formalism uses direct product $\otimes$ and direct sum $\oplus$ notation to decompose this nonet of nine mesons into a singlet and an octet

$$3 \otimes 3 = 1 \oplus 8 \tag{17}$$

and the octet is displayed in Fig. 26-3 on an S versus m_I diagram. Each particle in the diagram is labeled with its quark–antiquark pair combination. There are three possibilities for the point at the origin, namely uū, dd̄, and ss̄. The two particles of the octet that are located there, namely π^0 and η, plus the singlet particle η', are different linear combinations of the three $q\bar{q}$

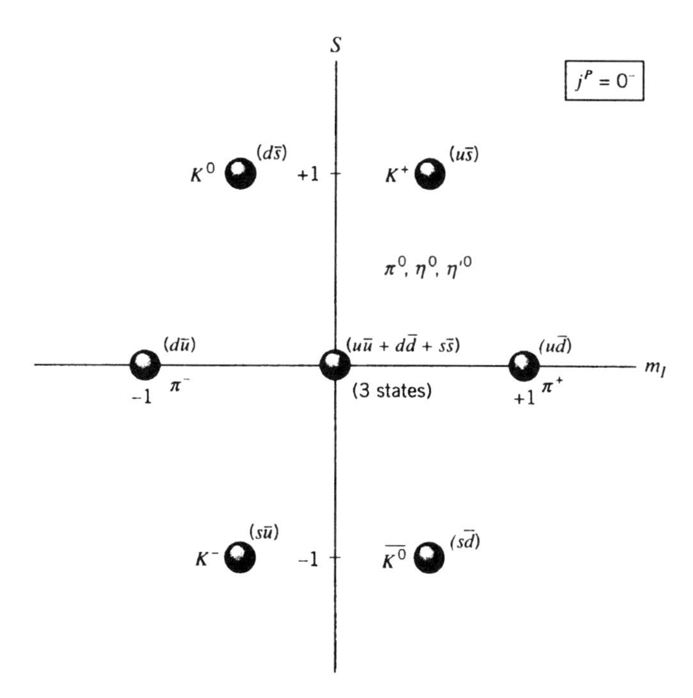

Fig. 26-3. S versus m_I plot for the $j^P = 0^-$ pseudoscalar meson octet. (From J. W. Rohlf, *Modern Physics from* α *to* Z^0, Wiley, New York, 1994, p. 484.)

TABLE 26-3. Characteristics of the octet of pseudoscalar mesons. These particles have spin $s = 0$, orbital angular momentum $\ell = 0$, total angular momentum $j = 0$, parity $P = -1$, baryon number $B = 0$, and charm $C = 0$. They are in the $j^P = 0^-$ state.

Property	π^+ $u\,\bar{d}$	π^0 $(u\,\bar{u},\, d\,\bar{d},\, s\,\bar{s})$	π^- $d\,\bar{u}$	K^0 $d\,\bar{s}$	K^+ $u\,\bar{s}$	K^- $s\,\bar{u}$	$\overline{K^0}$ $s\,\bar{d}$	η^0, η'^0 $(u\,\bar{u},\, d\,\bar{d},\, s\,\bar{s})$
Isospin, I	1	1	1	$\frac{1}{2}$	$\frac{1}{2}$	$\frac{1}{2}$	$\frac{1}{2}$	0
Isospin projections m_I	+1	0	−1	$-\frac{1}{2}$	$+\frac{1}{2}$	$-\frac{1}{2}$	$+\frac{1}{2}$	0
Charge, q/e	+1	0	−1	0	+1	−1	0	0
Strangeness, S	0	0	0	+1	+1	−1	−1	0
Mass, GeV/c^2	139.57	134.97	139.57	497.7	493.7	493.7	497.7	547, 958
Lifetime τ, s	2.6×10^{-8}	8.4×10^{-17}	2.6×10^{-8}	8.6×10^{-11}	1.2×10^{-8}	1.210^{-8}	8.6×10^{-11}	$10^{-19}, 10^{-21}$

possibilities at the origin. Particles on opposite sides of the octet, such as the pair $\pi+$, $\pi-$, and also the pair K^+, K^-, are antiparticles of each other.

Figure 26-4 shows, on the lower left, the masses of these nine mesons in the units GeV/c^2. We see from the figure that the singlet η' lies somewhat above the eight members of the octet. In this octet the spins of the quark and antiquark constituting each pair are antiparallel so the spin s of each meson is zero. In addition, the mesons have no angular momentum so $\ell = 0$, and hence the overall angular momentum $j = \ell + s = 0$. The parity of a meson in general is

$$P = P_q P_{aq}(-1)^{\ell} = (-1)^{\ell+1} \tag{18}$$

where the intrinsic quark, antiquark parities are $P_q = +1$ and $P_{aq} = -1$, respectively. For the present case $\ell = 0$ so the overall meson parity is -1. Since $j = 0$ each meson is looked upon as having a scalar angular momentum. A scalar with parity -1 is called a pseudoscalar so these mesons are

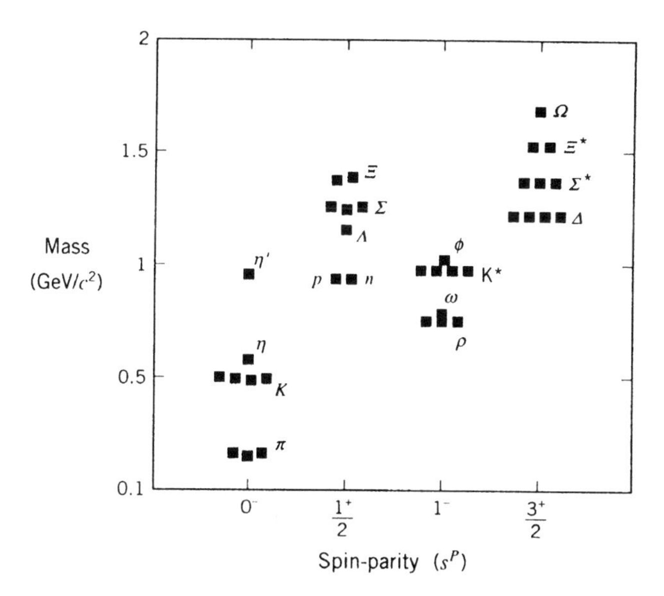

Fig. 26-4. Classification of the hadrons by their masses. From left to right we have the pseudoscalar meson octet (0^-), the ground state baryon octet ($\frac{1}{2}^+$), the vector meson octet (1^-), and the baryon decimet ($(\frac{3}{2})^+$). (From J. W. Rohlf, *Modern Physics from α to Z^0*, Wiley, New York, 1994, p. 482.)

called pseudoscalar types, and designated by $j^{P} = 0^{-}$ for short, as indicated along the absicissa under the pseudoscalar meson octet of Fig. 26-4.

There is another more massive excited state nonet of mesons of the type given in Fig. 26-3 in which the intrinsic quark–antiquark spins are parallel to each other for a total spin $s = 1$. Again we have $\ell = 0$, so for this case $j = \ell + s = 1$, and the mesons are called vector mesons, with the designation $j^{P} = 1^{-}$, as indicated in the figure. Additional excited state nonets have been found, some with $\ell > 1$.

The baryon particles are composed of three quarks each, for a total of $3^3 = 27$ baryons, and the SU(3) formalism decomposes this into a singlet, two octets, and one decimet (decuplet) as follows:

$$3 \otimes 3 \otimes 3 = 1 \oplus 8 \oplus 8 \oplus 10 \tag{19}$$

Figures 26-4, 26-5, and 26-6 show, respectively, the baryon mass diagrams and the S versus m_I plots for the ground state octet and for a decimet. The

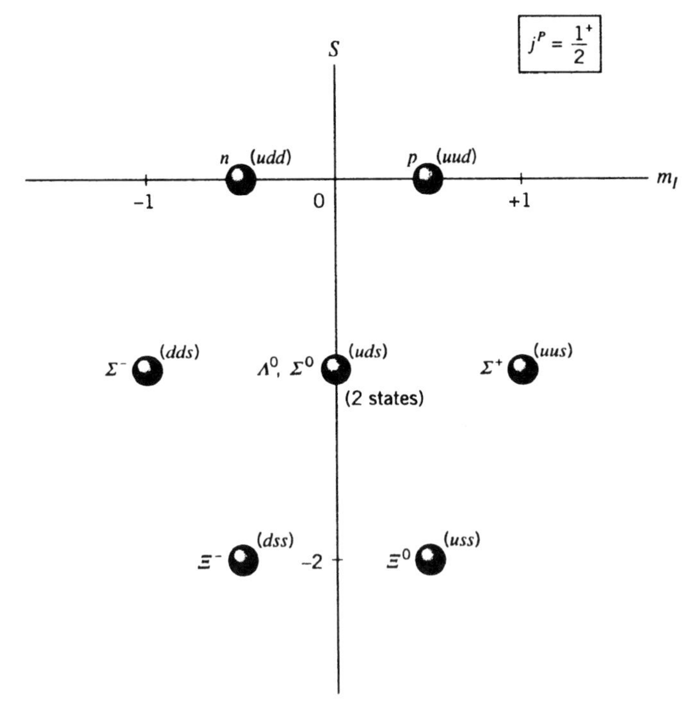

Fig. 26-5. S versus m_I plot for the $j^P = \frac{1}{2}^+$ ground state baryon octet. (From J. W. Rohlf, *Modern Physics from α to* Z^0, Wiley, New York, 1994, p. 487.)

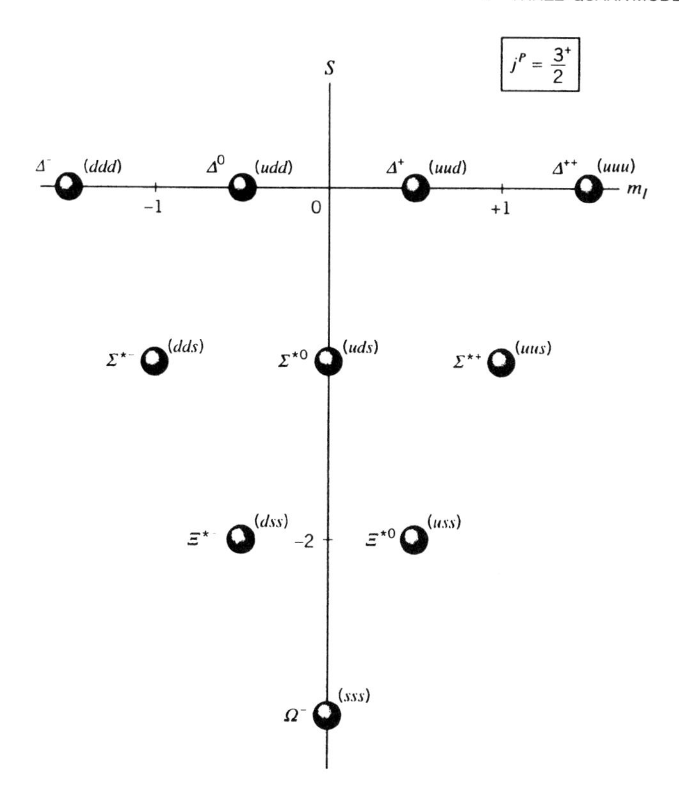

Fig. 26-6. S versus m_I plot for the $j^P = (\frac{3}{2})^+$ baryon decimet. (From J. W. Rohlf, *Modern Physics from α to Z⁰*, Wiley, New York, 1994, p. 485.)

ground state octet contains eight baryons with the properties given in Table 26-4 and Fig. 26-5. It is the most important multiplet because it contains the proton p and neutron n as the normal particles, i.e., as the zero strangeness particles. In this octet the three individual quark spins are not parallel so we have the net spin $s = \frac{1}{2}$. There is no orbital angular momentum, $\ell = 0$, so the total angular momentum $j = \frac{1}{2}$. For the bayron case of three quarks Eq. (18) for the parity becomes

$$P = P_{\mathrm{q}} P_{\mathrm{q}'} P_{\mathrm{q}''} (-1)^{\ell} = (-1)^{\ell} \tag{20}$$

where $P = +1$ for each of the three quarks. Since for the ground state octet $\ell = 0$, the overall parity is $+1$, giving $j^{\mathrm{P}} = \frac{1}{2}^{+}$. The decimet case shown in

TABLE 26-4. **Characteristics of the octet of ground state baryons. These particles have spin $s = \frac{1}{2}$, orbital angular momentum $\ell = 0$, total angular momentum $j = \frac{1}{2}$, parity $P = +1$, baryon number $B = +1$, and charm $C = 0$. They are in the $j^P = \frac{1}{2}^+$ state.**

Property	p uud	n udd	Σ^+ uus	Σ^0 uds	Σ^- dds	Ξ^0 uss	Ξ^- dss	Λ^0 uds
Isospin, I	$\frac{1}{2}$	$\frac{1}{2}$	1	1	1	$\frac{1}{2}$	$\frac{1}{2}$	0
Isospin projection, m_I	$+\frac{1}{2}$	$-\frac{1}{2}$	+1	0	−1	$+\frac{1}{2}$	$-\frac{1}{2}$	0
Charge, q/e	+1	0	+1	0	−1	0	−1	0
Strangeness, S	0	0	−1	−1	−1	−2	−2	0
mc^2, GeV	938.27	939.57	1189.4	1192.6	1197.4	1315	1321.3	1115.6
Lifetime τ, s	∞	0.89×10^{3}	8.0×10^{-11}	7.4×10^{-20}	1.5×10^{-10}	2.9×10^{-10}	1.6×10^{-10}	2.6×10^{-10}

Fig. 26-6 has three parallel quark spins so $s = \frac{3}{2}$ and no orbital angular momentum, $\ell = 0$, so for this case $j = \frac{3}{2}$ and $j^{\mathrm{P}} = \left(\frac{3}{2}\right)^{+}$. The masses for the particles of the $\frac{1}{2}^{+}$ and $\left(\frac{3}{2}\right)^{+}$ baryon multiplets are given in Fig. 26-4.

5 CONSERVATION LAWS AND REACTIONS

The reactions of the elementary particles, including their modes of decay, are restricted by conservation laws. There are, of course, the well-known conservation laws introduced earlier in the book, such as the conservation of linear and angular momentum, the conservation of energy, the conservation of charge, etc. In addition there are conservation laws regarding some of the new quantum numbers introduced in this chapter. For example, the overall baryon number is conserved in all interactions

$$B = \sum B_i = \text{const} \tag{21}$$

where B is the sum of the baryon numbers of all the particles that enter the reaction, and is likewise the sum of the baryon numbers of all the particles that exit from the reaction. In contrast to this the conservation of strangeness

$$S = \sum S_i = \text{const} \tag{22}$$

holds for the strong and for the electromagnetic interactions, but not for the weak interaction.

We saw in Section 2 that the strong, electromagnetic, and weak interactions have the respective time constants τ given by the values 10^{-23} s, 10^{-19} s, and 10^{-11} s. When a reaction can go by all three paths the overall decay rate is the sum of the three decay rates so the strong interaction dominates, with the other two being negligible. If the strong interaction is forbidden, then the decay will go by the electromagnetic mechanism, and only if both are forbidden will a particle decay via the weak interaction. For example a Δ^{++} baryon decays via the strong interaction process

$$\Delta^{++} \Rightarrow \mathrm{p} + \pi^{+} \tag{23}$$

with the strangeness change $\Delta S = 0$ since each particle has $S = 0$. The Ω^{-} baryon of the decimet with $S = -3$ cannot decay via a strong interaction because no combination of hadrons with a smaller mass than Ω^{-} has

$S = -3$, $q = -e$, and $B = +1$. It decays by the following sequence of three weak interactions with the indicated strangenesses:

$$\begin{aligned} \Omega^- &\Rightarrow \Xi^0 + \pi^- \\ (-3) &\Rightarrow (-2) + (0) \end{aligned} \tag{24a}$$

$$\begin{aligned} \Xi^0 &\Rightarrow \Lambda^0 + \pi^0 \\ (-2) &\Rightarrow (-1) + (0) \end{aligned} \tag{24b}$$

$$\begin{aligned} \Lambda^0 &\Rightarrow \mathrm{p} + \pi^- \\ (-1) &\Rightarrow (0) + (0) \end{aligned} \tag{24c}$$

corresponding to the overall reaction

$$\begin{aligned} \Omega^- &\Rightarrow \mathrm{p} + 2\pi^- + \pi^0 \\ (-3) &\Rightarrow (0) + (0) \quad (0) \end{aligned} \tag{25}$$

where the proton is a stable particle. The charged pions decay to leptons via weak interactions with a mean lifetime $\tau = 2.6 \times 10^{-8}$ s, and the neutral pion decays by the electromagnetic interaction in 8.7×10^{-17} s

$$\begin{aligned} \pi^+ &= \mu^+ + \nu_\mu \\ \pi^0 &= \gamma + \gamma \\ \pi^- &= \mu^- + \bar{\nu}_\mu \end{aligned} \tag{26}$$

The lifetimes are indicative of the interaction types.

Interactions involving the collision of two particles are used to produce other elementary particles. For example, the strong interaction ($\Delta S = 0$)

$$\begin{aligned} &\mathrm{K}^- + \mathrm{p} \Rightarrow \Omega^- + \mathrm{K}^+ + \mathrm{K}^0 \\ &(-1) \quad (0) \quad (-3) \quad (+1) \quad (+1) \end{aligned} \tag{27}$$

has been used to produce the Ω^- baryon that was discussed above.

6 LEPTONS

The leptons are relatively light particles, the best known examples being the electron e^- and its antiparticle, the positron e^+, which are stable, and the muons μ^- and μ^+, which behave like negative and positive electrons, respectively, except for their larger mass $m = 206.8\, m_e$, and their instability. The muon has a lifetime of 2.2 μs, and decays through the weak interaction

$$\mu^+ \Rightarrow e^+ + \bar{\nu}_\mu + \nu_e \tag{28a}$$

$$\mu^- \Rightarrow e^- + \nu_\mu + \bar{\nu}_e \tag{28b}$$

There is a heavy lepton called the tau-particle, τ, with a mass $m = 1777\, m_e$. Additional leptons are the massless electron, muon, and tau neutrinos ν_e, ν_μ, and ν_τ, together with their massless antineutrino counterparts $\bar{\nu}_e$, $\bar{\nu}_\mu$, and $\bar{\nu}_\tau$. The τ^- particle has a decay scheme similar to Eq. (28b) with the much shorter lifetime of 3.0 ps.

7 COLOR, CHARM, BEAUTY, AND TRUTH

In Section 4 we introduced the three-quark model and then proceeded to show how the baryons are built up by combinations of three quarks and the mesons are formed from quark–antiquark pairs. This model is adequate for explaining the properties of most of the known hadrons. In the theory of quantum chromodynamics (QCD) the quarks are assigned a strong charge, also called color, which comes in three varieties R, G, and B (red, green, blue). Each quark u, d, and s, can have any of these colors. Each of the three quarks in a baryon is a different color, so a baryon has zero net strong charge, and is called a color singlet. The same is true of antibaryons which contain three antiquarks, one of each anticolor, and hence are color singlets. Mesons, which are composed of a quark and an antiquark, are also zero color combinations.

In addition to the three fundamental quarks u, d, and s of this model, there are three additional much heavier quarks that have been discovered or postulated. The first of these is the charm quark, c, which has a rest energy of 1.5 GeV, and it is characterized by a special quantum number called charm, $C = +1$. Its characteristics are listed in Table 26-2. The three light quarks, u, d, and s, all have $C = 0$. There is a bound state of charm and anticharm called charmonium which has been observed.

We mentioned above that the simple unitary group SU(n) formalism has been used for the classification of elementary particles. SU(2) is used to classify isospin multiplets into their $2I+1$ states m_I along a straight line. In the next higher group SU(3) we introduce strangeness and project the classification into a second dimension S so the multiplets are singlets (η'), triplets (quarks), octets (mesons and baryons), and decimets (baryons), with representative diagrams in the S versus m_I plane appearing in Figs. 26-2, 26-3, 26-5, and 26-6. The next higher unitary group SU(4) introduces charm and projects SU(3) multiplets upward from the S versus M_I plane into a third dimension C, and this is illustrated in Fig. 26-7 for the four quarks u, d, s, and c pictured in the S, m_I, C space.

Baryons are composed of three quarks each, but now there are four choices for these three quarks. There are the usual uncharmed combinations, such as (uud) for a proton. In addition there are singly charmed baryons such as $\sum_c^+$ with the combination (udc), Ξ_{cc}^{++} formed from (ucc), and Ω_{ccc}^{++} with the constitution (ccc).

The total number of three quark ($qq'q''$) combinations with charm taken into account is $4^3 = 64$, and the SU(4) group theory decomposition gives one fourfold and three 20-fold supermultiplets

$$4 \otimes 4 \otimes 4 = 4 \oplus 20 \oplus 20 \oplus 20 \tag{29}$$

There is one tetrahedral configuration with uncharmed and singly charmed baryons, two 20-fold supermultiplets in the shape of an endekahedron (11-sided figure) with uncharmed, singly charmed, and doubly charmed bary-

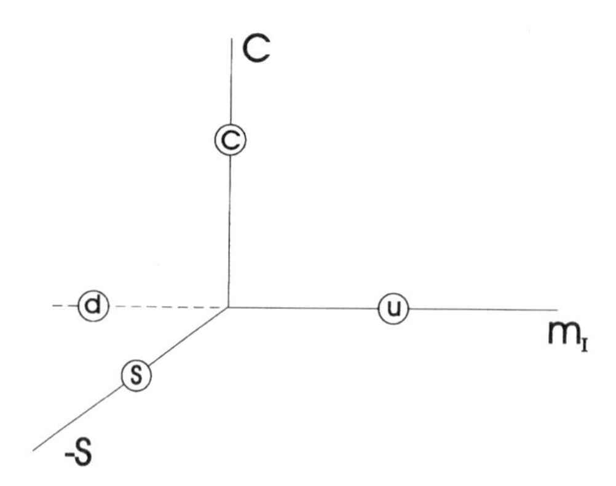

Fig. 26-7. The locations of the four quarks u, d, s, and c in the three-dimensional m_I, S, C space.

ons, and one pyramidal 20-fold configuration with uncharmed and singly, doubly, and triply (Ω_{ccc}^{++}) charmed baryons. The tetrahedral quartet contains the uncharmed (uds) and three singly charmed combinations (udc), (usc), and (dsc) which do not duplicate any quark. Figure 26-8 presents sketches of

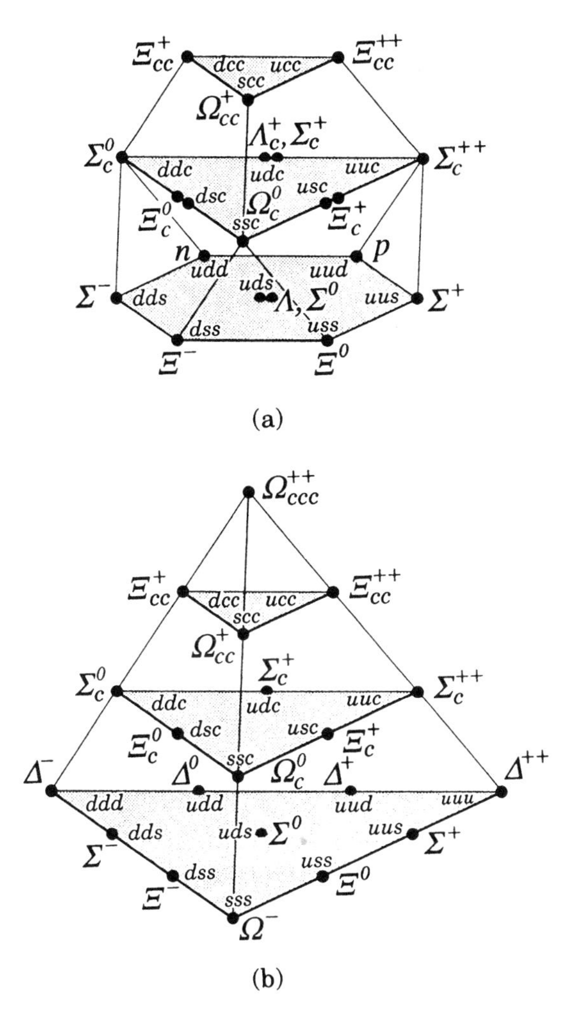

Fig. 26-8. Two of the 20-fold supermultiplets of the SU(4) classification of the baryons. Part (a) has the uncharmed ground state octet of Fig. 26-5 at the bottom, and part (b) has the uncharmed decimet of Fig. 26-6 at the bottom. (See *Phys. Rev.* **D54**, Part 1, 1996, p. 100 (Special journal issue, Review of Particle Physics); see also *Particle Properties Data Book*, AIP Press, New York, 1996.)

the pyramidal supermultiplets. We see from the figure that the ground state 20-fold has the decimet and the latter the ground state baryon octet at the bottom, namely at the lowest or uncharmed level. Some of the charmed particles in these supermultiplets have been observed.

Perhaps, for completeness, we should comment on the SU(2) decomposition for the baryons. In SU(2) there is no strangeness, i.e., $S = 0$, so there are only two quarks, u and d. The possible baryon quark combinations are (uuu) and (ddd) once each, and (uud) and (udd) three times each, for a total of eight. From another viewpoint the number of combinations is $2^3 = 8$. The SU(2) decomposition gives two doublets and one quartet

$$2 \otimes 2 \otimes 2 = 2 \oplus 2 \oplus 4 \tag{30}$$

so the only $S = 0$ multiplets that are allowed are those for which $2I + 1 = 2$ and 4, meaning $I = \frac{1}{2}$ and $\frac{3}{2}$. In other words, only doublet and quartet isospin multiplets are possible for baryons having $S = 0$. A check of Figures 26-2, 26-3, 26-5, and 26-8 indicates that this is the case. In Fig. 26-8 the uncharmed decimet contains the $S = 0$ quartet and the endekahedron, which occurs twice, contains the $S = 0$ (udd), (uud) doublet as part of the uncharmed octet, so Eq. (30) is satisfied.

In addition to charm, another heavy quark b with rest energy $\sim 5\,\text{GeV}$ and charge $-\frac{1}{3}e$ called the bottom quark, sometimes referred to as beauty, has been found. It is believed to be a doublet, but its counterpart top quark t, sometimes called truth, whose rest energy may exceed 100 GeV, has not yet been observed.

CHAPTER 27

MATHEMATICAL PHYSICS: TENSORS AND MATRICES

1 INTRODUCTION

In these final two chapters we cover the mathematical background needed for physics, and provide lists of the most commonly used mathematical expressions and equations. The present chapter deals with vector analysis, determinants, matrices, tensors, finite and infinite series, Fourier and integral transforms, integral equations, complex variables, group theory, and the Monte Carlo method. For the convenience of the reader we provide many of the mathematical functions that are frequently looked up in standard reference books, plus a number of additional formulae that are not so easy to find in these reference works. The next and final chapter of the book will cover additional mathematical material such as differential equations and their solutions, and some special topics such as gamma, delta, and Green's functions.

2 VECTOR RELATIONS

A vector is a quantity that has a magnitude and a direction. In this section we write vectors in bold face capital letters, and we write scalar functions, which have a magnitude but no direction, in lower case letters. Some fundamental vector relationships and a couple of tensor (**T**) relationships are

$$\mathbf{A}\cdot\mathbf{B}=\mathbf{B}\cdot\mathbf{A} \qquad \mathbf{A}\times\mathbf{B}=-\mathbf{B}\times\mathbf{A} \tag{1}$$

$$\begin{aligned}\mathbf{A}\cdot(\mathbf{B}\times\mathbf{C})&=(\mathbf{A}\times\mathbf{B})\cdot\mathbf{C}=-\mathbf{A}\cdot(\mathbf{C}\times\mathbf{B})=-(\mathbf{B}\times\mathbf{A})\cdot\mathbf{C}\\&=\text{volume of parallelepiped}\end{aligned} \tag{2}$$

$$\mathbf{A}\times(\mathbf{B}\times\mathbf{C})=\mathbf{B}(\mathbf{A}\cdot\mathbf{C})-\mathbf{C}(\mathbf{A}\cdot\mathbf{B}) \tag{3}$$

$$(\mathbf{A}\times\mathbf{B})\cdot(\mathbf{C}\times\mathbf{D})=\mathbf{A}\cdot[\mathbf{B}\times(\mathbf{C}\times\mathbf{D})]=(\mathbf{A}\cdot\mathbf{C})(\mathbf{B}\cdot\mathbf{D})-(\mathbf{A}\cdot\mathbf{D})(\mathbf{B}\cdot\mathbf{C}) \tag{4}$$

$$(\mathbf{A}\times\mathbf{B})\times(\mathbf{C}\times\mathbf{D})=\mathbf{A}[\mathbf{C}\cdot\mathbf{B}\times\mathbf{D})]+\mathbf{B}[\mathbf{D}\cdot(\mathbf{A}\times\mathbf{C})] \tag{5}$$

Relationships involving the gradient ∇, the divergence $\nabla\cdot$, and the curl $\nabla\times$ are as follows:

$$\text{if}\quad \nabla\cdot\mathbf{B}=0,\quad \text{then}\quad \mathbf{B}=\nabla\times\mathbf{A}\quad \text{and}\quad \mathbf{B}\text{ is a solenoidal vector} \tag{6}$$

$$\text{if}\quad \nabla\times\mathbf{B}=0,\quad \text{then}\quad \mathbf{B}=-\nabla f\quad \text{and}\quad \mathbf{B}\text{ is an irrotational vector} \tag{7}$$

$$\nabla(ab) = a\nabla b + b\nabla a \tag{8}$$

$$\nabla \cdot (a\mathbf{A}) = a\nabla \cdot \mathbf{A} + \mathbf{A} \cdot \nabla a \tag{9}$$

$$\nabla \times (a\mathbf{A}) = a\nabla \times \mathbf{A} + (\nabla a) \times \mathbf{A} \tag{10}$$

$$\nabla \cdot (\mathbf{A} \times \mathbf{B}) = (\nabla \times \mathbf{A})\mathbf{B} - (\nabla \times \mathbf{B})\mathbf{A} \tag{11}$$

$$\nabla \times (\mathbf{A} \times \mathbf{B}) = \mathbf{A}(\nabla \cdot \mathbf{B}) - \mathbf{B}(\nabla \cdot \mathbf{A}) - (\mathbf{A} \cdot \nabla)\mathbf{B} + (\mathbf{B} \cdot \nabla)\mathbf{A} \tag{12}$$

$$\mathbf{A} \times (\nabla \times \mathbf{B}) = (\nabla \mathbf{B}) \cdot \mathbf{A} - (\mathbf{A} \cdot \nabla)\mathbf{B} \tag{13}$$

$$\nabla(\mathbf{A} \cdot \mathbf{B}) = \mathbf{A} \times (\nabla \times \mathbf{B}) + \mathbf{B} \times (\nabla \times \mathbf{A}) + (\mathbf{A} \cdot \nabla)\mathbf{B} + (\mathbf{B} \cdot \nabla)\mathbf{A} \tag{14}$$

$$\nabla \cdot (\mathbf{AB}) = (\nabla \cdot \mathbf{A})\mathbf{B} + (\mathbf{A} \cdot \nabla)\mathbf{B} \tag{15}$$

$$\nabla \cdot (a\mathbf{T}) = a\nabla \cdot \mathbf{T} + (\nabla a) \cdot \mathbf{T} \quad \text{where } \mathbf{T} = \text{tensor} \tag{16}$$

Relationships involving successive differentiation operations such as the Laplacian ∇^2 over scalar functions f and vectors $\mathbf{A}$ are

$$\nabla^2 f = \nabla \cdot \nabla f \tag{17}$$

$$\nabla^2 \mathbf{A} = \nabla(\nabla \cdot \mathbf{A}) - \nabla \times (\nabla \times \mathbf{A}) \tag{18}$$

$$\nabla \times \nabla f = 0 \tag{19}$$

$$\nabla \cdot (\nabla \times \mathbf{A}) = 0 \tag{20}$$

Expressions involving the differentiation of the position vector $\mathbf{r}$ are

$$\nabla \cdot r = 3 \qquad \nabla \times \mathbf{r} = 0 \tag{21}$$

$$\nabla r = \mathbf{r}/r = \hat{\mathbf{r}} \qquad \nabla(1/r) = -\mathbf{r}/r^3 \tag{22}$$

$$\nabla^2(1/r) = -\nabla \cdot (\mathbf{r}/r^3) = 4\pi\delta(r) \tag{23}$$

There are several important relationships between an integral over a volume V and the related integral over the enclosing surface S, where $d\mathbf{S} = \hat{\mathbf{n}} dS$ and $\hat{\mathbf{n}}$ is a unit vector pointing outward from the surface

$$\iiint \nabla f \, dV = \iint \hat{\mathbf{n}} f \, dS \tag{24}$$

$$\iiint \nabla \cdot \mathbf{A}\, dV = \iint \hat{\mathbf{n}} \cdot \mathbf{A}\, dS \qquad \text{divergence theorem} \tag{25}$$

$$\iiint \nabla \cdot \mathbf{T}\, dV = \iint \hat{\mathbf{n}} \cdot \mathbf{T}\, dS \tag{26}$$

$$\iiint \nabla \times \mathbf{A}\, dV = \iint \hat{\mathbf{n}} \times \mathbf{A}\, dS \tag{27}$$

$$\iiint (f\nabla^2 g - g\nabla^2 f)\, dV = \iint \hat{\mathbf{n}} \cdot (f\nabla g - g\nabla f)\, dS \qquad \text{Green's theorem} \tag{28}$$

$$\iiint [\mathbf{A} \cdot (\nabla\times(\nabla \times \mathbf{B})) - \mathbf{B} \cdot (\nabla \times (\nabla \times \mathbf{A}))]\, dV$$
$$= \iint \hat{\mathbf{n}} \cdot [\mathbf{B} \times (\nabla \times \mathbf{A}) - \mathbf{A} \times (\nabla \times \mathbf{B})]\, dS \tag{29}$$

There are several important relationships between an integral over an open surface S and the related line integral along the contour C that bounds the surface, where ds is the line element

$$\iint \hat{\mathbf{n}} \times \nabla f\, dS = \oint f\,\mathbf{ds} \tag{30}$$

$$\iint (\nabla \times A) \cdot \hat{\mathbf{n}}\, dS = \oint \mathbf{A} \cdot \mathbf{ds} \qquad \text{Stokes' theorem} \tag{31}$$

$$\iint (\hat{\mathbf{n}} \times \nabla) \times \mathbf{A}\, dS = -\oint \mathbf{A} \cdot \mathbf{ds} \tag{32}$$

$$\int \hat{\mathbf{n}} \cdot (\nabla f \times \nabla g)\, dS = \oint f\nabla g \cdot \mathbf{ds} = -\oint g\nabla f \cdot \mathbf{ds} \tag{33}$$

3 COORDINATE SYSTEMS

Orthogonal curvilinear coordinates q_1, q_2, and q_3, and their associated orthogonal unit vectors $\hat{\mathbf{e}}_1, \hat{\mathbf{e}}_2$, and $\hat{\mathbf{e}}_3$ have the differential position vector

$$d\mathbf{r} = \hat{\mathbf{e}}_1 h_1\, dq_1 + \hat{\mathbf{e}}_2 h_2\, dq_2 + \hat{\mathbf{e}}_3 h_3\, dq_3 \tag{34}$$

The differential distance element ds is

$$ds = [(h_1\, dq_1)^2 + (h_2\, dq_2)^2 + (h_3\, dq_3)^2]^{1/2} \tag{35}$$

and the differential volume element $d\tau$ is

$$d\tau = h_1 h_2 h_3 \, dq_1 dq_2 \, dq_3 \tag{36}$$

where

$$h_1 = 1 \quad h_2 = 1 \quad h_3 = 1 \qquad \text{cartesian } x, y, z \tag{37a}$$
$$h_1 = 1 \quad h_2 = \rho \quad h_3 = 1 \qquad \text{cylindrical } \rho, \phi, z \tag{37b}$$
$$h_1 = 1 \quad h_2 = r \quad h_3 = r \sin\theta \quad \text{spherical } r, \theta, \phi \tag{37c}$$

for the coordinate systems of greatest interest.

In cartesian coordinates x, y, z we have

$$d\mathbf{r} = \hat{\mathbf{i}}\, dx + \hat{\mathbf{j}}\, dy + \hat{\mathbf{k}}\, dz \tag{38}$$
$$ds = [dx^2 + dy^2 + dz^2]^{1/2} \tag{39}$$
$$d\tau = dx\, dy\, dz \tag{40}$$

and the differential operators are given by

$$\nabla\Psi = \hat{\mathbf{i}}\frac{\partial\Psi}{\partial x} + \hat{\mathbf{j}}\frac{\partial\Psi}{\partial y} + \hat{\mathbf{k}}\frac{\partial\Psi}{\partial z} \tag{41}$$

$$\nabla \cdot V = \frac{\partial V_x}{\partial x} + \frac{\partial V_y}{\partial y} + \frac{\partial V_z}{\partial z} \tag{42}$$

$$\nabla \times V = \begin{vmatrix} \mathbf{i} & \hat{\mathbf{j}} & \hat{\mathbf{k}} \\ \dfrac{\partial}{\partial x} & \dfrac{\partial}{\partial y} & \dfrac{\partial}{\partial z} \\ V_x & V_y & V_z \end{vmatrix} \tag{43}$$

$$\nabla^2\Psi = \frac{\partial^2\Psi}{\partial x^2} + \frac{\partial^2\Psi}{\partial y^2} + \frac{\partial^2\Psi}{\partial z^2} \tag{44}$$

For cylindrical coordinates we have (see Fig. 27-1)

$$\begin{aligned} x &= \rho\cos\phi & \rho &= (x^2+y^2)^{1/2} \\ y &= \rho\sin\phi & \theta &= \tan^{-1} y/x \\ z &= z & z &= z \end{aligned} \tag{45}$$

$$\begin{aligned} \hat{e}_\rho &= \hat{\mathbf{i}}\cos\phi + \hat{\mathbf{j}}\sin\phi \\ \hat{\mathbf{e}}_\phi &= -\hat{\mathbf{i}}\sin\theta + \hat{\mathbf{j}}\cos\phi \end{aligned} \tag{46}$$

$$\begin{aligned} \hat{\mathbf{i}} &= \hat{\mathbf{e}}_\rho\cos\phi - \hat{\mathbf{e}}_\phi\sin\phi \\ \hat{\mathbf{j}} &= \hat{\mathbf{e}}_\rho\sin\phi + \hat{\mathbf{e}}_\phi\cos\phi \end{aligned} \tag{47}$$

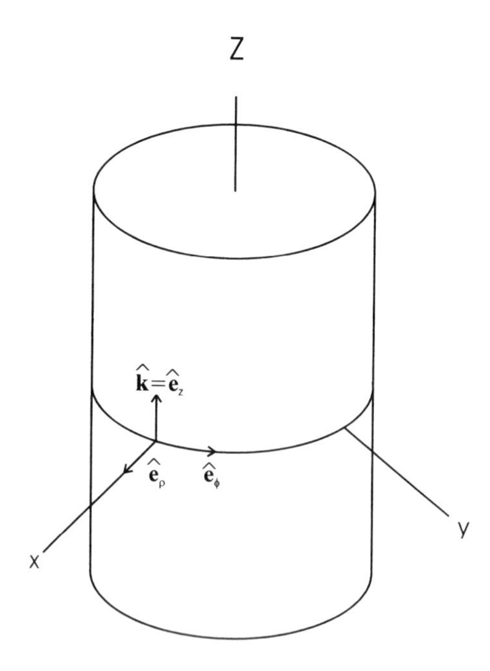

Fig. 27-1. The unit vectors $\mathbf{e}_\rho$, $\mathbf{e}_\phi$, and $\mathbf{e}_z = \mathbf{k}$ of cylindrical coordinates.

$$d\mathbf{r} = \hat{\mathbf{e}}_\rho d\rho + \hat{\mathbf{e}}_\phi \rho d\phi + \hat{\mathbf{k}} dz \tag{48}$$

$$ds = [d\rho^2 + \rho^2 d\phi^2 + dz^2]^{1/2} \tag{49}$$

$$d\tau = \rho \, d\rho \, d\phi \, dz \tag{50}$$

and the differential operators are given by

$$\nabla\Psi = \hat{\mathbf{e}}_\rho \frac{\partial\Psi}{\partial\rho} + \frac{\hat{\mathbf{e}}_\phi}{\rho} \frac{\partial\Psi}{\partial\phi} + \hat{\mathbf{k}} \frac{\partial\Psi}{\partial z} \tag{51}$$

$$\mathbf{\nabla} \cdot \mathbf{V} = \frac{1}{\rho} \frac{\partial}{\partial\rho}(\rho V_\rho) + \frac{1}{\rho} \frac{\partial V_\phi}{\partial\phi} + \frac{V_z}{\partial z} \tag{52}$$

$$\nabla \times V = \frac{1}{\rho} \begin{vmatrix} \hat{\mathbf{e}}_\rho & \rho\hat{\mathbf{e}}_\phi & \hat{\mathbf{k}} \\ \dfrac{\partial}{\partial\rho} & \dfrac{\partial}{\partial\phi} & \dfrac{\partial}{\partial z} \\ V_\rho & \rho V_\phi & V_z \end{vmatrix} \tag{53}$$

$$\nabla^2 V = \frac{1}{\rho} \frac{\partial}{\partial\rho}\left(\rho \frac{\partial\Psi}{\partial\rho}\right) + \frac{1}{\rho^2} \frac{\partial^2\Psi}{\partial\phi^2} + \frac{\partial^2 V_z}{\partial z^2} \tag{54}$$

For spherical polar coordinates we have (see Fig. 27-2)

$$\begin{aligned} x &= r\sin\theta\cos\phi \qquad & \rho &= (x^2 + y^2 + z^2)^{1/2} \\ y &= \sin\theta\sin\phi & \theta &= \cos^{-1} z/r \\ z &= \cos\theta & \phi &= \tan^{-1} y/x \end{aligned} \tag{55}$$

$$\begin{aligned} \hat{\mathbf{e}}_r &= \hat{\mathbf{i}}\sin\theta\cos\phi + \hat{\mathbf{j}}\sin\theta\sin\phi + \hat{\mathbf{k}}\cos\theta \\ \hat{\mathbf{e}}_\theta &= \hat{\mathbf{i}}\cos\theta\cos\phi + \hat{\mathbf{j}}\cos\theta\sin\phi - \hat{\mathbf{k}}\sin\theta \\ \hat{\mathbf{e}}_\phi &= -\hat{\mathbf{i}}\sin\phi + \hat{\mathbf{j}}\cos\theta \end{aligned} \tag{56}$$

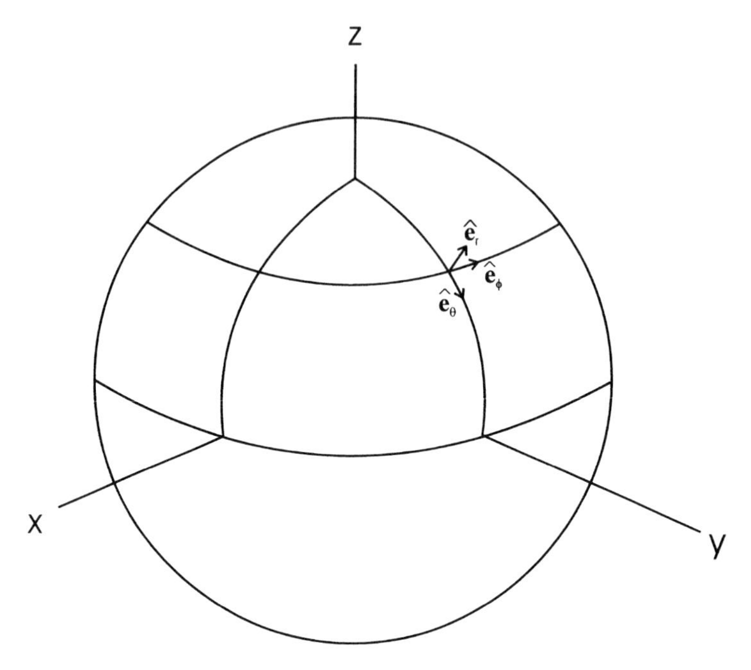

Fig. 27-2. The unit vectors $\mathbf{e}_r$, $\mathbf{e}_\theta$, and $\mathbf{e}_\phi$ of spherical polar coordinates.

$$\begin{aligned}
\hat{\mathbf{i}} &= \hat{\mathbf{e}}_r \sin\theta\cos\phi + \hat{\mathbf{e}}_\theta \cos\theta\cos\phi - \hat{\mathbf{e}}_\phi \sin\phi \\
\hat{\mathbf{j}} &= \hat{\mathbf{e}}_r \sin\theta\sin\phi + \hat{\mathbf{e}}_\theta \cos\theta\sin\phi + \hat{\mathbf{e}}_\phi \cos\phi \\
\hat{\mathbf{k}} &= \hat{\mathbf{e}}_r \cos\theta - \hat{\mathbf{e}}_\theta \sin\theta
\end{aligned} \tag{57}$$

$$d\mathbf{r} = \hat{\mathbf{e}}_r\, dr + \hat{\mathbf{e}}_\theta r\, d\theta + \hat{\mathbf{e}}_\phi r \sin\theta\, d\phi \tag{58}$$

$$ds = [dr^2 + r^2\, d\theta^2 + r^2 \sin^2\theta\, d\phi^2]^{1/2} \tag{59}$$

$$d\tau = r^2 \sin\theta\, dr\, d\theta\, d\phi \tag{60}$$

and the differential operators are given by

$$\nabla\Psi = \hat{\mathbf{e}}_r \frac{\partial\Psi}{\partial r} + \frac{\hat{\mathbf{e}}_\theta}{r}\frac{\partial\Psi}{\partial\theta} + \frac{\hat{\mathbf{e}}_\phi}{r\sin\theta}\frac{\partial\Psi}{\partial\phi} \tag{61}$$

$$\nabla \cdot V = \frac{1}{r^2}\frac{\partial}{\partial r}(r^2 V_r) + \frac{1}{r\sin\theta}\frac{\partial}{\partial\theta}(\sin\theta V_\theta) + \frac{1}{r\sin\theta}\frac{\partial V_\phi}{\partial\phi} \tag{62}$$

$$\nabla \times V = \frac{1}{r^2\sin\theta}\begin{vmatrix} \hat{\mathbf{e}}_r & r\hat{\mathbf{e}}_\phi & r\sin\theta\hat{\mathbf{e}}_\phi \\ \dfrac{\partial}{\partial r} & \dfrac{\partial}{\partial\theta} & \dfrac{\partial}{\partial\phi} \\ V_r & rV_\theta & r\sin\theta V_\phi \end{vmatrix} \tag{63}$$

$$\nabla^2 V = \frac{1}{r^2}\frac{\partial}{\partial r}\left(r^2\frac{\partial\Psi}{\partial r}\right) + \frac{1}{r^2\sin\theta}\frac{\partial}{\partial\theta}\left(\sin\theta\frac{\partial\Psi}{\partial\theta}\right) + \frac{1}{r^2\sin^2\theta}\frac{\partial^2 V_\phi}{\partial\phi^2} \tag{64}$$

4 DETERMINANTS

A 3×3 determinant has a value obtained by what is called expansion in minors

$$\begin{vmatrix} a_1 & b_1 & c_1 \\ a_2 & b_2 & c_2 \\ a_3 & b_3 & c_3 \end{vmatrix} = a_1(b_2c_3 - b_3c_2) - a_2(b_1c_3 - b_3c_1) + a_3(b_1c_2 - b_2c_1) \tag{65}$$

The value of a determinant is zero if:

1. two rows are equal or two columns are equal, and
2. each element in a row or each element in a column is zero.

The value of the determinant is unchanged if

1. a multiple of one row is added to another row, and
2. a multiple of a column is added to another column.

If each element in a row or each element in a column is multiplied by a constant, then the value of the determinant is multiplied by that constant.

5 MATRICES

A 2×2 matrix M written in its standard form

$$M = \begin{pmatrix} a & b \\ c & d \end{pmatrix} \tag{66}$$

has a transpose $\tilde{M}$

$$\tilde{M} = \begin{pmatrix} a & c \\ b & d \end{pmatrix} \tag{67}$$

a complex conjugate M^*

$$M = \begin{pmatrix} a^* & b^* \\ c^* & d^* \end{pmatrix} \tag{68}$$

an adjoint $M^\dagger$ or transpose conjugate $\tilde{M}^*$

$$M^\dagger = \tilde{M}^* = \begin{pmatrix} a^* & c^* \\ b^* & d^* \end{pmatrix} \tag{69}$$

and sometimes it has an inverse matrix M^{-1}

$$M^{-1} - = \begin{pmatrix} A & B \\ C & D \end{pmatrix} \tag{70}$$

defined by the relation $MM^{-1} = M^{-1}M = I$, where I is the unit matrix

$$MM^{-1} = \begin{pmatrix} a & b \\ c & d \end{pmatrix}\begin{pmatrix} A & B \\ C & D \end{pmatrix} = \begin{pmatrix} aA + bC & aB + bD \\ cA + dC & cB + dD \end{pmatrix} = \begin{pmatrix} 1 & 0 \\ 0 & 1 \end{pmatrix} \tag{71}$$

Three important matrix properties are matrix multiplication, defined in the previous equation, matrix addition, given by

$$\begin{pmatrix} 0 & 1 \\ 1 & 0 \end{pmatrix} + \begin{pmatrix} 0 & -i \\ i & 0 \end{pmatrix} = \begin{pmatrix} 0 & 1 - i \\ 1 + i & 0 \end{pmatrix} \tag{72}$$

and direct product expansion exemplified by

$$\begin{pmatrix} 0 & 1 \\ 1 & 0 \end{pmatrix} \otimes \begin{pmatrix} 0 & -i \\ i & 0 \end{pmatrix} = \begin{pmatrix} 0 & 0 & 0 & -i \\ 0 & 0 & i & 0 \\ 0 & -i & 0 & 0 \\ i & 0 & 0 & 0 \end{pmatrix} \tag{73}$$

Hamiltonian matrices $\mathcal{H}$ are Hermitian matrices, i.e., matrices with the property that

$$\mathcal{H}_{ij} = \mathcal{H}_{ji}^* \tag{74}$$

In other words the adjoint $\mathcal{H}^\dagger$ of a Hermitian matrix $\mathcal{H}$ is equal to the Hermitian matrix itself

$$\mathcal{H} = \mathcal{H}^\dagger \tag{75}$$

From Eq. (74) the diagonal elements of a Hermitian matrix are real, and its eigenvalues are real. A real Hermitian matrix is symmetric.

Transformation matrices are unitary matrices U, and a unitary matrix has the property that its inverse equals its adjoint

$$U^{-1} = U^\dagger \tag{76}$$

In other words, we have the pair

$$U = \begin{pmatrix} \alpha & \beta \\ \gamma & \delta \end{pmatrix} \qquad U^{-1} = \begin{pmatrix} \alpha^* & \gamma^* \\ \beta^* & \delta^* \end{pmatrix} \tag{77}$$

The rows and columns of unitary matrices obey the orthogonality conditions

$$\sum U_{ij} U_{ik}^* = \sum U_{ji} U_{ki}^* = \mathbb{I}\delta_{jk} \tag{78}$$

where $\mathbb{I}$ is a unit matrix, and from these conditions we can show that

$$\beta = -\gamma^* \tag{79}$$

so the most general 2×2 unitary matrix U with determinant $+1$ has the form

$$U = \begin{pmatrix} \alpha & -\beta^* \\ \beta & \alpha^* \end{pmatrix} \tag{80}$$

with the orthogonality condition

$$\alpha\alpha^* + \beta\beta^* = 1 \tag{81}$$

The elements of U can be expressed in terms of their real and imaginary parts e_j called Euler parameters (or sometimes Cayley Klein parameters)

$$\alpha = e_0 + ie_3 \tag{82}$$

$$\beta = e_2 + ie_1 \tag{83}$$

which satisfy the normalization condition

$$e_0^2 + e_1^2 + e_2^2 + e_3^2 = 1 \tag{84}$$

so three independent parameters must be specified to determine the matrix U.

A real unitary matrix is called an orthogonal matrix R, and it has the properties that its inverse equals its transpose

$$R^{-1} = \tilde{R} \tag{85}$$

and the rows and columns are orthogonal and normalized

$$\sum_i R_{ij} R_{ik} = \sum_i R_{ji} R_{ki} = \delta_{jk} \tag{86}$$

An example of R is a rotation matrix in three-dimensional cartesian space.

Conversions of real matrices M and Hermitian matrices $\mathcal{H}$ to their forms M' and $\mathcal{H}'$ in new systems are brought about by similarity transformations involving orthogonal (R) and unitary (U) matrices, respectively

$$R^{-1} M R = M' \tag{87}$$

$$U^{-1} \mathcal{H} U = \mathcal{H}' \tag{88}$$

which preserve the trace, Tr, or sum of the diagonal elements

$$\mathrm{Tr} = \sum M_{ii} = \sum M'_{ii} \tag{89}$$

$$\mathrm{Tr} = \sum \mathcal{H}_{ii} = \sum \mathcal{H}'_{ii} \tag{90}$$

These transformations (87) and (88) can be used to put their respective matrices in diagonal form, which displays their eigenvalues. The columns of the transformation matrix contain the eigenvectors of the transformation.

As an example consider the diagonalization of a 2×2 Hermitian matrix $\mathcal{H}$ by the transformation (88) using the unitary matrix (80)

$$\begin{pmatrix} \alpha^* & \beta^* \\ -\beta & \alpha \end{pmatrix} \begin{pmatrix} H_{11} & H_{12} \\ H_{12}^* & H_{22} \end{pmatrix} \begin{pmatrix} \alpha & -\beta^* \\ \beta & \alpha^* \end{pmatrix} = \begin{pmatrix} \lambda_1 & 0 \\ 0 & \lambda_2 \end{pmatrix} \tag{91}$$

The two column vectors

$$\begin{pmatrix} \alpha \\ \beta \end{pmatrix} \quad \text{and} \quad \begin{pmatrix} -\beta^* \\ \alpha^* \end{pmatrix} \tag{92}$$

are the eigenvectors of the original Hermitian matrix $\mathcal{H}$.

The eigenvalues λ_1 and λ_2 may be found from the secular equation determinant

$$\begin{vmatrix} H_{11} - \lambda & H_{12} \\ H_{12}^* & H_{22} - \lambda \end{vmatrix} = 0 \tag{93}$$

and this 2×2 case provides a quadratic equation with the solutions

$$\lambda = \tfrac{1}{2}(H_{11} + H_{22}) \pm \tfrac{1}{2}[(H_{11} - H_{22})^2 + 4|H_{12}|^2]^{1/2} \tag{94}$$

If the off-diagonal matrix element H_{12} is small compared with the diagonal elements H_{11} and H_{22}, and H_{12} is also small compared with their difference $H_{11} - H_{22} = \Delta$, then we obtain the following approximate values for the two energies $\lambda = E_1$ and E_2

$$E_1 = H_{11} + H_{12}^2/\Delta \tag{95}$$

$$E_2 = H_{22} - H_{12}^2/\Delta \tag{96}$$

6 PAULI MATRICES AND SPINORS

The Pauli spin matrices

$$\sigma_1 = \begin{pmatrix} 0 & 0 \\ 1 & 0 \end{pmatrix} \qquad \sigma_2 = \begin{pmatrix} 0 & -i \\ i & 0 \end{pmatrix} \qquad \sigma_3 = \begin{pmatrix} 1 & 0 \\ 0 & -1 \end{pmatrix} \tag{97}$$

have the rare distinction of being both unitary and Hermitian. They have the cyclic permutation property

$$\sigma_i \sigma_j = i\sigma_k \tag{98}$$

they anticommute

$$\sigma_i \sigma_j + \sigma_j \sigma_i = 2\mathbb{I}\delta_{ij} \tag{99}$$

and the square of each is the unit matrix

$$\sigma_i^2 = \mathbb{I} \tag{100}$$

The Pauli matrices constitute a representation of a spin $\frac{1}{2}$ vector

$$\mathbf{S} = \tfrac{1}{2}\sigma \qquad (S = \tfrac{1}{2}) \tag{101}$$

They provide a 2×2 representation of the coordinate vector $\mathbf{r}$

$$M_r = \begin{pmatrix} z & x + iy \\ x - iy & z \end{pmatrix} = x\sigma_1 + y\sigma_2 + z\sigma_3 \tag{102}$$

Half angle unitary matrices R such as

$$R_x = \begin{pmatrix} \cos\frac{1}{2}\theta & i\sin\frac{1}{2}\theta \\ i\sin\frac{1}{2}\theta & \cos\frac{1}{2}\theta \end{pmatrix} \tag{103}$$

bring about rotations of the matrix M_r via the similarity transformation

$$R_i^{-1} M_r R_i = M_r' \tag{104}$$

to the rotated matrix $M_r'(x', y', z')$.

Spinors are two-component complex vectors u, v in this 2×2 space called spinor space, and they transform to a new coordinate frame as vectors via a unitary transformation matrix (80)

$$\begin{pmatrix} \alpha & -\beta^* \\ \beta & \alpha^* \end{pmatrix} \begin{pmatrix} u \\ v \end{pmatrix} = \begin{pmatrix} u' \\ v' \end{pmatrix} \tag{105}$$

so

$$
\begin{aligned}
u' &= \alpha u - \beta^* v \\
v' &= \beta u + \alpha^* v
\end{aligned}
\tag{106}
$$

We see from Eq. (103) that the identity in this space is not 2π but rather 4π, since this is the angle of rotation that restores a spinor to its original configuration. This means that there is a homomorphism which, in this case, is a two to one mapping between the 2×2 unitary matrices U and associated 3×3 orthogonal rotation matrices R. In other words, the two angles θ and $\theta + 2\pi$ of $U(\theta)$ of Eq. (80) and R_x of Eq. (103) both map to θ of the 3×3 rotation matrix R of Eq. (3) in Chapter 3.

Spinors are important quantities in quantum mechanics since the two components u and v can represent the two spin states of an electron, namely $m = +\frac{1}{2}$ and $m = -\frac{1}{2}$.

7 TENSORS

Zero-, first-, and second-rank tensors are, respectively called scalars S, vectors V, and tensors T. They have pseudocounterparts S_p, V_p, and T_p, and the parity operation P which brings about the change $x \Rightarrow -x$, $y \Rightarrow -y$, $z \Rightarrow -z$ has the following effect on the various tensor types:

$$
\begin{aligned}
PS &= S & PS_p &= -S_p \\
P\mathbf{V} &= -\mathbf{V} & P\mathbf{V}_p &= \mathbf{V}_p \\
P\mathbf{T} &= \mathbf{T} & P\mathbf{T}_p &= -\mathbf{T}_p
\end{aligned}
\tag{107}
$$

and similarly for higher order tensors. An example of a pseudovector is the cross-product of two vectors.

A nine-component cartesian tensor $A_{xx}, A_{xy}, \ldots,$ can be decomposed into a one-component scalar S which is the trace

$$
S = A_{xx} + A_{yy} + A_{zz} \tag{108}
$$

a three-component (pseudo) vector part $\mathbf{V}$ which is the antisymmetric part of the tensor, where, for example, we have for the x component of the vector

$$
V_x = A_{yz} - A_{zy} \tag{109}
$$

and a zero trace tensor part having five independent components of the type

$$A_{xx} \quad \text{and} \quad A_{xy} + A_{yx} \tag{110}$$

which come from the symmetric part of the original cartesian tensor. The six components of Eq. (110) are subject to the zero trace condition, so only five of them are independent.

The three components of a vector can be expressed as a magnitude

$$V = (V_x^2 + V_y^2 + V_z^2)^{1/2} \tag{111}$$

plus two polar angles θ and ϕ which provide its direction in space. The five components of a tracelss second-rank tensor can be uniquely designated by the three Euler angles θ, ϕ, ψ that diagonalize it, by its largest magnitude element along the diagonal T_{zz}, and by the difference between the other two diagonal elements $T_{xx} - T_{yy}$ after diagonalization.

The scalar or dot product of two vectors **A** and **B** involves the contraction $\mathbf{A} \cdot \mathbf{B}$ of contravariant **A** and covariant **B** forms, or vice versa, to give

$$\mathbf{A} \cdot \mathbf{B} = A^1 B_1 + A^2 B_2 + A^3 B_3 = \mathbf{B} \cdot \mathbf{A} = B^1 A_1 + B^2 A_2 + B^3 A_3 \tag{112}$$

Unless covariant and contravarient vectors are identical, as in the case of cartesian coodinates, two covariant vectors should never be contracted, and two contravarient ones should never be contracted because the result could contain cross terms of the type $A_1 B_3$ or $A^2 B^1$, respectively. The reason for the presence of cross terms is that the basis vectors are, in general, not orthogonal to each other. These considerations generalize to higher rank tensors.

A contravariant vector with the components $V^i = V^1, V^2, \ldots$ in one coordinate system transforms to another coordinate system by the rule

$$V^{i\,\prime} = \sum_j \left(\frac{\partial x_i'}{\partial x_j} \right) V^j \tag{113}$$

and a covariant vector $V_i = V_1, V_2, \ldots$ transforms by the rule

$$V_i' = \sum_j \left(\frac{\partial x_j}{\partial x_i'} \right) V_j \tag{114}$$

Second-rank contravariant T^{ij}, mixed T^i_j, and covariant T_{ij} tensors transform in the following manners:

$$T^{ij\,\prime} = \sum_{kl} \left(\frac{\partial x_i' \partial x_j'}{\partial x_k \partial x_\ell} \right) T^{k\ell} \quad \text{contravariant} \tag{115}$$

$$T^{i\,\prime}{}_{j} = \sum_{k\ell} \left(\frac{\partial x_i' \partial x_\ell}{\partial x_k \partial x_j'} \right) T^k_\ell \quad \text{mixed} \tag{116}$$

$$T_{ij}' = \sum_{k\ell} \left(\frac{\partial x_k \partial x_\ell}{\partial x_i' \partial x_j'} \right) T_{k\ell} \quad \text{covariant} \tag{117}$$

These rules are easily generalized to higher rank tensors.

The distinction between contravariant and covariant is important for several situations in Physics: (i) in solid state physics oblique coordinates have the direct lattice and the reciprocal lattice as a contravariant–covariant pair; (ii) in quantum mechanics ket vectors $|i\rangle$ and bra vectors $\langle j|$ constitute such a pair; (iii) row and column vectors comprise such a covariant–contravariant pair; and (iv) special relativity is sometimes formulated using **r**, ct and **r**, $-ct$ as a covariant–contravariant pair. When relativity is formulated using the notation **r**, ict then contravariant and covariant vectors are identical to each other

8 INFINITE SERIES

In this section we summarize several of the relations and theorems involving infinite series. Consider the partial sum

$$s_i = \sum u_n \tag{118}$$

By the Cauchy criterion the partial sum converges to a limit S

$$\lim s_i = S \tag{119}$$

if for each $\varepsilon > 0$ there exists a fixed N such that

$$|S - s_i| < \varepsilon \quad \text{for } i > N \tag{120}$$

Two important comparison series are the geometric series which converges for $r < 1$ to the limit

$$\sum r^{n-1} = \frac{1}{1-r} \tag{121}$$

and diverges for $r \geq 1$, and the harmonic series

$$\sum n^{-1} = \tfrac{1}{2} + \tfrac{1}{3} + \tfrac{1}{4} + \ldots \tag{122}$$

which diverges.

If the terms u_n of a series vary in sign and $\sum u_n$ converges but $\sum |u_n|$ diverges, then the series converges conditionally. The harmonic series (122) converges conditionally. If $\sum |u_n|$ converges, then the series converges absolutely. Uniform convergence is a subtle concept which involves the dependence of the convergence of a series of functions $\sum u_n(x)$ on the value of the argument x.

The Euler–Mascheroni constant γ is defined by the limit $n \Rightarrow \infty$

$$\gamma = \lim_{n\to\infty} \left[\sum_{m=1}^{n} \frac{1}{m} - \ln(n) \right] = 0.577216\ldots \tag{123}$$

where both the partial sum $s_n = \sum m^{-1}$ and the natural logarithm of n diverge strongly, but the difference between them converges.

9 FOURIER SERIES

The expansion of a function $f(x)$ in a Fourier series is

$$f(x) = \tfrac{1}{2}a_0 + \sum_m a_m \cos mx + \sum_m b_m \sin mx \tag{124}$$

where m is a sequence of integers, and the coefficients of the trigonometric functions can be evaluated from the expressions

$$a_m = \frac{1}{\pi} \int_0^{2\pi} f(t) \cos mt \, dt \tag{125}$$

$$b_m = \frac{1}{\pi} \int_0^{2\pi} f(t) \sin mt \, dt \tag{126}$$

using the orthogonality relations

$$\int_0^{2\pi} \sin mx \sin px \, dx = \pi\delta_{mp} \quad m \neq 0$$
$$\int_0^{2\pi} \cos mx \cos px \, dx = \pi\delta_{mp} \quad m \neq 0 \tag{127}$$
$$\int_0^{2\pi} \sin mx \cos px \, dx = 0$$

An alternative Fourier series expansion is

$$f(x) = \sum_m c_n \, \mathrm{e}^{inx} \tag{128}$$

where expressions for c_n analogous to Eqs. (125)–(127) and relationships between the coefficients c_n, a_n, and b_n can easily be written down.

A Fourier series is, in general, a very good approximation to a function, but it can overshoot at the positions of discontinuities, as illustrated for the square wave in Fig. 27-3. This tendency to overshoot is called the Gibbs phenomenon. Table 27-1 gives examples of some common Fourier series.

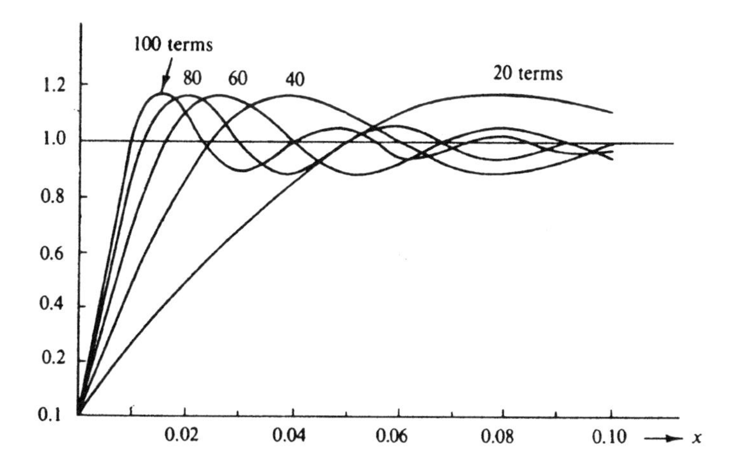

Fig. 27-3. Gibbs phenomenon or tendency of the Fourier series of a square wave to overshoot very close to the discontinuity. (From G. Arfken, and H. J. Weber, *Mathematical Methods for Physicists*, 4th ed., Academic Press, New York, 1995, p. 787.)

TABLE 27-1. Examples of some Fourier series

Name	Values of m	Series	Sketch
Sawtooth raised	1, 2, 3, ...	$\pi - 2\sum \frac{\sin mx}{m}$	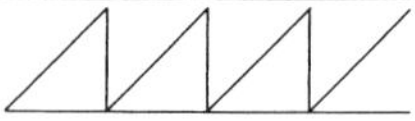
Sawtooth centered	1, 2, 3, ...	$2\sum (-1)^{m+1} \frac{\sin mx}{m}$	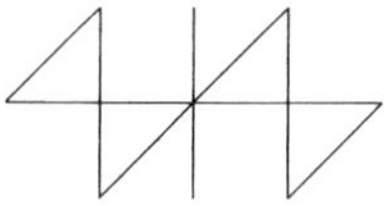
Triangular centered	1, 3, 5, ...	$\frac{4}{\pi}\sum (-1)^{(m+1)/2} \frac{\sin mx}{m^2}$	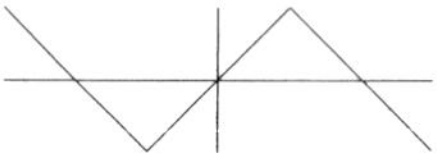
Square centered	1, 3, 5, ...	$\frac{4}{\pi}\sum \frac{\sin mx}{m}$	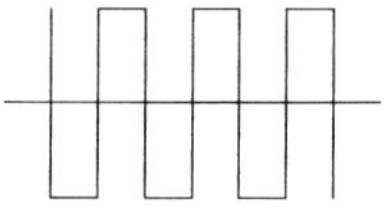
Full wave rectified	2, 4, 6, ...	$\frac{2}{\pi} - \frac{4}{\pi}\sum \frac{\cos mx}{m^2 - 1}$	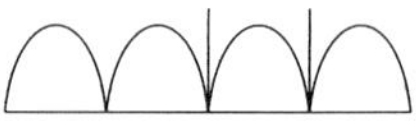

Source: See I. Bronshtein and K. Semendiaev, *Manual de Matemáticas*, Editorial, Mir Moscow, 1977, p. 638.

10 FOURIER TRANSFORMS

The Fourier series (124) is a useful way to represent a periodic function $f(x)$ by employing a series of terms $\sin mx$ and $\cos mx$, where m is an integer. If the function is not periodic, then we can replace the summation over m and n in Eqs. (124) and (128), respectively, by an integration over ω. For convenience we make use of Eq. (128) with the variable changed from x to t, to give for $f(t)$.

$$f(t) = \frac{1}{(2\pi)^{1/2}} \int_{-\infty}^{\infty} F(\omega)\, e^{-i\omega t}\, d\omega \tag{129}$$

where the discrete coefficients c_n of Eq. (128) with integral values of n are subsumed into a continuous function $F(\omega)$ of a variable ω which can take on all values from $-\infty$ to $+\infty$. The series representation (128) of $f(x)$ has been supplanted by the integral representation (129) of $f(t)$. We can also write the inverse transformation

$$F(\omega) = \frac{1}{(2\pi)^{1/2}} \int_{-\infty}^{\infty} f(\omega)\, e^{i\omega t}\, dt \tag{130}$$

The normalization factor $(2\pi)^{1/2}$ is included to make the reciprocal transforms $F(\omega)$ and $f(t)$ symmetrical with respect to each other, except for the sign in the exponential.

Equations (129) and (130) are applicable to general classes of non-periodic functions. If the original function $f(t)$ has even or odd symmetry under the parity operation $Pf(t) = f(-t)$

$$\begin{aligned} Pf_c(t) &= f_c(t) \quad \text{even} \\ Pf_s(t) &= -f_s(t) \quad \text{odd} \end{aligned} \tag{131}$$

then the exponential Fourier transform pair (129) and (130) can be replaced by cosine transform and sine transform counterparts, respectively. For even functions we have

$$f_c(t) = (2/\pi)^{1/2} \int_0^{\infty} F_c(\omega) \cos \omega t\, d\omega \tag{132a}$$

$$F_c(\omega) = (2/\pi)^{1/2} \int_0^{\infty} f_c(t) \cos \omega t\, dt \tag{132b}$$

and for odd functions

$$f_s(t) = (2/\pi)^{1/2} \int_0^{\infty} F_s(\omega) \sin \omega t\, d\omega \tag{133a}$$

$$F_s(\omega) = (2/\pi)^{1/2} \int_0^{\infty} f_s(t) \sin \omega t\, dt \tag{133b}$$

If the function $f(t)$ is localized over a range τ, then its transform $F(\omega)$ will be localized over the range $\Delta\omega$ which satisfies the uncertainty condition

$$\tau\, \Delta\omega \approx 1 \tag{134}$$

as in quantum mechanics.

It will be instructive to give several examples of Fourier transforms. If we select $e^{i\omega x}$ for $F(\omega)$ in Eq. (129) and renormalize slightly, then we obtain a representation for the delta function $\delta(t-x)$

$$\delta(t-x) = \frac{1}{2\pi}\int_{-\infty}^{\infty} e^{-i\omega(t-x)}\, d\omega \tag{135}$$

where

$$\delta(t-x) = \begin{cases} \infty & \text{for } t = x \\ 0 & \text{for } t \neq x \end{cases} \tag{136}$$

$$\int_{-\infty}^{\infty} \delta(t-x)\, dx = 1 \tag{137}$$

The delta function $\delta(t-x)$ is localized at a point and its Fourier transform $e^{-i\omega(t-x)}$ is totally delocalized, corresponding to the extreme case of the uncertainty relation (134).

In quantum mechanics the momentum space wavefunction $g(p)$ has a Fourier transform type relationship with the coordinate space wavefunction $\psi(x)$. As an example, for the ground state of the hydrogen atom we have

$$\psi(\mathbf{r}) = (1/\pi a_0)^{-1/2} e^{-r/a_0} \tag{138a}$$

$$g(\mathbf{p}) = \frac{1}{2\pi}\frac{(2\hbar/a_0)^{\frac{5}{2}}}{[p^2 + (\hbar/a_0)^2]^2} \tag{138b}$$

The uncertainty relation applied to these transforms

$$\Delta x\, \Delta p_x \approx \hbar \tag{139}$$

is, of course, the familiar one due to Heisenberg.

A finite wave train $f(t)$ N cycles long defined by the expression

$$f(t) = \begin{cases} \sin \omega_0 t & |t| < \tau \\ 0 & |t| > \tau \end{cases} \tag{140}$$

where $\tau = N\pi/\omega_0$ has a sine Fourier transform given by

$$F(\omega) = (2/\pi)^{1/2}\left\{\frac{\sin[(\omega_0 - \omega)\tau]}{2(\omega_0 - \omega)\tau} - \frac{\sin[(\omega_0 + \omega)\tau]}{2(\omega_0 + \omega)\tau}\right\} \tag{141}$$

A plot of the square of this function approximates the envelope of Fig. 14-7, in Chapter 14, which is the single wide slit diffraction pattern in optics, and also the Josephson junction Fraunhofer diffraction pattern in superconductivity.

In spectroscopy we utilize the fact that the cosine Fourier transform of a Lorentzian function $1/(1+y^2)$ is an exponential e^{-x}, as in the case of Eqs. (138). This means that a Lorentzian shaped line $Y(\omega)$ in the frequency domain of width $\Delta\omega$ centered at the frequency ω_0

$$Y(\omega) = \frac{1}{(\omega - \omega_0)^2 + (\Delta\omega)^2} \tag{142}$$

will exhibit oscillations in the time domain which are damped by an exponential decay factor $e^{-t/\tau}$. The classical way to do spectroscopy is to scan the frequency and obtain a spectrum of the type shown in Fig. 27-4(b). It is clear from this figure that most of the scanning time is spent between resonant lines when no useful information is being obtained. It is more efficient to irradiate the sample with successive pulses of radiation that contain a wide range of frequencies. The result is a time domain spectrum of the type shown in Fig. 27-4(a). This does not appear to contain useful information, but it can be Fourier transformed to provide the frequency domain spectrum shown at the bottom of the figure. There is a large gain in sensitivity because all of the irradiation time is spent gathering information about the spectrum. Currently almost all of infrared and nuclear magnetic resonance (NMR) spectroscopy is done the Fourier transform way.

A hundred years ago the main application of Fourier transform techniques in physics was in crystallography involving the determination of the atom positions in solids by Fourier transforming the experimentally measured distances between crystallographic planes. In band theory the properties of Brillouin zones in k-space (i.e., momentum space) are obtained by Fourier transforming crystal structure data from coordinate space.

11 INTEGRAL TRANSFORMS AND EQUATIONS

An integral transform relates pairs of functions $f(s)$ and $F(t)$ as follows:

$$f(s) = \int_a^b F(t)\, K(s, t)\, dt \tag{143}$$

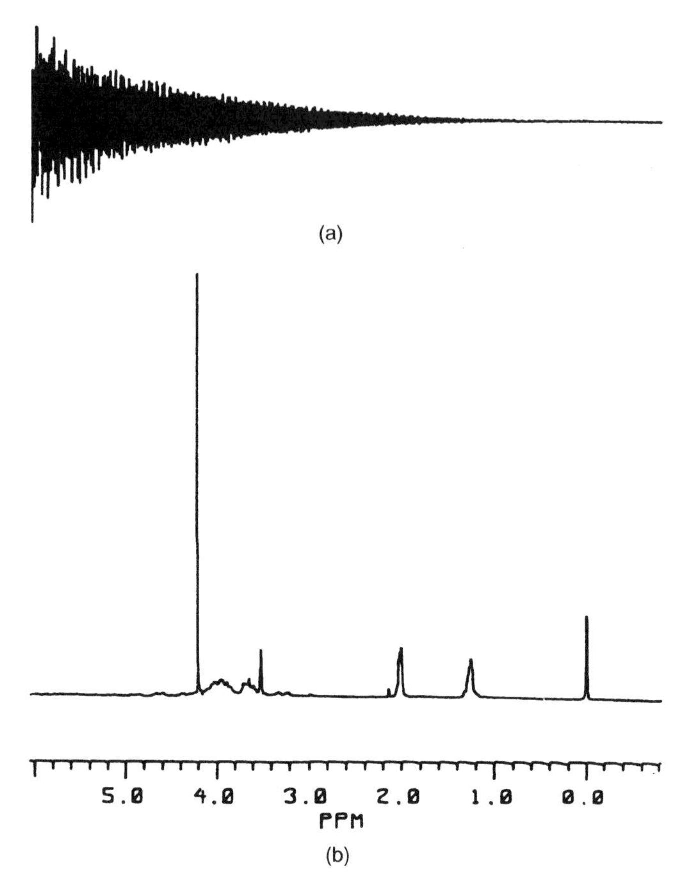

Fig. 27-4. Proton nuclear magnetic resonance (NMR) spectrum of a liposaccharide from the cell wall of Rhizobium showing (a) frequency domain, and (b) time domain scans. (From C. P. Poole, Jr. and H. A. Farach, *Theory of Magnetic Resonance*, Wiley, New York, 2nd Ed, 1987, p. 328; spectrum furnished by P. T. Ellis.)

where $f(s)$ is called the (integral) transform of $F(t)$, and $K(s, t)$ is the kernel of the transform. For the Fourier transform case (129, 130) the kernel is e^{ist}. Some additional examples are those due to Laplace $[K(s, t) = e^{st}]$, Mellin $[K(s, t) = s^{-t}]$, and Hankel $[K(s, t) = tJ_n(st)]$. The most widely used of these is the Laplace transform $\mathcal{L}\{F(t)\}$ defined as follows

$$\mathcal{L}\{F(t)\} = f(s) = \int_0^{\infty} e^{st} F(t)\, dt \tag{144}$$

Table 27-2 gives a list of some commonly used Laplace transforms.

The Laplace transforms of the first derivative $F'(t) = dF/dt$ and the second derivative $F''(t) = d^2F/dt^2$ are easily found by integration by parts

$$\mathcal{L}\{F'(t)\} = s\mathcal{L}\{F(t)\} - F(0) \tag{145}$$

$$\mathcal{L}\{F''(t)\} = s^2\mathcal{L}\{F(t)\} - sF(0) - F'(0) \tag{146}$$

with the derivatives taken with t coming from the positive side. These equations can be used to convert second-order differential equations to algebraic equations. For example, if we take the Laplace transform $f(s)$ of the harmonic oscillator equation by identifying $x(t) = F(t)$ and $f(s) = \mathcal{L}\{x(t)\}$

$$m\frac{d^2}{dt^2}x(t) + kx(t) = 0 \tag{147}$$

subject to the boundary conditions $F(0) = x(0) = x_0$ and $F'(0) = x'(0) = 0$ we obtain the algebraic equation

$$ms^2f(s) - msx_0 + kf(s) = 0 \tag{148}$$

Recalling that $k = m\omega_0^2$ we write for $f(s)$

$$f(s) = \frac{x_0 s}{s^2 + \omega_0^2} \tag{149}$$

The solution of the original Eq. (147) is now found by selecting from Table 27-2 the inverse Laplace transform (no. 8) of this function (149)

$$x(t) = x_0 \cos \omega_0 t \tag{150}$$

The general integral transform expression (143)

$$f(x) = \int_a^b F(t)\ K(x, t)\ dt \tag{151}$$

is called the Fredholm integral equation of the first kind if the function $f(x)$ is known and $F(t)$ is unknown. The Fredholm equation of the second kind is similar to this

$$F(x) = f(x) + \lambda \int_a^b F(t)\ K(x, t)\ dt \tag{152}$$

TABLE 27-2. Selected Laplace transforms

$f(s)$	$F(t)$	Limitation
1. 1	$\delta(t)$	Singularity at $+0$
2. $\frac{1}{s}$	1	$s > 0$
3. $\frac{n!}{s^{n+1}}$	t^n	$s > 0$ $n > -1$
4. $\frac{1}{s-k}$	e^{kt}	$s > k$
5. $\frac{1}{(s-k)^2}$	te^{kt}	$s > k$
6. $\frac{s}{s^2-k^2}$	$\cosh kt$	$s > k$
7. $\frac{k}{s^2-k^2}$	$\sinh kt$	$s > k$
8. $\frac{s}{s^2+k^2}$	$\cos kt$	$s > 0$
9. $\frac{k}{s^2+k^2}$	$\sin kt$	$s > 0$
10. $\frac{s-a}{(s-a)^2+k^2}$	$e^{at}\cos kt$	$s > a$
11. $\frac{k}{(s-a)^2+k^2}$	$e^{at}\sin kt$	$s > a$
12. $\frac{s^2-k^2}{(s^2+k^2)^2}$	$t\cos kt$	$s > 0$
13. $\frac{2ks}{(s^2+k^2)^2}$	$t\sin kt$	$s > 0$
14. $(s^2+a^2)^{-1/2}$	$J_0(at)$	$s > 0$
15. $(s^2-a^2)^{-1/2}$	$I_0(at)$	$s > a$
16. $\frac{1}{a}\cot^{-1}\left(\frac{s}{a}\right)$	$j_0(at)$	$s > 0$
17. $\left.\begin{array}{l}\frac{1}{2a}\ln\frac{s+a}{s-a}\\ \frac{1}{a}\coth^{-1}\left(\frac{s}{a}\right)\end{array}\right\}$	$i_0(at)$	$s > a$
18. $\frac{(s-a)^n}{s^{n+1}}$	$L_n(at)$	$s > 0$
19. $\frac{1}{s}\ln(s+1)$	$E_1(x) = -Ei(-x)$	$s > 0$
20. $\frac{\ln s}{s}$	$-\ln t - C$	$s > 0$

Source: G. Arfken, *Mathematical Methods for Physicists*, Academic Press, 1985, p. 863.

where again $f(x)$ is the known function. Two related integral equations are the Volterra equation of the first kind

$$f(x) = \int_a^x F(t)\, K(x, t)\, dt \tag{153}$$

and of the second kind

$$F(x) = f(x) + \int_a^x F(t)\, K(x, t)\, dt \tag{154}$$

In all four equations (151)–(154) the function $f(x)$ and the kernel $K(x, t)$ are the known quantities.

12 COMPLEX VARIABLES

Complex variables are widely used throughout physics, and in this section we review some of their salient features. The complex variable z can be expressed in terms of the cartesian coordinates x, y and polar coordinates r, θ

$$z = x + iy = re^{i\theta} \tag{155}$$

and the same is true of a function $w(z)$ of this complex variable

$$w(z) = u(x, y) + iv(x, y) = \rho e^{i\phi} \tag{156}$$

where we omitted the arguments for $\rho(r, \theta)$ and $\phi(r, \theta)$. The contour mapping of functions $w(z)$ between the u, v and x, y planes is angle preserving.

An analytic function is one whose derivative $df(z)/dz$ at a point x_0, y_0 is independent of the direction of approach to the point. The necessary and sufficient condition for analyticity is the satisfaction of the Cauchy–Riemann conditions

$$\frac{\partial u}{\partial x} = \frac{\partial v}{\partial y} \qquad \frac{\partial u}{\partial y} = -\frac{\partial v}{\partial x} \tag{157}$$

Both u and v satisfy Laplace's equation in two dimensions

$$\frac{\partial^2 u}{\partial x^2} + \frac{\partial^2 u}{\partial y^2} = 0 \qquad \frac{\partial^2 v}{\partial x^2} + \frac{\partial^2 v}{\partial y^2} = 0 \tag{158}$$

Cauchy's integral theorem states

$$\oint f(z)\, dz = 0 \tag{159}$$

where $f(z)$ is an analytical function and no singularities are inside the closed contour of integration C. If the contour encloses a point z_0 where the function is not analytic, then we have Cauchy's integral formula

$$\oint \frac{f(z)}{z - z_0}\, dz = 2\pi i f(z_0) \tag{160}$$

for a counter-clockwise contour, and also the expression

$$\oint f(z)\, dz = 2\pi i a_{-1} \tag{161}$$

where the residue a_{-1} at the point z_0 has the value

$$a_{-1} = [z - z_0) f(z)]_{z=z_0} \tag{162}$$

If the singularity is on the path, then the factor $2\pi i$ in Eq. (161) is replaced by $\pm\pi i$, where care must be exercised about the sign. If there are additional singular points z_0^n within the contour C, then we can write down the residue theorem

$$\oint f(z)\, dz = 2\pi i \sum a_{-1}^n \tag{163}$$

where each residue a_{-1}^n can be evaluated in the manner of Eq. (162).

The function $f(z)$ may be expanded about the point z_0 inside the contour C in a power series summed over powers of $(z - z_0)^n$ with the integer n going from zero to infinity

$$f(z) = \sum_{n=0}^{\infty} a_n (z - z_0)^n \tag{164}$$

where each coefficient a_n may be determined either from the nth derivative $f^{(n)}(x_0)$ evaluated at z_0 or in terms of an integral around the contour C

$$a_n = \frac{f^{(n)}(z_0)}{n!} = \frac{1}{2\pi i} \oint_c \frac{f(z')\, dz'}{(z' - z_0)^{n+1}} \tag{165}$$

If the range of analyticity is an annular region extending from an inner radius r_1 to an outer radius r_2, then the function $f(z)$ may be expanded in a Laurent series summed over powers of $(z - z_0)^n$, where the integer n goes from $-\infty$ to $+\infty$

$$f(z) = \sum_{n=-\infty}^{\infty} a_n (z - z_0)^n \tag{166}$$

with the coefficients a_n given by

$$a_n = \frac{1}{2\pi i} \oint_c \frac{f(z')\, dz'}{(z' - z_0)^{n+1}} \tag{167}$$

where C may be any contour of integration within the annular region $r_1 < |z - z_0| < r_2$ encircling z_0. The Laurent series (166) reduces to the Taylor series (164) in the limit $r_1 = 0$.

13 CONTOUR INTEGRATION

The residue theorem (163) can be used to evaluate definite integrals I of the type

$$I = \oint_A^B f(x)\, dx \tag{168}$$

by integrating the integrand written as the function $f(z)$ of the complex variable z around a closed contour which includes the path from A to B along the real or x axis. The value of the integral (163) from A to B is $2\pi i$ times the sum of the enclosed residues minus the value of the integral along the remainder of the path.

For example, consider the integral

$$I = \oint_{-\infty}^{\infty} f(x)\, dx \tag{169}$$

This can be written in the form

$$I = \oint_{-R}^{R} f(x)\, dx \tag{170}$$

in which we will take the limit of R going to infinity. The remainder of the path might be a semicircle of radius R in the upper half plane, as indicated in Fig. 27-5. If $f(z)$ vanishes as strongly as $1/z^2$ along this semicircle, then the integration of $f(z)$ along the semicircle will vanish when we take the limit of $R \Rightarrow \infty$, and we can write down from the residue theorem (163)

$$\oint f(x)\, dx = 2\pi i \sum_n a_{-1}^n \tag{171}$$

As an example consider the definite integral

$$I = \oint_{-\infty}^{\infty} \frac{dx}{1 + x^2} \tag{172}$$

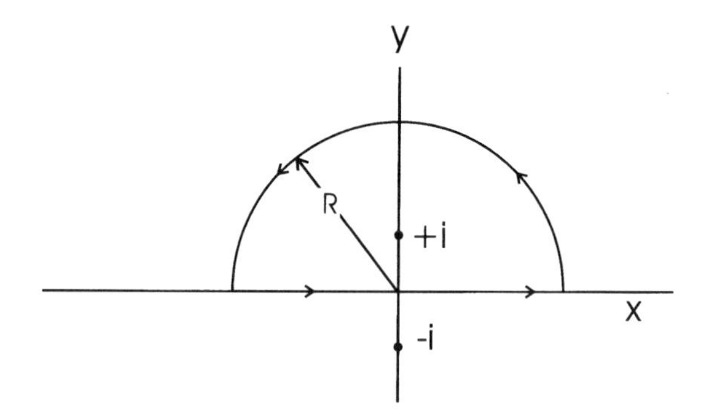

Fig. 27-5. Integration on the complex plane involving a path along the x axis followed by a path along a large semicircle of radius R in the upper half plane. The path encloses the singularity at $x = 0$, $y = +i$, but not the one at $x = 0$, $y = -i$. The limit $R \Rightarrow \infty$ is taken to evaluate the integral.

Extending the variable to the complex plane we can write the integral in the following way that displays its singularities:

$$I = \oint \frac{dz}{(z+i)(z-i)} \tag{173}$$

which lie along the y axis. The singularity $y = i$ in the upper half plane lies within the contour with the residue from Eq. (162) $a_{-1} = 1/2i$, while the other singularity $y = -i$ lies below outside the contour and so does not concern us. For the integration along the semicircle we use angular coordinates

$$z = Re^{i\theta} \qquad dz = iRe^{i\theta} d\theta \quad 0 \leq \theta \leq \pi \tag{174}$$

Inserting these expressions into Eq. (173) and assuming $R \gg 1$ we have for the path along the semicircle

$$I = \frac{1}{R} \oint_0^{\pi} ie^{-i\theta}\, d\theta \tag{175}$$

which vanishes in the limit $R \Rightarrow \infty$. Therefore, from Eq. (171) we obtain the value of the original integral

$$\oint_{\infty}^{\infty} f(x)\, dx = 2\pi i(1/2i) = \pi \tag{176}$$

Another type of integral which is often encountered has the form

$$I = \oint_0^{2\pi} f(\sin\theta, \cos\theta)\, d\theta \tag{177}$$

Using Eq. (174) we have

$$d\theta = -idz/z \tag{178}$$

and the integration is around the unit circle with $R = 1$. The value of the integral (177) is thus $2\pi i$ times the sum of the residues a_{-1} within the unit circle. These residues can be evaluated from the form of Eq. (177) using the exponential definitions of the sine and cosine where

$$\sin\theta = \frac{e^{i\theta} - e^{-i\theta}}{2i} = \frac{z - z^{-1}}{2i} \tag{179}$$

$$\cos\theta = \frac{e^{i\theta} + e^{-i\theta}}{2} = \frac{z + z^{-1}}{2} \tag{180}$$

14 GROUP THEORY

A group is a set of objects called elements with a product operation and the following properties:

1. Closure whereby the product of any two elements equals a third element in the group: $ab = c$.
2. Multiplication is associative: $a(b) = (ab)c$.
3. The group contains a unit element I called the identity with the property $aI = Ia = a$ for all elements a.
4. Each element a has an inverse element a^{-1} with the property $a\,a^{-1} = a^{-1}a = I$.

An abelian group is one in which the multiplication operation commutes, $ab = ba$, for all elements in the group. An example is the set of elements 1, -1, i, $-i$, where 1 is the identity, -1 is its own inverse, and $-i$ is the inverse of i. This group is cyclic, with the symbol C_4, and it has the generator i with the property that

$$i^2 = -1 \qquad i^3 = -i \qquad i^4 = 1 \tag{181}$$

A cyclic group with n elements has a generator g with $g^n = I$, and it is called C_n. The group multiplication table of the cyclic group of order 4 is

	1	-1	i	$-i$
1	1	-1	i	$-i$
-1	-1	1	$-i$	i
i	i	$-i$	-1	1
$-i$	$-i$	i	1	-1

Each element appears once and only once in each row and in each column of the multiplication table.

A subgroup is a collection of some of the elements of a larger group which, by themselves, form a smaller group. For example, the two elements 1 and -1 form a subgroup of the cyclic group C_4.

Two elements a and c belong to the same class if they are related by a similarity transformation

$$a\,b\,a^{-1} = c \tag{182}$$

and all possible similarity transformations divide the elements of a group into subsets called classes.

A representation is a set of matrices that satisfies the multiplication table of the group. An irreducible representation is such a set of matrices M_j that satisfies the multiplication table but cannot all be simultaneously reduced in order by the same similarity transformation $U\,M_j\,U^{-1}$

$$U\begin{pmatrix} a & b & c & d \\ e & f & g & h \\ i & j & k & l \\ m & n & o & p \end{pmatrix}U^{-1} = \begin{pmatrix} A & B & 0 & 0 \\ C & D & 0 & 0 \\ 0 & 0 & E & F \\ 0 & 0 & G & H \end{pmatrix} \tag{183}$$

Each group of order h has a number of irreducible representations formed from $n_i \times n_i$ matrices which satisfy the relation

$$\sum n_i^2 = h \tag{184}$$

where the summation is over the number of classes. An abelian group in which all elements commute with each other only has one-dimensional representations.

The properties of groups that we have introduced can be illustrated by considering the symmetry of an equilateral triangle. The triangle has six symmetry elements ($h = 6$), namely the identity I, rotations through 120° and 240°, and three reflection planes σ_i perpendicular to the plane of the triangle, as indicated in Fig. 27-6. The reflection operations do not commute with the rotations. The elements divide into three classes:

class 1: I

class 2: R^{120}, R^{240}

class 3: $\sigma_1, \sigma_2, \sigma_3$

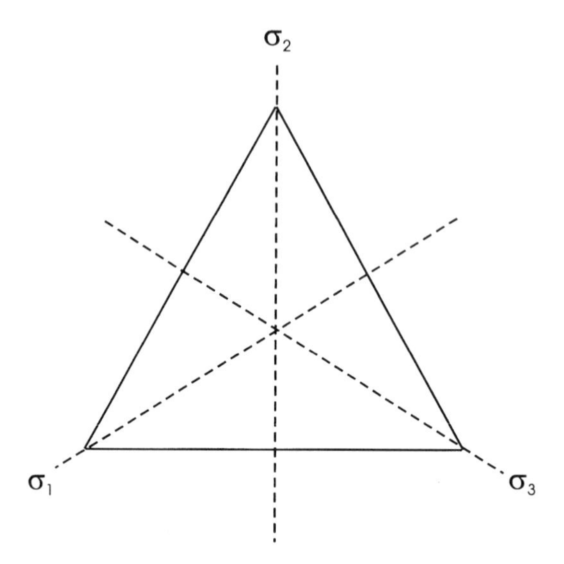

Fig. 27-6. The three symmetry planes σ_1, σ_2, and σ_3 of an equilateral triangle.

There are four subgroups:

subgroup 1: I, σ_1
subgroup 2: I, σ_2
subgroup 3: I, σ_3
subgroup 4: I, R^{120}, R^{240}

There are two one-dimensional representations and one two-dimensional representation, to give for Eq. (184)

$$1^2 + 1^2 + 2^2 = 6 \tag{185}$$

Group theory is important in crystallography. Table 27-3 gives the 11 proper point groups classified in the seven crystal systems, and lists the 10 derived and 11 inversion point groups associated with them. Most of these proper point groups are either cyclic, C_n, with an n-fold axis of symmetry, or dihedral, D_n, with an n-fold axis of symmetry plus a 2-fold rotation axis perpendicular to it. The cubic system has a tetrahedral (O) and an octahedral proper point group. A derived point group adds a horizontal (σ_h), vertical (σ_v) or diagonal (σ_d) reflection plane to the corresponding proper

TABLE 27-3. Relationships of 10 derived point groups and 11 inversion point groups with their corresponding 11 proper point group for the seven crystal systems. Some derived groups are obtained by the direct product operation denoted by $\otimes$ and some are obtained by adding the symmetry element given in parentheses to the indicated subgroup. Schoenflies symbols are used for the point groups with international symbols given in square brackets. Column 2 gives the relations of the lattice parameters and angles for the crystal systems. (Some alternative notations are $C_3 = C_{1h}$, $C_i = S_2$, $C_{3i} = S_6$, $D_2 = V$, $D_{2d} = D_{2v} = V_d$, $D_{2h} = V_h$, $D_{3d} = D_{3v}$, cubic = isometric.)

Crystal system	Lattice constants and angles	Proper point group	Order (n)	Derived point group (order n)	Inversion point group (order $2n$)
Triclinic	$a \neq b \neq c$	$C_1[1]$	1		$C_1 \otimes i = C_i[1]$
	$\alpha \neq \beta \neq \gamma$				
Monoclinic	$a \neq b \neq c$	$C_2[2]$	2	$C_1 \otimes \sigma_h = C_s[m]$	$C_2 \otimes i = C_{2h}[2/m]$
	$\alpha = \beta = 90^\circ \neq \gamma$				
Orthorhombic	$a \neq b \neq c$	$D_2[222]$	4	$C_2 \otimes \sigma_v = C_{2v}[2mm]$	$D_2 \otimes i = D_{2h}[2/mmm]$
	$\alpha = \beta = \gamma = 90^\circ$				
Trigonal	$a = b = c$	$C_3[3]$	3		$C_3 \otimes i = C_{3i}[3]$
	$\alpha = \beta = \gamma < 120^\circ$	$D_3[32]$	6	$C_3(+\sigma_v) = C_{3v}[3m]$	$D_3 \otimes i = D_{3d}[3m]$
	$\neq 90^\circ$				
Tetragonal	$a = b \neq c$	$C_4[4]$	4	$C_2(+S_4) = S_4[4]$	$C_4 \otimes i = C_{4h}[4/m]$
	$\alpha = \beta = \gamma = 90^\circ$	$D_4[422]$	8	$C_4(+\sigma_v) = C_{4v}[4mm]$	$D_4 \otimes i = D_{4h}[4/mmm]$
				$D_2(+\sigma_d) = D_{2d}[42m]$	
Hexagonal	$a = b \neq c$	$C_6[6]$	6	$C_3(+\sigma_h) = C_{3h}[6]$	$C_6 \otimes i = C_{6h}[6/m]$
	$\alpha = \beta = 90^\circ$				
	$\gamma = 120^\circ$	$D_6[622]$	12	$C_6(+\sigma_v) = C_{6v}[6mm]$	$D_6 \otimes i = D_{6h}[6/mmm]$
				$D_3 \otimes C_s = D_{3h}[6m2]$	
Cubic (isometric)	$a = b = c$	$T[23]$	12		$T \otimes i = T_h[m3]$
	$\alpha = \beta = \gamma = 90^\circ$	$O[432]$	24	$T(+\sigma_d) = T_d[43m]$	$O \otimes i = O_h[m3m]$

group. Adding the inversion operation i to a proper or derived point group generates one of the eleven inversion groups listed in the right-hand column of the table. Figure 27-7 shows how these various groups are generated from each other through descent of symmetry operations.

If symmetry operations involving translation are added to the 32 point groups then we obtain the 230 crystallographic space groups. There are 73 space groups called semidirect product (or symmorphic) which are derived from the point groups by adding only pure translation symmetry operations. The remaining 157 space groups contain glide planes and screw axes in addition to translations. Using pure group theory classification there are only 219 non-isomorphic space groups because 11 of them appear as mirror image or enantiomorphic pairs.

In addition to the 32 crystallographic point groups that are associated with crystal lattices there are some additional point groups, that reflect the symmetry of molecules. For example, a crystal lattice cannot have a fivefold (72°) rotation axis, but the molecule ferrocene $(C_5H_5)_2Fe$ has two pentagonal C_5H_5 rings aligned, one above and one below the Fe atom, staggered relative to each other, so there is a fivefold improper rotation symmetry operation S_5. The molecule belongs to the point group D_{5d}.

The groups that we have been discussing are finite groups. Examples of infinite groups are the set of all 3×3 rotation matrices that form the orthogonal group of order 3 with the symbol $O(3)$. The set of all $n \times n$ unitary matrices form the group $U(n)$. Of particular interest are the set of all 2×2 unitary matrices with determinant +1 that form the special or unimodular unitary group SU(2), and the set of all 3×3 unitary matrices with determinant +1 which form the special unitary group SU(3). Gell-Mann's eightfold way classification of the elementary particles in terms of SU(3), based on the up, down, and strange quarks (u, d, s), is exemplified by the baryon diagram of Fig. 26-5 in Chapter 26. Figures 26-7 and 26-8 show SU(4) classifications based on the addition of a fourth quark (c) called charm.

Before concluding we will say a word about Lie algebras. The elements τ_i of a Lie algebra have the property that the commutator of two elements

$$[\tau_i, \tau_j] = \tau_i\tau_j - \tau_j\tau_i \tag{186}$$

is a linear combination of other elements

$$[\tau_i, \tau_j] = -i \sum C_{ij}^k \tau_k \tag{187}$$

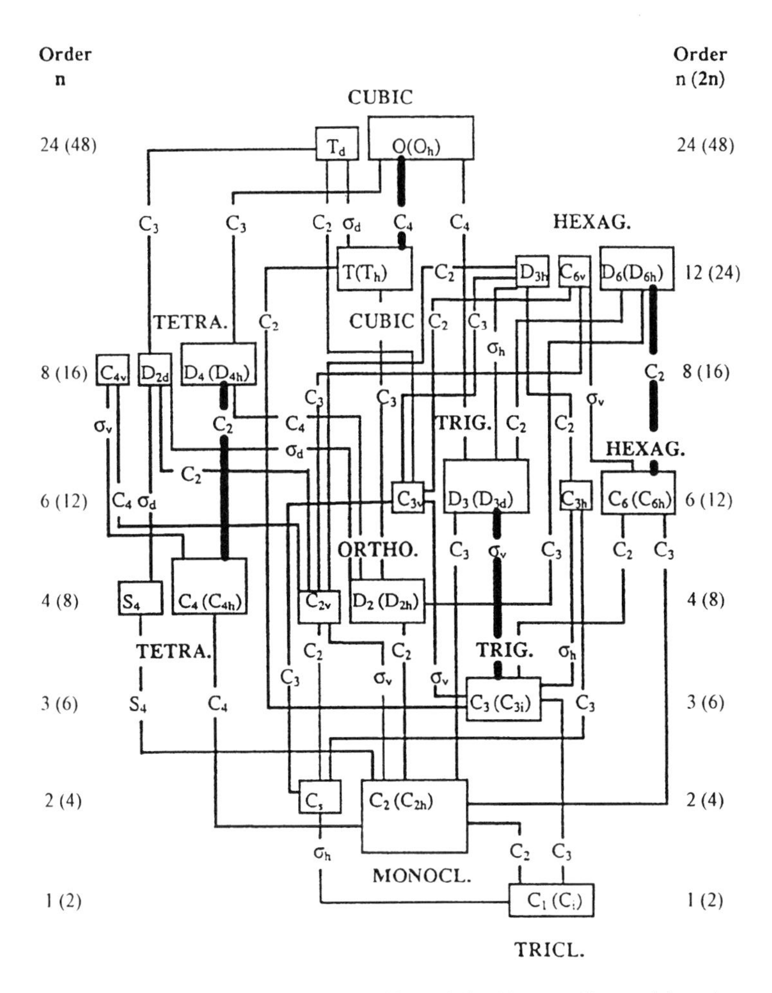

Fig. 27-7. Descent of symmetry relationships of the 32 crystallographic point groups with the crystal systems indicated. Each oblong box contains the symbol of a proper point group of order *n* on the left and that of the corresponding inversion point group of order 2*n* listed in parentheses on the right. Derived point groups are shown in smaller square boxes to the left of the larger box of their associated proper point group. The orders of the proper and derived groups are given by the scales on the left and right. Single lines connect derived groups with their subgroups below. Boxes of groups in the same crystal system are clustered below each other and connected by wide boldfaced lines. Shown also are the symmetry elements which must be added to a subgroup below to convert it to the group above. This figure should be compared with Table 27-3. (From S. Misra et al., *Applied Magnetic Resonance* **11**, 1996, 29.)

where the summation is over the k elements, and the real coefficients C_{ij}^k that define the algebra are called structure constants. Associated with the Lie algebra is a Lie group.

15 MONTE CARLO METHOD

Monte Carlo is a computational method based on the use of random variables. The method is difficult to define in general, so we illustrate it by the evaluation of the following definite integral

$$I = \int_a^b F(x)\, dx \tag{188}$$

where the integrand $F(x)$ satisfies the inequality

$$0 \leqslant F(x) \leqslant M \tag{189}$$

for values of x in the range $0 \leqslant a \leqslant x \leqslant b$.

Random number generators ordinarily provide output in the range $0 < x_i < 1$, and we begin by generating two sequences of N random numbers

$$\begin{aligned} &x_i', x_2', x_3', \ldots x_N' \\ &y_1', y_2', y_3', \ldots y_N' \end{aligned} \tag{190}$$

The transformations

$$\begin{aligned} x_i &= a + x_i' \;\; (b - a) \\ y_i &= y_i' \;\; M \end{aligned} \tag{191}$$

convert the original sets (190) to pairs of random numbers (x_i, y_i) that match the limits of the problem. For each value x_i the function $f(x_i)$ is compared with the associated y_i value and we count the number of times n for which $f(x_i)$ is less than y_i. The unlikely case $f(x_i) = y_i$ is counted as $\frac{1}{2}$. Finally the integral (188) has the value

$$I = \frac{nM}{N}(b - a) \tag{192}$$

Thus random numbers provide the solution to a problem (188) which has no intrinsic randomness.

CHAPTER 28

MATHEMATICAL PHYSICS: DIFFERENTIAL EQUATIONS AND ORTHOGONAL FUNCTIONS

1 INTRODUCTION

In the previous chapter we covered many miscellaneous topics in mathematical physics, including integral equations and their solutions. This chapter is a little more focused. It covers the various differential equations and special functions that are so widely used throughout physics. Most of these special functions, such as those of Bessel and Legendre, are solutions of particular second-order differential equations, although some, such as the gamma function, are not, and Green's functions are aids in solving differential equations. We discuss the equations themselves and the properties of their solutions. Tabulations are given of many of these special functions, including those with real and imaginary arguments, and in many cases asymptotic expressions are provided for the limiting cases of the argument approaching zero and approaching infinity.

2 FIRST-ORDER DIFFERENTIAL EQUATIONS

If the first-order differential equation

$$\frac{dy}{dx} = f(x, y) \tag{1}$$

can be written in the form

$$P(x, y)\,dx + Q(x, y)\,dy = 0 \tag{2}$$

with $f(x, y) = -P(x, y)/Q(x, y)$, and if a function $\phi(x, y)$ exists so that this equation can be expressed in terms of a differential $d\phi$

$$d\phi = \frac{\partial \phi}{\partial x}\,dx + \frac{\partial \phi}{\partial y}\,dy = 0 \tag{3}$$

where $P = \partial\phi/\partial x$ and $Q = \partial\phi/\partial y$, then the equation is said to be exact, and its solution is given by

$$\phi(x, y) = \text{const} \tag{4}$$

In thermodynamics, exact differential equations play a major role.

If the function $f(x, y)$ of Eq. (1) has the form $-p(x)y(x) + q(x)$, then we obtain the most general first-order differential equation

$$\frac{dy}{dx} + p(x)y = q(x) \tag{5}$$

If $q(x) = 0$, then the equation is said to be homogeneous, otherwise it is inhomogeneous. If an integrating factor $\alpha(x)$ can be found with the property

$$\frac{d}{dx}[\alpha(x)y] = \alpha(x)q(x) \tag{6}$$

then the equation can be integrated directly to give the solution. The integrating factor obeys the relation

$$\frac{d}{dx}[\alpha(x)] = \alpha(x)p(x) \tag{7}$$

which can be integrated

$$\alpha(x) = \alpha(x_0)\exp\left[\int p(x)\,dx\right] \tag{8}$$

to provide a closed form expression for $\alpha(x)$ as the exponential of the integral of $p(x)$ over x.

3 SECOND-ORDER DIFFERENTIAL EQUATIONS

Several homogeneous second-order differential equations, together with some of their features, are listed in Table 28-1, and the characteristics of some polynomial solutions are presented in Table 28-2. These equations are called homogeneous because each term is a product of a constant or a function $f(x)$ with either u'', u' or u. There are three common ways in which each equation can be written, and we will illustrate them for Bessel's equation:

$$x^2u'' + xu' + [a^2x^2 - n^2]u = 0 \quad \text{conventional form} \tag{9a}$$

$$xu'' + u' + [a^2x - (n^2/x)]u = 0 \quad \text{self-adjoint form} \tag{9b}$$

$$u'' + x^{-1}u' + [a^2 - (n^2/x^2)]u = 0 \quad \text{displaying singularities} \tag{9c}$$

The conventional form does not have any particular rationale, but is simply the way the equation is usually written. The self-adjoint form may be written in general as

$$pu'' + p'u' + qu + \lambda wu = 0 \tag{10}$$

TABLE 28-1. Characteristics of the principal second-order differential equations of interest for solving physical problems. The Bessel functions have an infinite number of terms, the other solutions are polynomials with finite numbers of terms. An asterisk (*) in column 7 denotes an irregular singularity. The equations are listed in column 1 in their conventional forms, while columns 3 to 6, respectively, give the functions p(x) and $q(x)$, the eigenvalue λ and the weighting function w(x) of their self-adjoint forms $p(x)u'' + p'(x)u' + q(x)u + \lambda u(x)u(x) = 0$.

Name and equation	Domain of x	Self-adjoint $p(x)$	Self-adjoint $q(x)$	Eigenvalue λ	Weighting function $w(x)$	Singularity	Parity
Legendre $(1-x^2)u'' - 2xu' + \ell(\ell+1)u = 0$	$-1 \le x \le 1$	$1-x^2$	0	$\ell(\ell+1)$	1	$-1, 1, \infty$	yes
Associated Legendre add term $-m^2u/(1-x^2)$	$-1 \le x \le 1$	$1-x^2$	$-m^2/(1-x^2)$	$\ell(\ell+1)$	1	$-1, 1, \infty$	yes
Bessel $x^2u'' + xu' + (x^2a^2 - n^2)u = 0$	$0 \le x \le \infty$	x	$-n^2/x$	a^2	x	$0, \infty^*$	yes
Laguerre $xu'' + (1-x)u' + nu = 0$	$0 \le x \le \infty$	$x\,e^{-x}$	0	n	e^{-x}	$0, \infty^*$	no
Associated Laguerre add term $+ku'$	$0 \le x \le \infty$	$x^{k+1}\,e^{-x}$	0	$n-k$	$x^k e^{-x}$	$0, \infty^*$	no
Hermite $u'' - 2xu' + 2nu = 0$	$-\infty \le x \le \infty$	e^{-x^2}	0	$2n$	e^{-x^2}	∞^*	yes

or alternatively

$$\frac{d}{dx}\left(p(x)\frac{du(x)}{dx}\right) + q(x)u(x) + \lambda w(x)u(x) = 0 \tag{11}$$

where the functions $p(x)$ and $q(x)$, the eigenvalue λ, and the density or weighting function $w(x)$ are listed in Table 28-1 for each equation.

The differential equation (10) has two independent solutions for each eigenvalue

$$u(x) = au_1(x) + bu_2(x) \tag{12}$$

and one of these can generally be found by Frobenius' method of assuming a power series solution

$$u(x) = x^k(a_0 + a_1x + a_2x^2 + a_3x^3 + \ldots) \tag{13}$$

This series is substituted into the differential equation (10) and the coefficient of each power of x in the resulting expression is set equal to zero. The lowest power of x provides what is called the indicial equation which furnishes the value of k, or sometimes two choices of values of k. The remaining terms provide recursion relations for the factors a_m in Eq. (13), to give, for example, the following expression for the Bessel equation:

$$a_{2m} = \frac{(-1)^m n! a_0}{2^{2m} m!(m+n)!} \tag{14}$$

where n is an integer which appears in the term $q(x) = -n^2/x$ in the differential equation itself, as shown in Table 28-1. Sometimes the series solution contains an infinite number of terms, as in the Bessel case, and sometimes it is cut off, and only has a finite number of terms, as in the Legendre case. Table 28-2 provides characteristics of several polynomial solutions, and Table 28-3 lists the lowest order terms of these polynomials.

There is another feature of the solutions of differential equations that should be mentioned, and that is their parity or lack of parity. Some equations, such as those of Bessel and Legendre, have solutions that contain only odd powers of x and others with only even powers of x, which means that these particular series solutions have parity such that

$$Pu(x) = u(-x) = \pm u(x) \tag{15}$$

TABLE 28.2. Characteristics of various differential equations

Polynomial	Singular points	Differential equation	Definition
Legendre $P_n(z)$ $-1 \le z \le 1$	$-1, +1, \infty$	$(1 - z^2)F'' - 2zF' + n(n+1)F = 0$	$P_n(z) = \frac{1}{2^n n!} \frac{d^n}{dz^n}(z^2 - 1)^n$
Hermite $H_n(z)$ $-\infty \le z \le \infty$	∞^*	$F'' - 2zF' + 2nF = 0$	$H_n(z) = (-1)^n e^{z^2} \frac{d^n}{dz^n}\left(e^{-z^2}\right)$
Laguerre $L_n(z)$ $0 \le z \le \infty$	$0, \infty^*$	$zF'' + (1 - z)F' + nF = 0$	$L_n(z) = \frac{e^z}{n!} \frac{d^n}{dz^n}(z^n e^{-z})$
Chebyshev Type I $T_n(z)$ $-1 \le z \le 1$	$-1, +1, \infty$	$(1 - z^2)F'' - zF' + n^2 F = 0$	$T_n(z) = \frac{(-1)^n \pi^{1/2}(1 - z^2)^{1/2}}{2^n(n - \frac{1}{2})!} \frac{d^n}{dz^n}(1 - z^2)^{n-1/2}$
Chebyshev Type II $U_n(z)$ $-1 \le z \le 1$	$-1, +1, \infty$	$(1 - z^2)F'' - 3zF' + n(n+2)F = 0$	$U_n(z) = \frac{(-1)^n (n+1)\pi^{1/2}}{2^{n+1}(n + \frac{1}{2})!(1 - z^2)^{1/2}} \frac{d^n}{dz^n}(1 - z^2)^{n+1/2}$

*Irregular singular point.

Generating function $g(z, t)$	Recursion relations	Orthogonality, normalization
$(1 - 2zt + t^2)^{-1/2} = \sum_{n=0}^{\infty} P_n(z)t^n$	$\left\{ \begin{array}{c} (2n+1)zP_n = (n+1)P_{n+1} + nP_{n-1} \\ P'_{n+1} + P'_{n-1} = 2zP'_n + P_n \\ P'_{n+1} - P'_{n-1} = (2n+1)P_n \end{array} \right\}$	$\int_{-1}^{1} P_n(z)P_m(z)\,dz = \frac{2\delta_{nm}}{2n+1}$
$e^{-t^2+2tz} = \sum_{n=0}^{\infty} H_n(z)\frac{t^n}{n!}$	$\left\{ \begin{array}{c} H_{n+1} = 2zH_n - 2nH_{n-1} \\ H'_n = 2nH_{n-1} \end{array} \right\}$	$\int_{-\infty}^{\infty} H_n(z)H_m(z)\,e^{-z^2}\,dz = 2^n\pi^{1/2}n!\delta_{nm}$
$\frac{e^{-tz/(1-t)}}{l-t} = \sum_{n=0}^{\infty} L_n(z)t^n$	$\left\{ \begin{array}{c} (n+1)L_{n+1} = \|2n+1-z\|L_n - nL_{n-1} \\ zL'_n = nL_n - nL_{n-1} \end{array} \right\}$	$\int_{0}^{\infty} L_n(z)L_m(z)e^{-z}dz = \delta_{nm}$
$\frac{1-t^2}{1-2zt+t^2} = T_0(z) + 2\sum_{n=0}^{\infty} T_n(z)t^n$	$\left\{ \begin{array}{c} T_{n+1} + T_{n-1} = 2zT_n \\ (1-z^2)T'_n = -nzT_n + nT_{n-1} \end{array} \right\}$	$\int_{-1}^{1} \frac{T_n(z)T_m(z)}{(1-z^2)^{1/2}}dz = \begin{cases} \frac{1}{2}\pi\,\delta_{nm} & n > 0 \\ \pi\delta_{nm} & n = 0 \end{cases}$
$\frac{1}{1-2zt+t^2} = \sum_{n=0}^{\infty} U_n(z)t^n$	$\left\{ \begin{array}{c} U_{n+l} + U_{n-1} = 2zU_n \\ (1-z^2)U_n = -nzU_n + (n+1)U_{n-1} \end{array} \right\}$	$\int_{-1}^{1} U_n(z)U_m(z)(1-z^2)^{1/2}dz = \frac{\pi}{2}\delta_{nm}$

TABLE 28.2. (*continued*)

Parity	Special values	Associate polynomial
$P_n(-z) = (-1)^n P_n(z)$	$\left\{ \begin{array}{c} P_n(1) = 1 \\ P_n(-1) = (-1)^{-n} \end{array} \right\}$	$P_n^m(z) = (1 - z^2)^{m/2} \frac{d^m}{dz^m} P_n(z)$
$H_n(-z) = (-1)^n H_n(z)$	$\left\{ \begin{array}{c} H_{2n}(0) = (-1)^n \frac{(2n)!}{n!} \\ H_{2n+1}(0) = 0 \end{array} \right\}$	None
No parity (z cannot be negative)	$L_n(0) = 1$	$L_n^m(z) = (-1)^m \frac{d^m}{dz^m} L_{n+m}(z)$
$T_n(-z) = (-1)^n T_n(z)$	$\left\{ \begin{array}{c} T_n(1) = 1 \\ T_n(-1) = (-1)^n \\ T_{2n}(0) = (-1)^n \\ T_{2n+1}(0) = 0 \end{array} \right\}$	None
$U_n(-z) = (-1)^n U_n(z)$	$\left\{ \begin{array}{c} U_n(1) = n + 1 \\ U_n(-1) = (-1)^n (n + 1) \\ U_{2n}(0) = (-1)^n \\ U_{2n+1}(0) = 0 \end{array} \right\}$	None

The parity of a particular solution $u_n(x)$ of this type alternates between even and odd for successive values of the integer n. Solutions of other equations, such as Laguerre polynomials, contain both odd and even powers of x so they have no parity. We see from Table 28-3 that the polynomials P_n, H_n, T_n, and U_n alternate in parity, while each L_n polynomial contains both odd and even powers of x, and so the Laguerre polynomials L_n lack parity.

It is not so easy to find the second independent solution $u_2(x)$ of Eq. (12). Since this second solution is orthogonal to the first, its Wronskian W must not vanish

$$W = \begin{vmatrix} u_1 & u_2 \\ u_1' & u_2' \end{vmatrix} = u_1 u_2' - u_2 u_1' \neq 0 \tag{16}$$

Sometimes the two solutions have different ranges of applicability in practical problems. For example, the Bessel functions $J_n(x)$ are well behaved at the origin $x \Rightarrow 0$, whereas the second solution, called a Neumann function $N_n(x)$, blows up at the origin.

The solutions $u_i(x)$ of a second-order differential equation for different eigenvalues λ are linearly independent of each other, and sometimes they constitute an orthogonal set capable of being normalized to form a set of functions $\phi_n(x)$ that satisfy the orthonormality condition

$$\int \phi_i^*(x)\phi_j(x)w(x)dx = \delta_{ij} \tag{17}$$

where $w(x)$ is the weighting function of Eqs. (10) and (11), and the integration is carried out over the domain of the variable. Such orthonormal solutions are called eigenfunctions.

If the solutions are linearly independent but not orthogonal, then the Gram–Schmidt procedure can be used to convert them to a set of orthonormal eigenfunctions ϕ_n. Consider a linearly independent set of solutions $u_0, u_1, u_2, \ldots$. Select the first one u_0 as an unnormalized eigenfunction, write $\phi_0(x) = N_0 u_0$, and normalize it as follows:

$$\phi_0(x) = \frac{u_0(x)}{\left(\int u_0^2(x)w(x)\,dx\right)^{1/2}} \tag{18}$$

so $\phi_0(x)$ becomes the first of our orthonormal set of eigenfunctions. Then write the second eigenfunction $\phi_1(x)$ in the form

$$\phi_1(x) = N_1[u_1(x) + a\phi_0] \tag{19}$$

TABLE 28-3. Lowest order polynomials of the polynomial types defined in Table 28-2

Legendre polynomials

$P_0(z) = 1$

$P_1(z) = z = \cos\theta$

$P_2(z) = \frac{1}{2}(3z^2 - 1) = \frac{1}{2}(3\cos^2\theta - 1)$

$P_3(z) = \frac{1}{2}(5z^3 - 3z) = \frac{1}{2}(5\cos^3\theta - 3\cos\theta)$

$P_4(z) = \frac{1}{8}(35z^4 - 30z^2 + 3) = \frac{1}{8}(35\cos^4\theta - 30\cos^2\theta + 3)$

Associated Legendre polynomials $P_n^m(z)$, where $m \leq n$, and $P_n^0(z) \equiv P_n(z)$

$P_1^0(z) = z = \cos\theta$

$P_1^1(z) = (1 - z^2)^{1/2} = \sin\theta$

$P_2^0(z) = \frac{1}{2}(3z^2 - 1) = \frac{1}{2}(3\cos^2\theta - 1)$

$P_2^1(z) = 3z(1 - z^2)^{1/2} = 3\cos\theta\sin\theta$

$P_2^2(z) = 3(1 - z^2) = 3\sin^2\theta$

$P_3^0(z) = \frac{1}{2}(5z^3 - 3z) = \frac{1}{2}(5\cos^3\theta - 3\cos\theta)$

$P_3^1(z) = \frac{3}{2}(5z^2 - 1)(1 - z^2)^{1/2} = \frac{3}{2}(5\cos^2\theta - 1)\sin\theta$

$P_3^2(z) = 15z(1 - z^2) = 15\cos\theta\sin^2\theta$

$P_3^3(z) = 15(1 - z^2)^{3/2} = 15\sin^3\theta$

Hermite polynomials

$H_0(z) = 1$

$H_1(z) = 2z$

$H_2(z) = 2(2z^2 - 1)$

$H_3(z) = 4(2z^3 - 3z)$

$H_4(z) = 4(4z^4 - 12z^2 + 3)$

Laguerre polynomials

$L_0(z) = 1$

$L_1(z) = -z + 1$

$L_2(z) = (z^2 - 4z + 2)/2!$

$L_3(z) = (-z^3 + 9z^2 - 18z + 6)/3!$

$L_4(z) = (z^4 - 16z^3 + 72z^2 - 96z + 24)/4!$

Type I Chebyshev polynomials

$T_0(z) = 1$

$T_1(z) = z$

$T_2(z) = 2z^2 - 1$

$T_3(z) = 4z^3 - 3z$

$T_4(z) = 8z^4 - 8z^2 + 1$

Type II Chebyshev polynomials

$U_0(z) = 1$

$U_1(z) = 2z$

$U_2(z) = 4z^2 - 1$

$U_3(z) = 8z^3 - 4z$

$U_4(z) = 16z^4 - 12z^2 + 1$

and evaluate the parameter a by the requirement that ϕ_1 be orthogonal to ϕ_0. The normalization constant N_1 is evaluated in the manner of Eq. (18). The next eigenfunction ϕ_2 is then written

$$\phi_2(x) = N_2[u_2(x) + b\phi_1 + c\phi_0] \tag{20}$$

where b and c are evaluated from orthogonality of ϕ_2 with ϕ_1 and ϕ_0, and so forth. This procedure can be followed to convert all of the solutions u_i to orthonormal eigenfunctions ϕ_i.

The eigenfunctions ϕ_n for the different eigenvalues λ_n form a complete set in the sense that any well-behaved function $F(x)$ on the domain of x can be expressed as a series

$$F(x) = \sum a_n \phi_n(x) \tag{21}$$

to any desired degree of accuracy. This is the property of completeness.

4 LAPLACE AND HELMHOLTZ EQUATIONS

Laplace's equation

$$\nabla^2 \Psi = 0 \tag{22}$$

is perhaps the most important equation in mathematical physics. When we solve Schrödinger's equation with a radial potential the angular parts are the same as those of Laplace's equation.

Consider the Cartesian case where the equation is easily expressed in separable form

$$\frac{1}{X(x)} \frac{\partial^2 X(x)}{\partial x^2} + \frac{1}{Y(y)} \frac{\partial^2 Y(y)}{\partial y^2} + \frac{1}{Z(z)} \frac{\partial^2 Z(z)}{\partial z^2} = 0 \tag{23}$$

by writing the solution as a product function

$$\Psi(x, y, z) = X(x)Y(y)Z(z) \tag{24}$$

The individual component solutions can be harmonic or growth–decay types

$$\begin{aligned} \text{harmonic:} &\quad \sin kx,\ \cos kx,\ \mathrm{e}^{ikx},\ \mathrm{e}^{-ikx} \\ \text{growth–decay:} &\quad \sin \kappa x,\ \cosh \kappa x,\ \mathrm{e}^{\kappa x},\ \mathrm{e}^{-\kappa x} \end{aligned} \tag{25}$$

and it is the nature of the equation that one of the solutions must be harmonic, one must be growth–decay, and the third can be either type. In other words, a possible solution is

$$\Psi(x, y, z) = \sin k_x x \cos k_y y \sinh \kappa_z z \tag{26}$$

Inserting this into Laplace's equation (23) gives

$$-k_x^2 - k_y^2 + \kappa_z^2 = 0 \tag{27}$$

and we see immediately why we need at least one harmonic and at least one growth–decay component of the solution. This statement applies to all coordinate systems where the equation is separable. It is not true for the wave equation

$$\nabla^2 \Psi - \frac{1}{c^2} \frac{\partial^2 \Psi}{\partial t^2} = 0 \tag{28}$$

which, with a harmonic time dependence $\Psi(x, y, z, t) = \Psi(x, y, z)\mathrm{e}^{i\omega t}$ becomes Hemholtz's equation

$$\nabla^2 \Psi + k^2 \Psi = 0 \tag{29}$$

where $k = \omega/c$, and it is possible for all the space components to be harmonic. The requirement remains, however, that at least one space component must be harmonic.

In cylindrical coordinates we have for Laplace's equation in separable form

$$\frac{1}{\rho R} \frac{\partial}{\partial \rho}\left(\rho \frac{\partial R}{\partial \rho}\right) + \frac{1}{\rho^2 \Phi}\left(\frac{\partial^2 \Phi}{\partial \phi^2}\right) + \frac{1}{Z} \frac{\partial^2 Z}{\partial z^2} = 0 \tag{30}$$

where

$$\Psi(\rho, \phi, z) = R(\rho)\Phi(\phi)Z(z) \tag{31}$$

Selecting $\Phi(\phi)$ to be harmonic and $Z(z)$ to be growth–decay

$$\Psi(\rho, \phi, z) = R(\rho)\, e^{im\phi} \sinh \kappa z \tag{32}$$

gives Bessel's equation

$$\rho \frac{\partial}{\partial \rho}\left(\rho \frac{\partial R}{\partial \rho}\right) + (\kappa^2 \rho^2 - m^2) R = 0 \tag{33}$$

and the solutions are Bessel functions $R(\rho) = J_m(\kappa\rho)$ which are of the harmonic type. If we had selected both $\Phi(\phi)$ and $Z(z)$ to be harmonic, then we would have obtained the modified Bessel equation

$$\rho \frac{\partial}{\partial \rho}\left(\rho \frac{\partial R}{\partial \rho}\right) - (\kappa^2 \rho^2 + m^2) R = 0 \tag{34}$$

with the growth–decay solutions such as $R(\rho) = K_n(\kappa\rho)$ which are called modified Bessel functions. The ordinary and modified Bessel functions will be discussed in the next section.

In spherical coordinates Helmholtz's equation (29) is separable

$$\frac{1}{Rr^2} \frac{\partial}{\partial r}\left(r^2 \frac{\partial R}{\partial r}\right) + \frac{1}{\Theta r^2 \sin\theta} \frac{\partial}{\partial \theta}\left(\sin\theta \frac{\partial \Theta}{\partial \theta}\right) + \frac{1}{\Phi r^2 \sin^2\theta} \frac{\partial^2 \Phi}{\partial \phi^2} + k^2 = 0 \tag{35}$$

where

$$\Psi(r, \theta, \phi) = R(r)\Theta(\theta)\ \Phi(\phi) \tag{36}$$

Recalling that the angular solution is a spherical harmonic $Y_{LM}(\theta, \phi)$ to be discussed in Section 7

$$\Psi(r, \theta, \phi) = R(r) Y_{LM}(\theta, \phi) \tag{37}$$

gives the radial equation

$$r^2 \frac{d^2 R}{dr^2} + 2r \frac{dR}{dr} + [k^2 r^2 - n(n+1)] R = 0 \tag{38}$$

where we write $n(n+1)$ instead of $L(L+1)$. The change of variable

$$R(r) = P(kr)/(kr)^{1/2} \tag{39}$$

converts Eq. (38) to Bessel's equation for half integer orders

$$r^2 \frac{d^2 P}{dr^2} + r \frac{dP}{dr} + [k^2 r^2 - (n + \tfrac{1}{2})^2] P = 0 \tag{40}$$

and the solution $P(r)$ to the radial equation (40) is a spherical Bessel function $j_n(kr)$ to be discussed in the next section, where

$$j_n(kr) = (\pi/2kr)^{1/2} J_{n+1/2}(kr) \tag{41}$$

The next three sections treat, in turn, the Bessel functions, the Legendre polynomials, and then the spherical harmonics.

5 BESSEL FUNCTIONS

Bessel's equation (33), which can be written in the form

$$z^2 \frac{d^2 J_n(z)}{dz^2} + z \frac{dJ_n(z)}{dz} + (z^2 - n^2) J_n(z) = 0 \tag{42}$$

has solutions called Bessel functions $J_n(z)$ which have an infinite series represention

$$J_n(z) = \sum_{m=0}^{\infty} \left(\frac{(-1)^m}{m!(m+n)!} \right) \left(\frac{z}{2} \right)^{n+2m} \tag{43}$$

an integral representation for $|\arg z| < \pi/2$

$$J_n(z) = \frac{1}{\pi} \int_0^{\pi} \cos(n\theta - z \sin \theta)\, d\theta \tag{44}$$

with the special case

$$J_0(z) = \frac{1}{2\pi} \int_0^{2\pi} e^{iz \sin \theta} d\theta = \frac{1}{2\pi} \int_0^{2\pi} e^{iz \cos \theta}\, d\theta \tag{45}$$

a generating function representation

$$e^{z(t^2-1)/2t} = \sum_{n=-\infty}^{\infty} J_n(z) t^n \tag{46}$$

and several others. Figure 28-1 shows how the Bessel function $J_n(z)$ varies with both the order n and the argument z. Table 28-4 gives roots, i.e., values of the argument z, where the Bessel functions $J_n(z)$ and their derivatives

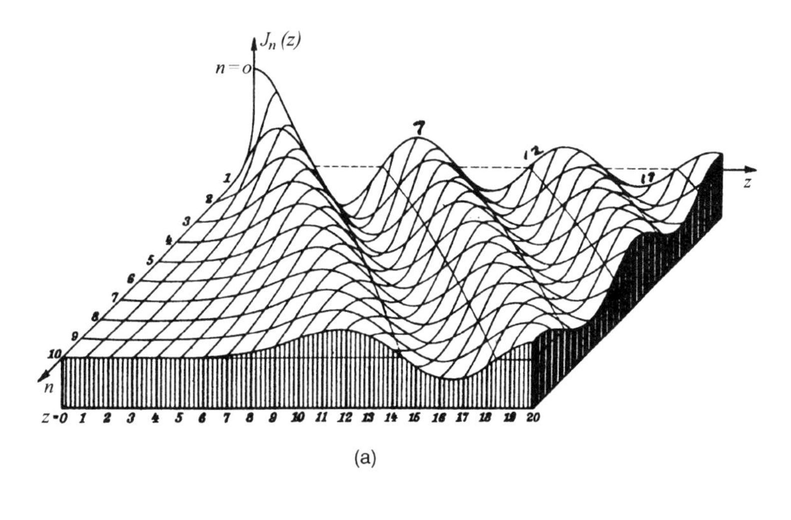

(a)

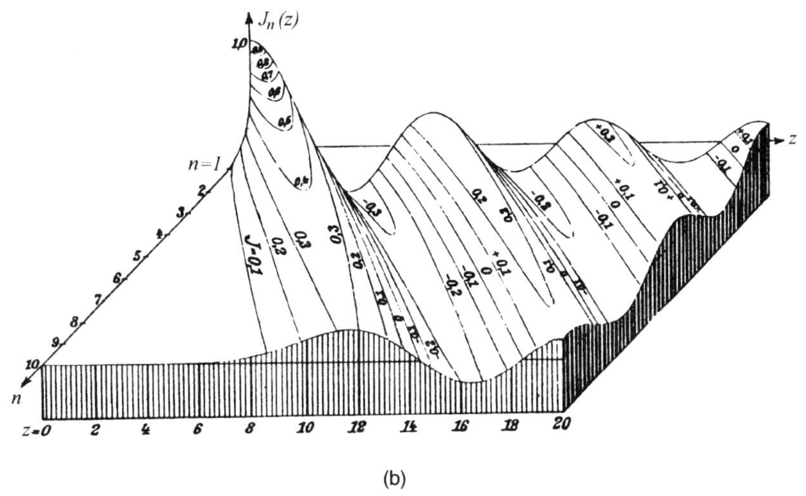

(b)

Fig. 28-1. Dependence of the Bessel function $J_n(z)$ on the order n and on real values of their argument z, where plot (a) shows contours of constant n and constant z, plus contours for which $J_n(z) = 0$, and plot (b) shows contours of constant amplitude. (From E. Jahnke and F. Emde, *Tables of Functions*, Dover, New York, 1945, pp. 127(a), 153(b).)

TABLE 28-4. Roots of Bessel functions ($J_n(z) = 0$) and their derivatives $J_n'(z) = 0$)

Root	$J_0(z)$	$J_1(z)$	$J_2(z)$	$J_3(z)$
1	2.4048	3.8317	5.1356	6.3802
2	5.5201	7.0156	8.4172	9.7610
3	8.6537	10.1735	11.6198	13.0152
4	11.7915	13.3237	14.7960	16.2235

Root	$J_0'(z)$	$J'(z)$	$J_2'(z)$	$J_3'(z)$
1	3.8317	1.8412	3.0542	4.2012
2	7.0156	5.3314	6.7061	8.0152
3	10.1735	8.5363	9.9695	11.3459

$J_n'(z) = dJ_n/dz$ are zero. Table 28-5 gives asymptotic expressions for very small and very large z.

The Bessel functions for positive and negative n are linearly independent when n is not an integer, but when n is an integer they are not independent, but rather obey the relation

$$J_{-n}(z) = (-1)^n J_n(z) \tag{47}$$

The Bessel functions for positive n are well behaved at the origin ($z \Rightarrow 0$). A second linearly independent solution to Bessel's equation is the Neumann function $N_n(z)$

$$N_n(z) = \frac{J_n(z)\cos n\pi - J_{-n}(z)}{\sin n\pi} \tag{48}$$

which has an infinity at the origin where $z \Rightarrow 0$. Linear combinations of $J_n(z)$ and $N_n(z)$ provide the Hankel functions of the first kind, $H_n^{(1)}(z)$, and the second kind, $H_n^{(2)}(z)$

$$H_n^{(1)}(z) = J_n(z) + iN_n(z) \tag{49a}$$

$$H_n^{(2)}(z) = J_n(z) - iN_n(z) \tag{49b}$$

TABLE 28-5. Asymptotic expressions of various varieties of Bessel functions for small and large values of the arguments

Function	Limit for $z \ll 1$	Limit for $z \gg 1$
$J_n(z)$	$\begin{Bmatrix} 1 - z^2/4 & n = 0 \\ z^n/2^n n! & n > 0 \end{Bmatrix}$	$\sqrt{\frac{2}{\pi z}} \cos\left[z - \left(n + \frac{1}{2}\right)\frac{\pi}{2}\right]$
$N_n(z)$	$\begin{Bmatrix} \frac{2}{\pi}(\ln z + \gamma - \ln 2) & n = 0 \\ \frac{-(n-1)!}{\pi}\left(\frac{2}{z}\right)^n & n > 0 \end{Bmatrix}$	$\sqrt{\frac{2}{\pi z}} \sin\left[z - \left(n + \frac{1}{2}\right)\frac{\pi}{2}\right]$
$H_n^{(1)}(z)$	$\begin{Bmatrix} 2i \ln z/\pi & n = 0 \\ \frac{-(n-1)!i}{\pi}\left(\frac{2}{z}\right)^n & n > 0 \end{Bmatrix}$	$\sqrt{\frac{2}{\pi z}} \exp i\left[z - \left(n + \frac{1}{2}\right)\frac{\pi}{2}\right]$
$H_n^{(2)}(z)$	$\begin{Bmatrix} -2i \ln z/\pi & n = 0 \\ \frac{(n-1)!i}{\pi}\left(\frac{2}{z}\right)^n & n > 0 \end{Bmatrix}$	$\sqrt{\frac{2}{\pi z}} \exp -i\left[z - \left(n + \frac{1}{2}\right)\frac{\pi}{2}\right]$
$I_n(z)$	$\begin{Bmatrix} 1 + z^2/4 & n = 0 \\ z^n/2^n n! & n > 0 \end{Bmatrix}$	$\frac{e^z}{\sqrt{2\pi z}}$
$K_n(z)$	$\begin{Bmatrix} -\ln z & n = 0 \\ 2^{n-1}(n-1)!z^{-n} & n > 0 \end{Bmatrix}$	$\sqrt{\frac{\pi}{2z}} e^{-z}$
$j_n(z)$	$\frac{z^n}{(2n+1)!!}$	$\frac{1}{z} \sin\left(z - \frac{n\pi}{2}\right)$
$n_n(z)$	$-(2n-1)!!z^{-n-1}$	$-\frac{1}{z} \cos\left(z - \frac{n\pi}{2}\right)$
$h_n^{(1)}(z)$	$-i(2n-1)!!z^{-n-1}$	$(-i)^{n+1} \frac{e^{iz}}{z}$
$h_n^{(2)}(z)$	$i(2n-1)!!z^{-n-1}$	$i^{n+1} \frac{e^{-iz}}{z}$
$i_n(z)$	$\frac{z^n}{(2n+1)!!}$	$\frac{e^z}{2z}$
$k_n(z)$	$\frac{(2n-1)!!}{z^{n+1}}$	$\frac{e^{-z}}{z}$

All four of these functions (47)–(49b) obey the following recursion relations:

$$J_{n-1}(z) + J_{n+1}(z) = \frac{2n}{z} J_n(z) \tag{50a}$$

$$J_{n-1}(z) - J_{n+1}(z) = 2\frac{dJ_n(z)}{dz} \tag{50b}$$

The Bessel functions obey the orthogonality relation for $p \neq m$

$$\int_0^a J_n(\alpha_{nm}\rho/a)J_n(\alpha_{np}\rho/a)\rho\, d\rho = 0 \tag{51}$$

where the α_{nm} are the $J_n(z)$ roots listed in the upper part of Table 28-4. For $m = p$ we obtain the normalization condition

$$\int_0^\infty [J_n(\alpha_{nm}\rho/a)]^2 \rho\, d\rho = \tfrac{1}{2}a^2[J_{n+1}(\alpha_{nm})]^2 \tag{52}$$

The modified Bessel equation (34) has a negative sign in front of both the z^2 and the n^2 terms

$$z^2\frac{d^2J_n(z)}{dz^2} + z\frac{dJ_n(z)}{dz} - (z^2 + n^2)J_n(z) = 0 \tag{53}$$

and its solutions are Bessel functions of imaginary argument $J_n(iz)$ called modified Bessel functions $I_n(z)$ which are defined by the expression

$$I_n(z) = i^{-n}J_n(iz) \tag{54}$$

The second linearly independent modified Bessel function is chosen to be

$$K_n(z) = \frac{\pi}{2} i^{n+1} H_n^{(1)}(iz) \tag{55}$$

They, of course, have infinite series representations and obey recursion relations. Figure 28-2 shows how I_0, I_1, K_0, and K_1 depend on z.

In the previous section we showed how the radial part (38) of the Helmholtz equation (35) in spherical coordinates can be converted to Bessel's equation (40) for half integral orders. The solutions are spherical Bessel and spherical Neumann functions

$$j_n(z) = (\pi/2z)^{1/2} J_{n+1/2}(z) \tag{56}$$

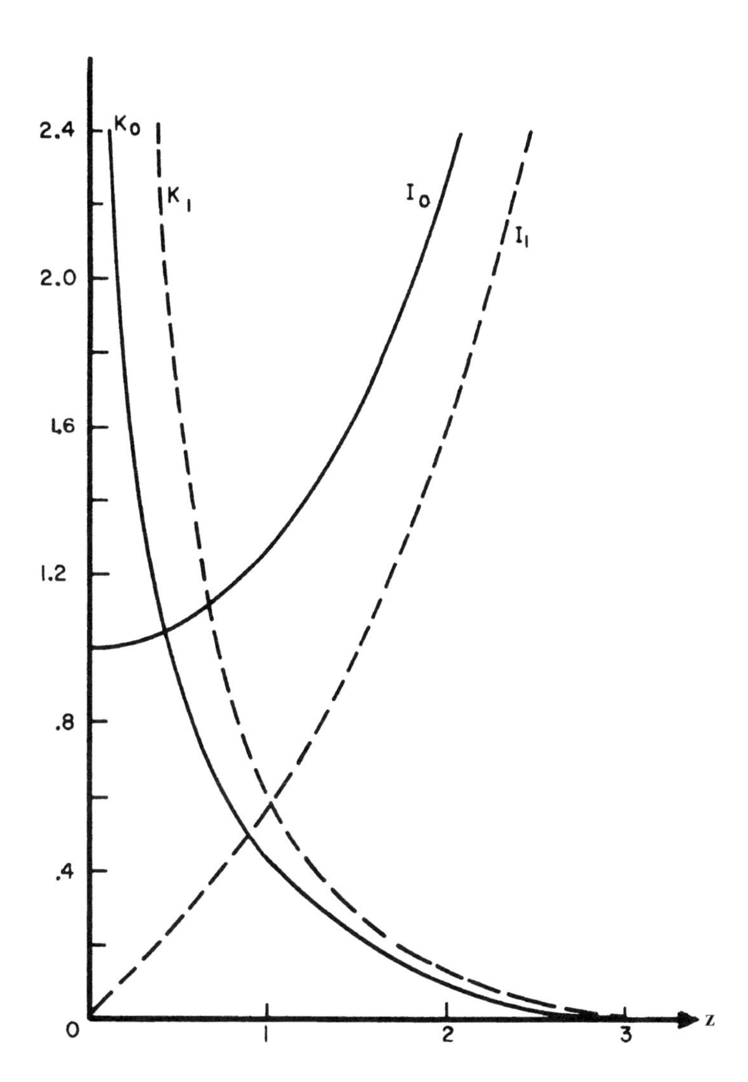

Fig. 28-2. Dependence of the modified Bessel functions $I_0(z)$ and $I_1(z)$, and the modified Hankel functions $K_0(z)$ and $K_1(z)$, on the argument z. (From M. Abramowitz and I. A. Stegun, eds., *Handbook of Mathematical Functions*, National Bureau of Standards (NIST), 1970, Maryland, p. 375.)

$$n_n(z) = (\pi/2z)^{1/2} N_{n+1/2}(z) = (-1)^{n+1}(\pi/2z)^{1/2} J_{-n-1/2}(z) \tag{57}$$

and their spherical Hankel function counterparts

$$h_n^{(1)}(z) = j_n(z) + in_n(z) \tag{58a}$$

$$h_n^{(2)}(z) = j_n(z) - in_n(z) \tag{58b}$$

When the infinite series for these expressions are written out we can show that the lowest order solutions have the following forms:

$$j_0(z) = \frac{\sin z}{z} \tag{59}$$

$$j_1(z) = \frac{\sin z}{z^2} - \frac{\cos z}{z} \tag{60}$$

$$n_0(z) = -\frac{\cos z}{z} \tag{61}$$

$$n_1(z) = -\frac{\cos z}{z^2} - \frac{\sin z}{z} \tag{62}$$

$$h_0^{(1)}(z) = -\frac{i\,\mathrm{e}^{iz}}{z} \tag{63}$$

$$h_1^{(1)} = \mathrm{e}^{iz}\left(-\frac{1}{z} - \frac{i}{z^2}\right) \tag{64}$$

$$h_0^{(2)}(z) = \frac{i\,\mathrm{e}^{-iz}}{z} \tag{65}$$

$$h_1^{(2)}(z) = \mathrm{e}^{-iz}\left(-\frac{1}{z} + \frac{i}{z^2}\right) \tag{66}$$

These expressions satisfy recurrence and orthogonality relations. Their limiting values for small and large z are given in Table 28-5.

Modified spherical Bessel functions can also be defined

$$i_n(z) = (\pi/2z)^{1/2} I_{n+1/2}(z) \tag{67a}$$

$$k_n(z) = (2/\pi z)^{1/2} K_{n+1/2}(z) \tag{67b}$$

with the special cases

$$i_0(z) = \frac{\sinh z}{z} \tag{68}$$

$$k_0(z) = \frac{e^{-z}}{z} \tag{69}$$

and the asymptotic forms for small and large z are presented in Table 28-5.

6 LEGENDRE POLYNOMIALS

The solutions of Legendre's differential equation

$$(1 - z^2)\frac{d^2 P_n(z)}{dz^2} - 2z\frac{dP_n(z)}{dz} + n(n+1)P_n(z) = 0 \tag{70}$$

are the Legendre polynomials listed in Tables 28-2 and 28-3. They are defined by the generating function $(1 - 2zt + t^2)^{-1/2}$

$$\frac{1}{(1 - 2zt + t^2)^{1/2}} = \sum_{n=0}^{\infty} P_n(z)t^n \qquad |t| < 1 \tag{71}$$

which arises from the expansion of $1/r_{12}$ in spherical coordinates

$$\frac{1}{|\mathbf{r}_1 - \mathbf{r}_2|} = \frac{1}{r_>}\sum_{n=0}^{\infty}\left(\frac{r_<}{r_>}\right)^n P_n(\cos\theta) \tag{72}$$

where $r_{12} = |\mathbf{r}_1 - \mathbf{r}_2|$, $t = r_</r_>$, and $z = \cos\theta$. This is a special case of the ultraspherical or Gegenbauer polynomials $C_n^\alpha(z)$ that are obtained from the more general generating function

$$\frac{1}{(1 - 2zt + t^2)^\alpha} = \sum_{n=0}^{\infty} C_n^\alpha(z)t^n \tag{73}$$

where $\alpha = \frac{1}{2}$ for the Legendre case, $\alpha = 1$ for type II Chebyshev polynomials, and $\alpha = 0$ for the type I Chebyshev case

Legendre polynomials can also be calculated using Rodrigues' formula

$$P_n(z) = \frac{1}{2^n n!}\left(\frac{d}{dz}\right)^n (z^2 - 1)^n \tag{74}$$

These polynomials obey a number of recurrence relations, such as

$$(2n+1)zP_n(z) = (n+1)P_{n+1}(z) + nP_{n-1}(z) \tag{75a}$$

$$P'_{n+1}(z) + P'_{n-1}(z) = 2zP'_n(z) + P_n(z) \tag{75b}$$

$$P'_{n+1}(z) - P'_{n-1}(z) = (2n+1)P_n(z) \tag{75c}$$

where $n = 1, 2, 3\ldots,$ and $P'_n(z) = dP_n/dz$ is the derivative of $P_n(z)$. The polynomials satisfy the parity relation

$$P_n(-z) = (-1)^n P_n(z) \tag{76}$$

they have the special values

$$P_n(1) = 1 \tag{77}$$

$$P_n(-1) = (-1)^n \tag{78}$$

and they comply with the orthogonality relation

$$\int_{-1}^{1} P_n(z)P_n(z)\,dz = \frac{2\delta_{mn}}{2n+1} \tag{79}$$

which is equivalent to

$$\int_{0}^{\pi} P_m(\cos\theta)P_n(\cos\theta)\sin\theta\,d\theta = \frac{2\delta_{mn}}{2n+1} \tag{80}$$

since $z = \cos\theta$.

The associated Legendre equation

$$\frac{1}{\sin\theta}\frac{d}{d\theta}\left(\sin\theta\frac{dP_n^m(\cos\theta)}{d\theta}\right) + \left[n(n+1) - \frac{m^2}{\sin^2\theta}\right]P_n^m(\cos\theta) = 0 \tag{81}$$

which may be written in a somewhat simpler form by the change of variable $z = \cos\theta$

$$(1-z^2)\frac{d^2P_n^m(z)}{dz^2} - 2z\frac{dP_n^m(z)}{dz} + \left[n(n+1) - \frac{m^2}{1-z^2}\right]P_n^m(z) = 0 \tag{82}$$

is solved by the associated Legendre polynomials $P_n^m(z)$, which are obtained by differentiating Legendre polynomials

$$P_n^m(z) = (1 - z^2)^{m/2} \frac{d^m P_n(z)}{dz^m} \tag{83}$$

where, by definition, $P_n^0(z) = P_n(z)$. The lower order associated Legendre polynomials are listed in Table 28-3.

There is parity relation

$$P_n^m(-z) = (-1)^{m+n} P_n^m(z) \tag{84}$$

and the end point values

$$P_n^m(\pm 1) = 0 \qquad m \neq 0 \tag{85a}$$
$$P_n^m(\pm 1) = (-1)^n \quad m = 0 \tag{85b}$$

The orthogonality relations are

$$\int_{-1}^{1} P_n^m(z) P_p^m(z)\, dz = \frac{2(n+m)!\delta_{np}}{(2n+1)(n-m)!} \tag{86}$$

There are several other polynomials that commonly occur in physical problems, and Tables 28-2 and 28-3 summarize their properties. They do not occur often enough to merit the more extensive treatment given here to Legendre polynomials.

7 SPHERICAL HARMONICS

The spherical harmonics $Y_{LM}(\theta, \phi)$ are normalized forms of the product $P_L^M(\cos\theta)\,\mathrm{e}^{iM\phi}$ of an associated Legendre polynomial P_L^M and the factor $\mathrm{e}^{iM\phi}$

$$Y_{LM}(\theta, \phi) = (-1)^M \left(\frac{2L+1}{4\pi} \frac{(L-M)!}{(L+M)!}\right)^{1/2} P_L^M(\cos\theta)\mathrm{e}^{iM\phi} \tag{87}$$

They have important cartesian and polar coordinate representations, and we give these here for $L = 0, 1, 2$ without including the normalization factors, and with initial minus signs ignored. Table 28-6 provides normalized spherical harmonics for $L = 0, 1, 2, 3$. Recall that $e^{\pm i\phi} = \cos\phi \pm i \sin\phi$.

TABLE 28.6 Spherical harmonics $Y_{LM}(\theta, \phi)$ for $L = 0, 1, 2, 3$

$$Y_{00} = \frac{1}{\sqrt{4\pi}}$$

$$Y_{11} = -\sqrt{\frac{3}{8\pi}} \sin\theta \, e^{i\phi}$$

$$Y_{10} = \sqrt{\frac{3}{4\pi}} \cos\theta$$

$$Y_{1-1} = +\sqrt{\frac{3}{8\pi}} \sin\theta \, e^{-i\varphi}$$

$$Y_{22} = \sqrt{\frac{15}{32\pi}} \sin^2\theta \, e^{2i\phi}$$

$$Y_{21} = -\sqrt{\frac{15}{8\pi}} \sin\theta \cos\theta \, e^{i\phi}$$

$$Y_{20} = \sqrt{\frac{5}{16\pi}} \left(3\cos^3\theta - 1\right)$$

$$Y_{2-1} = +\sqrt{\frac{15}{8\pi}} \sin\theta \cos\theta \, e^{-i\varphi}$$

$$Y_{2-2} = \sqrt{\frac{15}{32\pi}} \sin^2\theta \, e^{-2i\varphi}$$

$$Y_{33} = -\sqrt{\frac{35}{64\pi}} \sin^3\theta \, e^{3i\phi}$$

$$Y_{32} = \sqrt{\frac{105}{32\pi}} \sin^2\theta \cos\theta \, e^{2i\phi}$$

$$Y_{31} = -\sqrt{\frac{21}{64\pi}} \sin\theta (5\cos^2 - 1) \, e^{i\phi}$$

$$Y_{30} = \sqrt{\frac{7}{16\pi}} (5\cos^3\theta - 3\cos\theta)$$

$$Y_{3-1} = +\sqrt{\frac{21}{64\pi}} \sin\theta (5\cos^2\theta - 1) \, e^{-i\phi}$$

$$Y_{3-2} = \sqrt{\frac{105}{32\pi}} \sin^2\theta \cos\theta \, e^{-2i\phi}$$

$$Y_{3-3} = +\sqrt{\frac{35}{64\pi}} \sin^3\theta \, e^{-3i\phi}$$

$$Y_{00} = 1 \tag{88}$$

$$\begin{aligned} Y_{1-1} &= \frac{x - iy}{r} = \sin\theta\,\mathrm{e}^{-i\phi} \\ Y_{10} &= z/r = \cos\theta \\ Y_{11} &= \frac{x + iy}{r} = \sin\theta\,\mathrm{e}^{i\phi} \end{aligned} \tag{89}$$

$$\begin{aligned} Y_{2-2} &= \frac{(x - iy)^2}{r^2} = \sin^2\theta\,\mathrm{e}^{-2i\phi} \\ Y_{2-1} &= \frac{z(x - iy)}{r^2} = \cos\theta\sin\theta\,\mathrm{e}^{-i\phi} \\ Y_{20} &= \frac{3z^2 - r^2}{r^2} = 3\cos^2\theta - 1 \\ Y_{21} &= \frac{z(x + iy)}{r^2} = \cos\theta\sin\theta\,\mathrm{e}^{i\phi} \\ Y_{22} &= \frac{(x + iy)^2}{r^2} = \sin^2\theta\,\mathrm{e}^{2i\phi} \end{aligned} \tag{90}$$

The spherical harmonics $Y_{LL}(\theta, \phi)$ and $Y_{LL-1}(\theta, \phi)$ are always of the form

$$Y_{LL} = \frac{(x + iy)^L}{r^L} = \sin^L\theta\,\mathrm{e}^{Li\phi} \tag{91}$$

$$Y_{LL-1} = \frac{z(x + iy)^{L-1}}{r^L} = \cos\theta\sin^{L-1}\theta\mathrm{e}^{(L-1)i\phi} \tag{92}$$

and analogously for the $M = -L$ and $M = -L + 1$ cases.

The spherical harmonics satisfy the orthogonality relation

$$\int_{\phi=0}^{2\pi}\int_{\theta=0}^{\pi} Y_{L,M}^*(\theta, \phi)Y_{L',M'}(\theta, \phi)\sin\theta\,d\theta\,d\phi = \delta_{L,L'}\delta_{M,M'} \tag{93}$$

and the sum rule

$$\sum_{M=-L}^{L} |Y_{LM}(\theta, \phi)|^2 = \frac{2L + 1}{4\pi} \tag{94}$$

The spherical harmonic addition theorem is

$$P_L(\cos\gamma) = \frac{4\pi}{2L+1} \sum_{M=-L}^{L} (-1)^M Y_{LM}(\theta_1\phi_1) Y_{L-M}(\theta_2\phi_2) \tag{95}$$

where γ is the angle between the directions specified by the angles θ_1, ϕ_1 and θ_2, ϕ_2 [$\cos\gamma = \cos\theta_1 \cos\theta_2 + \sin\theta_1 \sin\theta_2 \cos(\phi_1 - \phi_2)$].

Tesseral harmonics $Z_{LM}^{S,C}(\theta, \phi)$ are real forms of the spherical harmonics obtained by taking the linear combinations $Y_{LM}(\theta, \phi) \pm Y_{L-M}(\theta, \phi)$. Their unnormalized forms are tabulated in Chapter 10 for $L = 0, 1, 2$, and normalized forms are listed in Table 28-7. For $M = 0$ we always have $Y_{00} = Z_{00}$. The tesseral harmonics provide real solutions to potential problems in spherical coordinates because the potential is a real quantity.

8 GAMMA AND RELATED FUNCTIONS

A gamma function $\Gamma(z)$ satisfies the equation

$$\Gamma(z+1) = z\Gamma(z) \tag{96}$$

where in general $z = x + iy$ is a complex number. If z is an integer, then we have the following relationship to the factorial:

$$\Gamma(n+1) = n! = 1 \cdot 2 \cdot 3 \cdot 4 \cdot \ldots (n-1)n \tag{97}$$

with the following special values:

$$\Gamma(2) = \Gamma(1) = 1! = 0! = 1 \tag{98}$$

$$\Gamma\left(\tfrac{1}{2}\right) = \sqrt{\pi} \tag{99}$$

Gamma functions for negative values of z can be obtained from the relation

$$\Gamma(z)\Gamma(1-z) = \frac{\pi}{\sin \pi z} \tag{100}$$

The Euler definition of $\Gamma(z)$ is the following infinite limit:

$$\Gamma(z) = \lim_{n \Rightarrow \infty} \frac{1 \cdot 2 \cdot 3 \ldots (n-1)n(n^z)}{z(z+1)(z+2)\ldots(z+n)} \qquad z \neq 0, -1, -2, \ldots \tag{101}$$

TABLE 28-7. Tesseral harmonics $Z^{C,S}_{LM}(\theta, \phi)$ for $L = 0, 1, 2, 3$

Z_{00}	$\frac{1}{2\sqrt{\pi}}$	$\frac{1}{2\sqrt{\pi}}$
Z_{10}	$\frac{1}{2}\sqrt{\frac{3}{\pi}}\left[\frac{z}{r}\right]$	$\frac{1}{2}\sqrt{\frac{3}{\pi}}[\cos\theta]$
Z^C_{11}	$\frac{1}{2}\sqrt{\frac{3}{\pi}}\left[\frac{x}{r}\right]$	$\frac{1}{2}\sqrt{\frac{3}{\pi}}[\sin\theta\cos\phi]$
Z^S_{11}	$\frac{1}{2}\sqrt{\frac{3}{\pi}}\left[\frac{y}{r}\right]$	$\frac{1}{2}\sqrt{\frac{3}{\pi}}[\sin\theta\sin\phi]$
Z_{20}	$\frac{1}{4}\sqrt{\frac{5}{\pi}}\left[\frac{3z^2 - r^2}{r^2}\right]$	$\frac{1}{4}\sqrt{\frac{5}{\pi}}[3\cos^2\theta - 1]$
Z^C_{21}	$\frac{1}{2}\sqrt{\frac{15}{\pi}}\left[\frac{zx}{r^2}\right]$	$\frac{1}{2}\sqrt{\frac{15}{\pi}}[\cos\theta\sin\theta\cos\phi]$
Z^S_{21}	$\frac{1}{2}\sqrt{\frac{15}{\pi}}\left[\frac{yz}{r^2}\right]$	$\frac{1}{2}\sqrt{\frac{15}{\pi}}[\cos\theta\sin\theta\sin\phi]$
Z^C_{22}	$\frac{1}{4}\sqrt{\frac{15}{\pi}}\left[\frac{x^2 - y^2}{r^2}\right]$	$\frac{1}{4}\sqrt{\frac{15}{\pi}}[\sin^2\theta\cos 2\phi]$
Z^S_{22}	$\frac{1}{4}\sqrt{\frac{15}{\pi}}\left[\frac{xy}{r^2}\right]$	$\frac{1}{4}\sqrt{\frac{15}{\pi}}[\sin^2\theta\sin 2\phi]$
Z_{30}	$\frac{1}{4}\sqrt{\frac{7}{\pi}}\left[\frac{5z^3 - 3z}{r^3}\right]$	$\frac{1}{4}\sqrt{\frac{7}{\pi}}[(5\cos^2\theta - 3)\cos\theta]$
Z^C_{31}	$\frac{1}{4}\sqrt{\frac{21}{2\pi}}\left[\frac{x(5z^2 - r^2)}{r^3}\right]$	$\frac{1}{4}\sqrt{\frac{21}{2\pi}}[(5\cos^2\theta - 1)\sin\theta\cos\phi]$
Z^S_{31}	$\frac{1}{4}\sqrt{\frac{21}{2\pi}}\left[\frac{y(5z^2 - r^2)}{r^3}\right]$	$\frac{1}{4}\sqrt{\frac{21}{2\pi}}[(5\cos^2\theta - 1)\sin\theta\sin\phi]$
Z^C_{32}	$\frac{1}{4}\sqrt{\frac{105}{\pi}}\left[\frac{(x^2 - y^2)z}{r^3}\right]$	$\frac{1}{4}\sqrt{\frac{105}{\pi}}[\sin^2\theta\cos\theta\cos 2\phi]$
Z^S_{32}	$\frac{1}{4}\sqrt{\frac{105}{\pi}}\left[\frac{xyz}{r^3}\right]$	$\frac{1}{4}\sqrt{\frac{105}{\pi}}[\sin^2\theta\cos\theta\sin 2\phi]$
Z^C_{33}	$\frac{1}{4}\sqrt{\frac{35}{2\pi}}\left[\frac{x(x^2 - y^2)}{r^3}\right]$	$\frac{1}{4}\sqrt{\frac{35}{2\pi}}[\sin^3\theta\cos 3\phi]$
Z^S_{33}	$\frac{1}{4}\sqrt{\frac{35}{2\pi}}\left[\frac{y(x^2 - y^2)}{r^3}\right]$	$\frac{1}{4}\sqrt{\frac{35}{2\pi}}[\sin^3\theta\sin 3\phi]$

The Weierstrass definition is

$$\frac{1}{\Gamma(z)} = z\,\mathrm{e}^{\gamma z} \prod_{n=1}^{\infty} \left(1 + \frac{z}{n}\right) \mathrm{e}^{-z/n} \tag{102}$$

where the value of the Euler Mascheroni constant γ is obtained from the following double limit:

$$\gamma = \lim_{n \Rightarrow \infty} \left[\sum_{m=1}^{n} \frac{1}{m} - \ln(n) \right] = 0.5772156649\ldots \tag{103}$$

Three Euler integral definitions valid for Re $z > 0$ are

$$\Gamma(z) = \int_0^{\infty} \mathrm{e}^{-t} t^{z-1} dt = 2 \int_0^{\infty} \mathrm{e}^{-t^2} t^{2z-1}\, dt = \int_0^1 [\ln(1/t)]^{z-1}\, dt \tag{104}$$

Table 28-8 gives explicit values for some gamma functions and Fig. 28-3 plots $\Gamma(z)$ and its reciprocal $1/\Gamma(z)$ versus z. Incomplete gamma functions correspond to $\Gamma(\frac{1}{2})$, with a finite upper or lower limit on the integration and they provide the error integrals

$$\mathrm{erf}(z) = \frac{2}{\sqrt{\pi}} \int_0^z \mathrm{e}^{-t^2} dt \tag{105}$$

$$\mathrm{erfc}(z) = 1 - \mathrm{erf}(z) = \frac{2}{\sqrt{\pi}} \int_z^{\infty} \mathrm{e}^{-t^2} dt \tag{106}$$

where

$$\mathrm{erf}(\infty) = 1 \tag{107}$$

The Riemann zeta function $\zeta(z)$ is defined by

$$\zeta(z) = \sum_{m=1}^{\infty} \frac{1}{m^z} \tag{108}$$

and one can show that

$$\ln(z!) = -\gamma z + \sum_{n=2}^{\infty} (-1)^n \frac{z^n}{n} \zeta(n) \tag{109}$$

TABLE 28-8. Values of gamma functions of several selected arguments

Function	Value
$\Gamma_{(1)} = 0!$	1
$\Gamma_{(2)} = 1!$	1
$\Gamma_{(3)} = 2!$	2
$\Gamma_{(4)} = 3!$	6
$\Gamma_{(5)} = 4!$	24
$\Gamma_{(1/2)}$	$\sqrt{\pi} = 1.772$
$\Gamma_{(3/2)}$	$\sqrt{\pi}/2 = 0.886$
$\Gamma_{(5/2)}$	$3\sqrt{\pi}/4 = 1.329$
$\Gamma_{(7/2)}$	$15\sqrt{\pi}/8 = 3.323$
$\Gamma_{(9/2)}$	$105\sqrt{\pi}/16 = 11.63$
$\Gamma_{(1/4)}$	3.626
$\Gamma_{(1/3)}$	2.679
$\Gamma_{(2/3)}$	1.354
$\Gamma_{(3/4)}$	1.225
$\Gamma_{(-1/2)}$	$-2\sqrt{\pi} = -3.545$
$\Gamma_{(-3/2)}$	$4\sqrt{\pi}/3 = 2.363$
$\Gamma_{(-5/2)}$	$-8\sqrt{\pi}/15 = -0.945$
$\Gamma_{(-7/2)}$	$16\sqrt{\pi}/105 = 0.270$

A better approximation for the factorial of large numbers, $z \gg 1$, is Stirling's series

$$\ln(z!) = \tfrac{1}{2}\ln 2\pi + (z + \tfrac{1}{2})\ln z - z + \frac{1}{12z} - \frac{1}{360z^3} \tag{110}$$

where the last two terms on the right-hand side are negligible for large z.

The digamma function $F(z)$ is

$$F(z) = -\gamma + \sum_{m=1}^{\infty}\left(\frac{1}{m} - \frac{1}{(z+m)}\right) \tag{111}$$

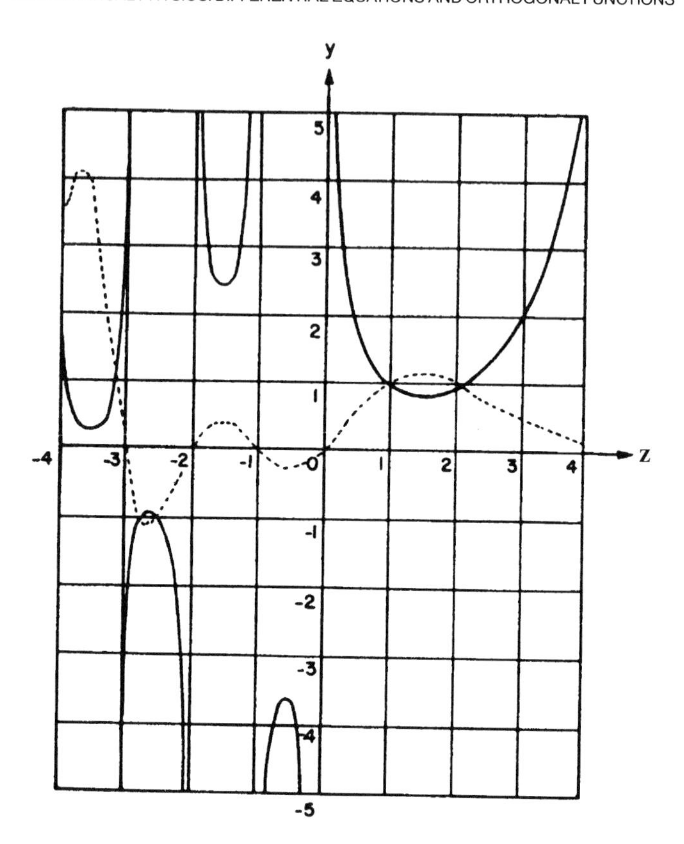

Fig. 28-3. Dependence of the gamma function $y = \Gamma(z)$ (———) and the inverse gamma function $y = 1/\Gamma(x)$ (- - - - -) on the argument for real values of z. (From M. Abramowitz and I. A. Stegun, eds, *Handbook of Mathematical Functions*, National Bureau of Standards (NIST), Maryland, 1970, p. 255.)

where the Euler–Mascheroni constant γ is given by Eq. (103).

The beta function $B(m, n)$ is

$$B(m, n) = \frac{\Gamma(m)\Gamma(n)}{\Gamma(m+n)} = \frac{(m-1)!(n-1)!}{(m+n-1)!} \tag{112}$$

The beta function has the following two integral representations:

$$B(m+1, n+1) = \int_0^1 t^m (1-t)^n dt = \int_0^\infty \frac{u^m du}{(1+u)^{m+n+2}} \tag{113}$$

and also the following definite integral obtained by setting $t = \cos\theta$ in the previous equation:

$$B(m+1, n+1) = 2\int_0^{\pi/2} \cos^{2m+1}\theta \sin^{2n+1}\theta \, d\theta \tag{114}$$

There are several factorial and double factorial expressions that are of interest. The first is the Legendre duplication formula

$$z!(z-\tfrac{1}{2})! = 2^{-2z}\pi^{1/2}(2z)! \tag{115}$$

and for integer z this becomes

$$(n+\tfrac{1}{2})! = \frac{\pi^{1/2}(2n+1)!!}{2^{n+1}} \tag{116}$$

where the double factorial notation means

$$(2n+1)!! = (2n+1)(2n-1)\ldots 5\cdot 3\cdot 1 \tag{117}$$

$$(-1)!! = 1 \tag{118}$$

$$(2n)!! = (2n)(2n-2)\ldots 4\cdot 2 \tag{119}$$

$$0!! = 1 \tag{120}$$

The double factorial is related to the regular factorial through the expressions

$$(2n+1)!! = \frac{(2n+1)!}{2^n n!} \tag{121}$$

$$(2n)!! = 2^n n! \tag{122}$$

9 DELTA FUNCTIONS

An ideal delta function $\delta(x-a)$ is infinite for $x = a$, is zero for $x \neq a$, and integrates to a unit area

$$\delta(x-a) = 0 \quad x \neq a \tag{123}$$

$$\int \delta(x-a)\,dx = 1 \tag{124}$$

An arbitrary function $f(x)$ has the following value $f(a)$ at the point $x = a$:

$$f(a) = \int f(x)\delta(x-a)\,dx \tag{125}$$

A delta function is symmetric in its argument

$$\delta(x-y) = \delta(y-x) \tag{126}$$

It is related to the Laplacian associated with a point charge

$$\nabla^2 \frac{1}{|r_1 - r_2|} = -4\pi\delta(r_1 - r_2) \tag{127}$$

and to the Laplacian of a Green's function $G(x, y)$

$$\nabla^2 G(x-y) = -4\pi\delta(x-y) \tag{128}$$

A delta function is often defined in terms of the limit

$$\delta(x) = \lim_{n \Rightarrow \infty} \delta_n(x) \tag{129}$$

and some examples of $\delta_n(x)$ that satisfy Eq. (129) are

$$\delta_n(x) = \frac{\sin nx}{\pi x} \qquad \text{Dirichlet form} \tag{130}$$

$$\delta_n(x) = (n/\sqrt{\pi})\exp(-n^2x^2) \qquad \text{Gaussian form} \tag{131}$$

$$\delta_n(x) = \frac{n/\pi}{1 + n^2x^2} \qquad \text{Lorentzian form} \tag{132}$$

$$\delta(x) = \frac{1}{2\pi} \lim_{n \Rightarrow \infty} \int_{-n}^{n} e^{inx}\,dx \qquad \text{Fourier form} \tag{133}$$

10 GREEN'S FUNCTIONS

To simplify the mathematics we describe Green's functions in one dimension. Assume that we wish to solve the inhomogeneous Helmholtz equation

$$\Psi'' + k^2\Psi = -f(x) \qquad 0 \leq x \leq a \tag{134}$$

subject to the boundary conditions

$$\Psi(0) = \Psi(a) = 0 \tag{135}$$

and we already know the two solutions $\phi_1(x) = \sin kx$ and $\phi_2(x) = \cos kx$ of the corresponding homogeneous equation

$$\phi'' + k^2\phi = 0 \tag{136}$$

A Green's function $G(x{:}y)$ can be constructed by assuming a solution of (134) of the form

$$\Psi(x) = A(x)\sin kx + B(x)\cos kx \tag{137}$$

and after some mathematical manipulations to evaluate $A(x)$ and $B(x)$ (Dettman, McGraw Hill, N.Y., 1952, p. 190) we find the Green's function

$$G(x{:}y) = \frac{\sin ky \sin k(a-x)}{k \sin ka} \qquad 0 \leq y \leq x \tag{138a}$$

$$G(x{:}y) = \frac{\sin kx \sin k(a-y)}{k \sin ka} \qquad x \leq y \leq a \tag{138b}$$

which is a solution to the delta function equation

$$G'' + k^2 G = -\delta(y-x) \tag{139}$$

The solution of the original Helmholtz equation (134) is given by

$$\Psi(x) = \int_0^a f(y)\, G(x{:}y)\, dy \tag{140}$$

where, of course, $f(y)$ is assumed to be known. Analogous procedures apply in two and three dimensions.

The Green's function has five important properties:

1. The Green's function satisfies the homogeneous equation (136) in each interval $0 \leq y < x$ and $x < y \leq a$, but not at the point $y = x$.
2. The Green's function is continuous at the point $y = x$

$$G(x{:}x^+) = G(x{:}x^-) \tag{141}$$

   where x^+ and x^- denote y approaching x from the right and from the left, respectively.
3. The derivative of the Green's function is discontinuous at $y = x$

$$G'(x{:}x^+) - G'(x{:}x^-) = -1 \tag{142}$$

   and because G' is discontinuous the second derivative G'' does not exist.
4. The Green's function satisfies the boundary conditions

$$G(x;0) = G(x;a) \tag{143}$$

5. The Green's function is symmetric under the interchange of x and y

$$G(x;y) = G(y;x) \tag{144}$$

A knowledge of these properties can be an aid in the construction of Green's function.

BIBLIOGRAPHY

MECHANICS—CHAPTERS 1 TO 5

V. I. Arnold, *Mathematical Methods of Classical Mechanics*, Springer Verlag, Berlin, 2d ed., 1989.

A. L. Fetter and J. D. Walecka, *Theoretical Foundations of Particles and Continuous Media*, McGraw-Hill, New York, 1980.

H. Goldstein, *Classical Mechanics*, Addison-Wesley, 2d ed., Reading, MA, 1980.

D. Hestenes, *New Foundations for Classical Mechanics*, Reidel, Dordrecht, Holland, 1986.

L. D. Landau and E. M. Lifshitz, *Mechanics*, 3d ed., Pergamon, New York, 1982.

NON-LINEAR DYNAMICS AND CHAOS—CHAPTER 6

A. B. Çambel, *Applied Chaos Theory*, Academic Press, New York, 1993.

R. J. Creswick, H. A. Farach, and C. P. Poole, Jr., *Introduction to Renormalization Group Methods in Physics*, Wiley, New York, 1992.

M. Hénon, *Numerical Exploration of Hamiltonian Systems*, Course 2 in *Chaotic Behaviour of Deterministic Systems*, Les Houches École de Physique Théoretique, 1981.

E. A. Jackson, *Perspectives of Nonlinear Dynamics*, 2 Vols, Cambridge University Press, Cambridge, 1989.

H. O. Peitgen, H. Jürgens, and D. Saupe, *Chaos and Fractals, New Frontier of Science*, Springer Verlag, 1992.

RELATIVITY—CHAPTER 7

L. D. Landau, *The Classical Theory of Fields*, Pergamon, New York, 1975.

H. M. Schwartz, *Introduction to Special Relativity*, McGraw-Hill, New York, 1968.

B. F. Schultz, *A First Course in General Relativity*, Cambridge, 1985.

S. Weinberg, *Gravitation and Cosmology*, Wiley, New York, 1972.

THERMODYNAMICS AND STATISTICAL MECHANICS—CHAPTERS 8 AND 9

H. B. Callen, *Thermodynamics and an Introduction the Thermostatics*, Wiley, New York, 1985.

K. Huang, *Statistical Mechanics*, Wiley, New York, 1963.

C. Kittel and H. Kroemer, *Thermal Physics*, Freeman, 1980.

ELECTRODYNAMICS—CHAPTERS 10 TO 12, 15, AND 16

J. D. Jackson, *Classical Electrodynamics*, 2d ed., Wiley, New York, 1975.

L. D. Landau and E. M. Lifshitz, *Electrodynamics of Continuous Media*, Pergamon, New York, 1984.

H. C. Ohanian, *Classical Electrodynamics*, Allyn-Bacon, Boston, 1988.

A. M. Portis, *Electromagnetic Fields*, Wiley, New York, 1978.

E. M. Purcell, *Electricity and Magnetism*, Berkeley Series, vol. 2, McGraw-Hill, New York, 1985.

OPTICS—CHAPTERS 13 AND 14

E. Hecht and A. Zajac, *Optics*, 3d ed., Addison-Wesley, Reading, MA, 1998.

S. G. Lipson, H. Lipson, and D. S. Tannhauser, *Optical Physics*, Cambridge, UK, 1995.

J. R. Meyer-Arendt, *Introduction to Classical and Modern Optics*, Prentice-Hall, New Jersey, 1995.

F. L. Pedrotti and L. S. Pedrotti, *Introduction to Optics*, 2d ed., Prentice-Hall, New Jersey, 1993.

QUANTUM MECHANICS—CHAPTERS 17 TO 22

G. Baym, *Lectures on Quantum Mechanics*, Addison-Wesley, Redwood City, CA, 1990.

C. Cohen-Tannoudji, B.Diu, and F. Laloe, *Quantum Mechanics*, Wiley, New York, 1977.

A. Messiah, *Quantum Mechanics*, 2 vols, North Holland, Amsterdam, 1968.

J. J. Sakurai, *Modern Quantum Mechanics*, Addison Wesley, 1994.

CONDENSED MATTER—CHAPTERS 23 AND 24

N. W. Ashcroft and N. D. Mermin, *Solid State Physics*, Saunders, Philadelphia, 1976.

G. Burns, *Solid State Physics*, Academic Press, New York, 1985.

C. Kittel, *Solid State Physics*, 7th ed., Wiley, New York, 1996.

NUCLEAR PHYSICS—CHAPTERS 25 AND 26

J. W. Rohlf, *Modern Physics from α to Z^0*, Wiley, New York, 1994.

D. Griffiths, *Introduction to Elementary Particles*, Harper and Row, New York, 1987.

MATHEMATICAL PHYSICS—CHAPTERS 27 AND 28

G. Arfken and H. J. Weber, *Mathematical Methods for Physicists*, 4th ed, Academic Press, New York, 1995.

R. Courant and D. Hilbert, *Methods of Mathematical Physics*, 2 vols, Wiley, New York, 1953.

S. Hassani, *Foundations of Mathematical Physics*, Allyn and Bacon, Boston, 1991.

H. S. Jeffreys and B. S. Jefferies, *Methods of Mathematical Physics*, Cambridge, 1956.

P. M. Morse and H. Feshbach, *Methods of Theoretical Physics*, 2 vols., McGraw-Hill, New York, 1953.

E. T. Whittaker and G. N. Watson, *A Course of Modern Analysis* 4th ed., Cambridge, 1952.

HANDBOOKS—CHAPTERS 27 AND 28

M. Abramowitz and I. A. Stegun, *Handbook of Mathematical Functions*, Dover, New York, 1972.

H. L. Anderson, Ed., *A Physicist's Desk Reference* (formerly *Physics Vade Mecum*), American Institute of Physics Press, New York, 1989.

E. R. Cohen, *Physics Quick Reference Guide*, American Institute of Physics Press, New York, 1996.

E. Jahnke and F. Emde, *Tables of Functions*, Dover, New York, 1945; McGraw-Hill, New York, 1960.

R. W. Weast and D. R. Lide, eds., *Handbook of Chemistry and Physics,* CRC Press, Boca Raton, FL, 1997–1998.

INDEX